The Poison Trials

The Poison Trials

Wonder Drugs, Experiment, and the Battle for Authority in Renaissance Science

ALISHA RANKIN

The University of Chicago Press   Chicago and London

The University of Chicago Press, Chicago 60637
The University of Chicago Press, Ltd., London

Published 2021
Printed in the United States of America

30 29 28 27 26 25 24 23 22 21 21 1 2 3 4 5

ISBN-13: 978-0-226-74471-1 (cloth)
ISBN-13: 978-0-226-74485-8 (paper)
ISBN-13: 978-0-226-74499-5 (e-book)
DOI: https://doi.org/10.7208/chicago/9780226744995.001.0001

Library of Congress Cataloging-in-Publication Data

Names: Rankin, Alisha Michelle, author.
Title: The poison trials : wonder drugs, experiment, and the battle for authority in Renaissance science / Alisha Rankin.
Other titles: Synthesis (University of Chicago Press)
Description: Chicago : The University of Chicago Press, 2020. | Series: Synthesis | Includes bibliographical references and index.
Identifiers: LCCN 2020026506 | ISBN 9780226744711 (cloth) | ISBN 9780226744858 (paperback) | ISBN 9780226744995 (ebook)
Subjects: LCSH: Poisons—History. | Toxicology, Experimental. | Toxicity testing—History. | Science—History.
Classification: LCC RA1221 .R355 2020 | DDC 615.9—dc23
LC record available at https://lccn.loc.gov/2020026506

⊗ This paper meets the requirements of ANSI/NISO Z39.48-1992 (Permanence of Paper).

For John, Nico, and Eli, who know more about poison than any family should

Contents

INTRODUCTION

Caravita's Oil

In the heat of a Roman August in 1524, an agent called L'Abbatino wrote to his patron, Federico II Gonzaga of Mantua, with a bold pronouncement. When he returned home to Mantua, he would bring the prince "the most precious thing one could have in the world." This marvelous item was not a rare gem, nor a priceless metal, nor a weapon of war, but instead a medicinal oil that had been proven to work against poison. The antidote had been created by a surgeon named Gregorio Caravita, who offered it to Pope Clement VII with the assurance that it would cure any poison taken into the body. Far from ignoring Caravita's claims, Pope Clement decided to have the oil tested. He commanded his medical personnel to try it on two criminals who had been condemned to death.[1]

Led by the pope's personal physician, Paolo Giovio, the doctors gave both prisoners a good quantity of a deadly aconite called napellus, enough to kill "not merely two men, but one hundred." As the poison took effect, the prisoners started to gesticulate wildly and cry out from the pain in their hearts. Immediately, Caravita anointed one of them with some of the oil, and the man's heart and pulse quickly returned to normal. The other prisoner, who was given no antidote, died in great agony. A second test of the oil, on a Mantuan criminal poisoned with arsenic, yielded similar results. L'Abbatino boasted that Giovio had promised him a sample of the antidote to bring back to Mantua. He also noted that Caravita knew how to make the oil and might be willing to pass on that knowledge, for a price.[2]

As gruesome as these tests may seem, the awestruck tone of L'Abbatino's letter hints at their impact at the time. Less than two weeks later, a four-page pamphlet appeared in print, described as a "Testimony of the most true and admirable virtues of a composite oil against plague and all poison, with which an experiment was conducted by distinguished men, at the command of the Supreme Pontiff Clement VII, in the Roman Capitoline edifice." Written in Latin, the pamphlet was addressed "to all good mortals" and signed by the physician Giovio, the papal pharmacist Tomasso Bigliotti, and the Roman senator Pietro Borghese. Pope Clement VII himself had ordered its publication.[3]

From the pamphlet we learn that the first test of Caravita's oil involved two Corsicans named Gianfrancesco and Ambrogio. The medics gave them special marzipan cakes crafted out of almond paste and deadly aconite from the Apennine Mountains. Caravita anointed Gianfrancesco with the antidote, on his temples, arteries, and heart. Ambrogio was given no antidote because he was the more "savage" of the two, and he suffered for four hours before he died. Gianfrancesco, after his impressive survival, was sent to the slave galleys instead of being executed, intended as a reward. The testers then wished to see if the antidote also worked against other poisons. They obtained permission to conduct a test on a Mantuan man named Antonio, convicted of murder, who drank a swill of raw eggs mixed with sugar and arsenic. This time, Giovio, Bigliotti, and Borghese administered the poison and the antidote "with our own hands," to make sure that Caravita was not tricking them. The second test, too, was successful, and, like Gianfrancesco, Antonio was sent off to "perpetual rowing" in the galleys. The pontiff then granted Caravita a substantial sum of money as a reward.[4]

These two accounts leave little doubt that in early August 1524, Pope Clement VII (1478–1534) and his personal physician devised a test using deadly poison on human beings, to see if they could be cured. The event appears to have all the quintessential elements of the darker side of Renaissance Italy: poison, violent prisoners, cruel popes, greedy princes, and Mantuan spies. We might be tempted to relegate it to the files of historical oddities, as some historians have already done.[5] Yet the test of Caravita's oil had a significance that went far beyond Pope Clement's fear of poison and his power over those condemned to die. It provides very early evidence of contrived experiments conducted on human test subjects by respected medical professionals. Giovio, Bigliotti, and Borghese not only signed the public pamphlet announcing the success but also specifically claimed the report was true and presented it as a significant—and valid—medical finding. The test was also witnessed by a soon-to-be-famous young physician, Pietro Andrea Mattioli (1501–77), who was studying under Caravita at the time. Two decades later, Mattioli included a description of this trial in his *Commentaries on Dioscorides* (1544), one of the most influential medical books of the sixteenth century. In his section on aconites, under the entry for napellus, Mattioli used the trial of Caravita's oil as evidence of the herb's toxicity.[6] Its appearance in the work of the respected Mattioli spread word of Clement VII's antidote test far and wide and reinforced its medical legitimacy, especially after Mattioli's book was translated into Latin in 1554.

By 1540, a few copycat trials had already sprung up in other parts of Italy, all testing derivatives of Caravita's oil. After the publication of Mattioli's book, poison trials spread around Europe and expanded to include other antidotes. Between 1524 and 1600, over a dozen documented poison trials on condemned criminals took place in Italy, France, and the Holy Roman Empire, with allusions to many others. The trials varied widely in the antidotes tested and the treatment of the prisoners, but they all took place at the behest of a powerful prince or church prelate, and they all involved physicians, often very prestigious doctors, or other courtly medical personnel. There were prominent ancient precedents for this practice. King Mithridates VI of Pontus (135–63 BCE) had famously tested poisons and antidotes on condemned criminals and thereby developed his namesake antidote mithridatium. The influential Greek physician Galen (129–c. 216 CE) mentioned these tests, and a work attributed to him included a poison trial on roosters. Medieval Islamic and European texts occasionally referenced Galen's rooster trials, but only in the sixteenth century was the practice of using condemned criminals for fatal tests revived.

Poison Trials

This book tells the forgotten story of these poison trials. It focuses on antidote tests both as a mirror of Renaissance medical practice and as an important crucible for ideas about evidence, authority, and proof. At a time when poison was widely feared as a harming agent and as the root of deadly diseases like plague, the urgent need for effective cures provoked intense excitement about promising drugs like Caravita's oil. Poison trials arose out of a fascination with new wonder drugs that cut across aristocratic, scholarly, and common spheres. Historically, many poison antidotes had a special status as near cure-alls, recommended as plague cures in particular. Unlike most drugs, moreover, antidotes could be tested relatively easily by giving a test subject poison. All poison trials on condemned criminals took place at princely courts, backed by prominent monarchs. This setting meant that the tests were never just about validating or rejecting a particular cure: whatever their medical intent, they also had a deeper symbolic value. A proven antidote would be a "most precious thing" indeed.

Poison trials on condemned criminals represented a new kind of experiment, developed by elite physicians and powerful princes in the context of a rapidly shifting medical and material landscape in the Renaissance.[7] With the rediscovery of many Greek and Roman medical texts in the fifteenth century, humanist physicians had a heady sense of their own authority, steeped in Galen's humoral medical theory. Mattioli boasted proudly of "our current most flourishing age," in which erudite doctors recouped ancient wisdom.[8] Yet university-trained physicians made up only a tiny proportion of healers in Renaissance Europe. The vibrant, pluralistic medical culture included surgeons, barbers, midwives, bonesetters, tooth pullers, and herb sellers, as well as unlicensed generalists who acted as doctors.[9] Physicians derided this last group, in particular, as dangerous empirics, but they also lived among these other practitioners and often treated the same patients, and they were not immune to the lure of promising cures.[10] Poison antidotes were particularly enticing because of the expansive healing powers associated with them—and because powerful aristocrats, many of whom acted as physicians' patrons, valued them so highly. Throughout this book, we will see physicians both eager to embrace wonder drugs and anxious to separate themselves from what they viewed as the riffraff. These competing impulses raised questions that are central to the history of pharmacy and medicine even today and feel eerily familiar in the era of the COVID-19 pandemic. What counts as evidence? Who has the authority to make claims about efficacy? How can one be sure a drug is authentic? What method should be used for testing drugs? What constitutes success?

This book makes three overarching arguments, one about experiment, one about authority, and one about medical ethics. I contend, first, that in the process of developing poison trials, physicians did far more than merely make a statement about particular cures. They devoted careful attention to method, wrote detailed experimental reports, and engaged with the problem of using human subjects for fatal tests. Antidote trials thus generated extensive engagement with "experimental thinking" in the century before the great seventeenth-century experimental boom.[11] Ever since the time of Galen, learned medical writings had portrayed poison as particularly testable, but Renaissance poison trials vastly expanded the scope of that testing. The scholarly conversations that sprung up around the tests raised increasingly complex questions about what constituted reliable proof and evidence in medicine. In making this argument, I join historians like Paula Findlen, Michael McVaugh, and Evan Ragland, who have shown that significant medical experimentation took place in the Middle Ages and Renaissance, and Harold Cook, who has demonstrated the enduring centrality of medicine to experimental thinking in the seventeenth century.[12] This perspective expands the purview of experiment beyond conflicts over Aristotelian natural philosophy, a project long undertaken by historians of alchemy.[13] It also puts scholarly medicine at the center of the drive to develop experimental conventions. If there was a Scientific Revolution, the revolt took hold among some of the most traditional physicians at a very early date.

Yet learned doctors did not create poison trials in a scholarly vacuum, a point that forms the basis of my second main argument. When Caravita's oil was tested in 1524, Renaissance Italians would have been familiar with a very different kind of poison trial: vibrant marketplace shows by mountebanks and charlatans who sold antidotes and cure-alls through dramatic displays of self-poisoning or poison demonstrations on animals.[14] Physicians developed the poison trial as an explicit, scholarly contrast to these marketplace shows. At the same time, they could never escape the market. Some physicians, like Giovio, tested drugs hawked by lower-class empirics. Others, like Mattioli, created and sold their own alchemical cures. Many others eagerly sought (and sometimes even advertised) new wonder drugs with claims of success that, to our ears, sound far-fetched. This book aims to blur the lines that Galenic physicians tried—and failed—to draw around their practice. I argue that the contested status of medical professionals left physicians open to empirics' ideas. Significant research over the past two decades has shown the importance of alchemists, artisans, charlatans, and craftsmen, as well as women and the domestic sphere, in developing scientific theories and practices.[15] I take these ideas a step further by arguing that empirical practitioners had a direct effect on the interests and methodologies of traditional Galenic doctors. Advancements in experiment went hand in hand with broader cultural forces, and, in fact, these trends helped drive experimental thinking.

Finally, I argue, the experimenters were forced to take into account the limitations of using humans for deadly tests, a kind of early medical ethics centered on cultural and religious norms. Previous historical scholarship has given little attention to humans as experimental subjects before the eighteenth century.[16] Philosopher Michel Foucault helped drive this perception with his focus on the later seventeenth century as a turning point, or rupture, in historical conceptions of the human condition.[17] Yet the evidence I have found shows that sixteenth-century individuals were well aware of the problems inherent in using human test subjects. The Renaissance use of condemned criminals to test poison, I argue, grew out of the existing practice of using the cadavers of executed criminals for anatomical investigation, a convention that had long been hamstrung by resistance from local populaces.[18] Using living criminals to test poison was even more complicated than using dead ones for dissection, and testers used a variety of devices to make the practice acceptable. These strategies focused on procedure rather than human rights. There was little concern, as in the modern perspective, with the exploitation of human bodies for experimental purposes or with the gruesome effects of the test itself. Nevertheless, the testers devoted surprising attention to the narratives surrounding human experimentation.

The Poison Trials shows the intertwined nature of medical innovations, professional rivalries, and political power. It is at once a story of top-down efforts by Europe's powerful princes to find remedies that might protect them, no matter how cruel the means, and of a bottom-up attempt by scrappy empirical healers like Caravita to sell their remedies. While isolated poison trials have been mentioned in previous scholarship, particularly in the work of Alessandro Pastore and Sheila Barker, no modern-day historian has put them at the center of a historical study.[19] Historians have noted the important role of poison research in the history of experiment, but the story usually starts in the 1660s, when the Italian physician Francesco Redi conducted hundreds of experiments on viper venom and an exotic antidote called snakestone.[20] Jutta Schickore's masterful book *About Method* (2017) put snake venom at the very center of the development of experimental methodologies from the seventeenth to the twentieth century, with Redi as her first main example.[21] Without a doubt, Redi's obsession with repetition added an entirely new scale to experiments, but many aspects of his method had already been developed in the sixteenth century, and the concept of poison as a particularly testable substance dated back to antiquity. Redi himself invoked that long history, as Jay Tribby has argued.[22] Like earlier physicians, Redi cannot be seen as a neutral hunter of experimental truths. To use Martha Baldwin's words, he also aimed to stake a claim to "who had the authority to pronounce medical facts and control medical knowledge."[23] That ulterior motive was a strong theme of sixteenth-century poison trials, and it is at the center of the story I tell with this book.

Poison in the Renaissance

A few years back, friends of mine decided to concoct their own herbal bitters out of Queen Anne's lace while on holiday in Norway. After ingesting it, they belatedly realized that a nearly identical-looking plant (*Selsnepe*, or water hemlock) was a deadly poison with no antidote. A panicked call to poison control revealed that if they had consumed a lethal dose of the poisonous herb, they probably would be dead before they could reach a hospital. Happily they were fine, but their experience points to our enduring anxieties about poison. Today, poison evokes a sense of something sinister and shadowy, a toxin hiding in everyday life that can take us by surprise. Although rare in most places, poisoning cases cause a special kind of fear and panic when they occur. The toxic levels of lead tragically discovered in the drinking water in Flint, Michigan, in 2014 demonstrate how quickly poison can turn the basic necessities of everyday life into a malicious enemy. Our occasional grocery store food poisoning epidemics recast ordinary lettuce or chicken as deadly antagonists. Meanwhile, any parent whose child has accidentally consumed a potentially poisonous substance knows the utter terror of that experience.

In the Renaissance, that fear was ever-present. Poison lurked everywhere. Especially in Southern Europe, poisonous snakes and spiders abounded, and a variety of poisonous herbs and fungi could be found throughout the continent. Authors wrote of waters infested with poisonous worms (a nod to the dangers of an untreated drinking supply), and even the air sometimes appeared poisonous. Ever since the Black Death of 1347–51, poison had been connected to plague and other pestilential diseases, which many physicians viewed as caused by the spread of invasive poisons in the body via venoms in the air. During times of epidemic disease, any bad smell could point to the presence of poison.[24] In his definitive book on poison in the Middle Ages and Renaissance, historian Frederick Gibbs has traced a pattern of linking ever more diseases to poisonous causes.[25] Cures for poison therefore could also be cures for illnesses. Caravita first came to Pope Clement's attention because he claimed to have successfully cured many plague patients in the hospital of San Giovanni Laterano during the terrible epidemic of 1523, under Clement's predecessor Adrian VI. Caravita's oil, proven in a seemingly irrefutable trial, offered the possibility of protection both from poison and from deadly disease.

Accidental poisoning and poisoned air represented only part of the threat. Rumors of poison plots ran rampant in the Renaissance, in all areas of society. In the middling and lower classes, suspected poisoning cases prompted vicious gossip and sometimes sensationalistic news pamphlets. Arsenic was regularly found in homes as a means against household pests such as rats and lice and was easily accessible. With rumors flying about successful, attempted, or suspected poisonings at various princely courts and aristocratic manor houses, the power brokers of Europe faced a constant fear of being poisoned. It is impossible to know how many actual poisoning deaths occurred among the nobility. Poison mimicked many other ailments and, significantly, often failed to be fatal. Whatever the reality, the *threat* of attempted poisoning was real, and poison was used as a tactic in political maneuvering. In 1547, Duke Cosimo I de' Medici of Tuscany (d. 1574) and Duke Ferrante I Gonzaga (d. 1557) of Mantua conspired to poison the military leader Piero Strozzi. Their plot never succeeded, and Strozzi lived another decade, but the letters of the Medici family reveal its existence.[26] Pope Clement VII's two direct predecessors, Adrian VI (d. 1523) and Leo X (d. 1521), were rumored to have died of poison.

The threat of poison diminished the power of the prince; it made him vulnerable to unseen outside forces. Concerns about this appearance of weakness prompted an increased attention to punishing the crime of poisoning in the Middle Ages, as Franck Collard has noted.[27] Although princely power had increased by the early sixteenth century, so too had rumors of poisoning attempts. The search for antidotes was therefore tied up in an age-old effort to maintain both the prince's health and his air of invincibility. Knowledge of poison was powerful, and physicians who held this power had been highly sought at aristocratic courts for millennia. The fear of poisoning remained a constant across rising and falling empires, as did the rewards kings and princes bestowed on physicians thought to have special knowledge of how to prevent and cure it. Anyone with an effective antidote, proven in a seemingly irrefutable test, would possess a protection against poison and plague and a demonstration of princely indomitability. In the era of Iberian colonial expansion and the Protestant Reformation, with all of their destabilizing effects, princes had urgent reasons to seek new ways to project power.

Drug Testing and Medical Theory

Poison was an unusual substance in medieval and Renaissance medicine because it *was* viewed as so testable. In the prevailing Galenic medical theory, it was difficult to determine the efficacy of a drug (that is, how well it worked in a general sense) through empirical tests. Incorporating ideas from Hippocrates and Aristotle, Galen theorized that the body was made up of four humors: blood, phlegm, red or yellow bile, and black bile. Each of these humors had a corresponding "complexion" composed of pairings of the qualities hot, cold, wet, and dry. Every person's particular balance of humors was different; your body was individual, special. Illness occurred when your humors got out of the balance that was healthy for you. All ingestible substances—food, drink, and drugs—had their own complexions and could be used to keep your body in its proper humoral balance or to correct imbalance.[28]

This theory made drug therapy extremely complex. Not only did physicians have to know all of the properties of simple and compound drugs, they also had to judge how those drugs would work for every individual's distinct humoral composition. Arabic philosopher-physicians such as Avicenna (ibn Sina, 980–1037) and Averroes (ibn Rushd, 1126–98) tried to rationalize this complicated system by quantifying the properties of drugs. As Greek and Arabic medicine filtered into Western Europe in the Middle Ages, many physicians remained skeptical that one could ever definitively establish drugs' effectiveness, given the large number of variables.[29] Drug testing did occur despite these concerns. As a number of scholars have noted, Avicenna developed clear rules for testing drugs, which medieval physicians adapted and expanded.[30] The primary purpose of these tests was not, however, to discover whether a particular drug "worked" against a particular disease, but rather to determine its properties (i.e., its humoral complexion).[31] This distinction helps explain why modern randomized clinical trials (RCTs) took so long to develop.[32] It was not because physicians did not think rationally about questions of testing, but because the individualized concept of the body meant that drug tests appeared to be of limited use.

Poison presented a major exception to this rule. It did not fit very well into the Galenic humoral system. Not only could a small amount of certain substances cause intense harm, it caused this harm nearly universally, no matter the humoral complexion. From the time of Galen through the sixteenth century, physicians slowly consolidated a concept of "poison" and developed a theory of its actions in the body, the subject of Gibbs's study. One line of thought stayed within the boundaries of Galenic theory and described poisons as substances that could radically imbalance the humors. Increasingly, however, physicians posited a theory of poison that circumvented humoral theory altogether. Poison, they reasoned, could harm a person's body through its entire substance, a theory they called "total substance" or "specific form". This mode of action was occult, or hidden, and it focused on poison's natural attraction to the heart rather than a humoral imbalance.[33]

If a single substance could cause nearly universal harm, it stood to reason that a single cure could counteract that injury for just about everybody. The key was to find an antidote that would impede this harming agent. There was no unified view of how, exactly, antidotes would counteract poisons, although the concept of specific form provided an explanatory framework. Many authors used ideas of sympathy or antipathy to explain how the specific form of antidotes neutralized, drew out, or broke down poisons. As poison also came to be seen as a disease agent, the same antidotes recommended for poison tended to be recommended for diseases like plague and syphilis.[34] This expansive role for antidotes, combined with the notion that poison worked similarly on nearly everyone, made the prospect of testing antidotes particularly appealing.

Contrived Trials and Clinical Encounters

Unlike most substances, poison gave testers the advantage of control: they could pick the time, place, and conditions to create the symptoms that they wanted to cure. Physicians and other healers usually assessed the efficacy of remedies by trying them on sick patients, and in most cases they did not expect drugs to work the same for all people. Poison trials, in contrast, were deliberately contrived events that allowed physicians to make a broader statement about an antidote's efficacy.[35] That possibility helps explain why Renaissance physicians were so eager to test poison on condemned criminals. Notions of drug testing dating back to Avicenna had held that tests were most reliable if done on humans rather than animals. In most cases, however, it was not acceptable to use humans to test dangerous substances.[36] The revival of trials on the condemned allowed physicians to get around this restriction.

Poison trials did not appear out of thin air. They fit neatly into a gradually increasing interest in the clinical encounter as a forum for new medical knowledge. In general, the status of information gained from observation and experience began to rise precipitously in the later fifteenth century, as scholars such as Gianna Pomata, Nancy Siraisi, Brian Ogilvie, and Pamela Smith have established definitively.[37] This new focus on experience came, on one hand, from artisans and other empirical practitioners, who had always relied on experiential knowledge and who gained a stronger voice with the new medium of print.[38] At the same time, it also became a strong interest among scholarly physicians, who had once used their book learning to distance themselves from empirics. Physicians increasingly began to incorporate experiential knowledge into their writings, especially in the fields of anatomy and botany, although they emphasized that it had to be informed by the proper medical theory.[39] The medical case study also took on new life in this context and became, as Pomata has shown, an important new "epistemic genre" in scholarly printed texts.[40] Specific patients' illnesses and outcomes attracted increasing attention among learned physicians.

Poison trials were part of this general trend. Indeed, Renaissance physicians did not necessarily distinguish between contrived trials and evidence gathered through clinical experience. The learned surgeon Claudius Richardus, for example, described four instances of successfully using bezoar in the 1560s: two poison trials on condemned criminals and two cases of sick patients. For both patients and prisoners, Richardus aimed to demonstrate how the bezoar purged poison from the body; in his account, all four cases presented a unified body of evidence. Richardus gave no indication that deliberately poisoning a criminal was inherently different from observing the progression of a poisonous disease in a sick patient.[41]

Even if sixteenth-century physicians did not neatly divide clinical experience and contrived trials, they did convey the sense of the poison trial as manufactured and deliberate. The most common verb they used to describe the trials was "to make"—to make a trial, to make an experience, to make an experiment. Ragland has noted the use of this same language in other kinds of contrived trials, such as anatomical experiments or tests of mineral waters.[42] The extensive, detailed reports on the poison trials that physicians produced, however, went beyond the usual genre of patient case studies. Poison trials remained too rare to become an "epistemic" genre in their own right; they appeared in a variety of texts such as herbals, drug treatises, and plague pamphlets, which is one reason why they have been overlooked. Contrived trials on humans represented exceptional cases, but these exceptions throw light on a number of nascent trends both in Renaissance medicine and in Renaissance culture more broadly.

Wonder Drugs and Panaceas

All of the poison antidotes discussed in this book also functioned as wonder drugs and near panaceas. The increasingly expansive concept of poison meant that nearly any poison antidote could also potentially cure plague, syphilis, epilepsy, internal bleeding, coughs, the common cold, and even headaches. This development occurred just at a time when a host of new substances were hailed as wonder drugs. With the Spanish colonization of the Americas and Portuguese expansion into portions of Southeast Asia, hundreds of potentially useful drugs poured into Europe and led to an unprecedented collaboration between physicians and merchants.[43] Some of these drugs were remedies for specific diseases or specific purposes, such as guaiacum (for syphilis) or mechoachan root (a purgative). Many others, however, were touted as near panaceas that could, among other things, cure poison. The most famous example is tobacco, which was described as a "panacea herb" in a Dutch publication from 1587 and counted antipoison properties among its many virtues.[44]

New cure-alls also emerged out of a separate tradition, the practice of alchemical medicine and distillation. A flourishing interest in the techniques of distillation (if not alchemical theory) led to a boom in new cures from the stills of noblewomen, pharmacists, monks, nuns, lower-class town distillers, and physicians.[45] Many of these drugs were viewed as near cure-alls, and like poison antidotes, they were assumed to work universally on all bodies. The writings of the famous Swiss alchemist Paracelsus, which became popular at the princely courts from the 1560s, helped solidify the interest in alchemical cures, the notion of universal applicability, and the role of poison in disease.[46] Although the Galenic physicians in this book either ignored or rejected Paracelsian medicine, almost all of them embraced distillation and its potential to produce cures for poison and disease. The blurred lines between poison antidotes and panaceas are central to understanding why antidotes were so highly sought. They may also help explain the decline of poison trials in the late sixteenth century, as miraculous tales of curing a variety of diseases became commonplace.

A Story in Eight Antidotes

This book traces poison trials and their wider medical meaning over the *longue durée*, from ancient Greece into the seventeenth century, with a focus on the sixteenth century. I take a pan-European approach, drawing on examples from around Europe and its colonies, but the bulk of evidence comes from Renaissance Italy and the Holy Roman Empire. I have broken the book into three sections. The first ("Authorities") focuses on precedents for the poison trials; the second ("Experiments") examines their apogee in the mid-sixteenth century; and the third ("Wonder Drugs") explores the connection between poison antidotes and cure-alls. Each chapter highlights an antidote that illustrates the chapter's themes.

Poison trials dated back to antiquity, when Greek, Persian, and Egyptian kings used condemned criminals in their efforts to develop effective antidotes. Ever since that time, poison was viewed as a particularly testable substance. Chapter 1, "Poison on Trial," uses the ancient antidote theriac to trace learned poison trials back to their origins in the Greek medical literature, particularly the writings of Galen, which described testing theriac through a poison trial on roosters. This precedent was picked up in the literature of Arabic physicians, who translated and augmented Greek medical works in the great intellectual powerhouse of the Islamic caliphates. From there poison trials found their way into Latin literature in the European Middle Ages, as doctors trained at the newly founded universities began to assimilate the writings of Galen and the Arabs. Poison as a substance fascinated medieval physicians, and they both explored it as an intellectual topic and wrote practical advice for avoiding it. This interest intensified when poison became associated with the Black Death, the deadly plague that swept through Europe from 1347 to 1351. Through a millennium of medical history, I argue, poison was thought to lend itself unusually well to testing. From Galen's time through the Middle Ages, however, humans were used only for nonfatal tests, and only animals were used to try poison.

Pope Clement VII changed these long-standing practices by reviving the practice of testing poison on condemned criminals. Chapter 2, "Condemned Bodies," revisits the trial of Caravita's oil to explore how Pope Clement and his physician Paolo Giovio (1483–1552) came to test a new poison antidote on condemned criminals in 1524. Most influentially, the study of human anatomy at universities grew by leaps and bounds from the late fifteenth century, and university dissections typically used the bodies of condemned criminals. The concept of the "condemned body" as an experimental subject had been well established by the 1520s. Even so, the shift from dissecting dead criminals to poisoning living ones was quite a leap, and the actors involved with the poison trials knew it. They thus made every effort to portray it as consistent with existing medical, cultural, and religious traditions and expectations. The poison trials were introduced not as brash innovation, I argue, but rather as a conventional event. Caravita's oil later morphed into the "Oleum Clementis," said to be Pope Clement's marvelous antidote and plague cure.

Also present at the trial of Caravita's oil was Pietro Andrea Mattioli, a young doctor from northern Italy who had come to Rome to study surgery with Caravita. Two decades later, Mattioli unexpectedly became famous in European medical circles, and his continued interest in poison trials led to their heyday in the 1560s–1580s. Chapter 3, "Experimenting with Drugs," examines physicians' efforts, led by Mattioli, to portray poison trials as learned medical experiments conducted in accord with venerated medical tradition. In so doing, they deliberately contrasted their efforts with the dramatic marketplace shows conducted by mountebanks and other empirical practitioners, who publicly tested their own antidotes using poison, often through self-poisoning. To avoid being conflated with charlatans, physicians at princely courts developed a model for a learned poison trial that involved both methods of testing and strategies of communication. They laid out the exact circumstances of the tests on criminals in excruciating detail, describing the symptoms of poisoning and the effects of the antidote in ways that conformed with accepted medical theory, and they recorded and shared the results. Yet they could never quite escape the overlap with empirics' practices. Mattioli himself paired his complaints about charlatans with a description of his own marvelous "scorpion oil."

In their trial reports, physicians dispassionately described the horrific effects of poison on the criminals who acted as test subjects. However, the use of condemned criminals for fatal tests was hardly the straightforward process that it seemed. Chapter 4, "To Cure a Thief," provides a case study of one criminal, a thief from southwest Germany named Wendel Tümler, to demonstrate the many challenges inherent in testing poison on humans. After an enticing new drug called Silesian terra sigillata became a much-discussed antidote and cure-all in 1580, Count Wolfgang II of Hohenlohe (1546–1610) wanted to test it on a human. When Tümler was apprehended for a petty theft, he appeared to have potential as a test subject. Yet the archival records show that Count Wolfgang's councilors were very concerned about the ramifications of using Tümler to test poison. They were worried, in particular, about what would happen if Tümler survived. To avoid social unrest, they recommended a course of action that would make it clear to all that he had agreed to be a test subject "out of his own free will, without any tricks or force." I argue that this emphasis on prisoner consent frequently appeared in accounts of poison trials, and many trial reports also emphasized that the criminal would go free if he survived. Such assurances about the treatment of test subjects did not signify medical ethics as we know them today. Instead, they reflected the interruption that poison trials threatened to the regular process of execution, an important cultural ritual in Renaissance Europe and one that was intended to guide the penitent criminal to a good, pious death. These difficulties in using human subjects for poison trials, I suggest, were one reason why they petered out toward the end of the sixteenth century, in favor of tests on animals.

Poison antidotes did not just cure poison. They were also seen as valuable wonder drugs, useful against a variety of illnesses. Chapter 5, "Powerful and Artful Substances," explores the status of antidotes as wonder drugs at the intersection of medicine and commerce. In the second half of the sixteenth century, a new genre of physicians' treatises touting the wondrous effects of drugs, many of them from Spanish and Portuguese colonies, began to make waves in print culture. All of the new wonder drug treatises included dramatic anecdotes of healing drawn from experience as proof of the medicaments' powers. German physician Johann Wittich (1537–96) proclaimed an explicit need for "powerful and artful substances that have been proven adequately in a trial."[47] The value of exotic wonder drugs like bezoar stone made them precious items in princely treasuries and cabinets of curiosity (*Kunstkammer*), and that value imbued human poison trials with a deeper meaning, one that reflected on the monarch himself. Only powerful princes had condemned criminals at their disposal, however, meaning that very few medical practitioners could actually use them as evidence. Healers situated outside of the princely courts thus tended to use anecdotes drawn from clinical experience in cases of illness, rather than contrived poison trials on criminals, to emphasize the efficacy of antidote cure-alls. At the same time, they frequently included brief poison trials using animals as proof that a marvelous drug was authentic, a sign of the deep suspicions about fraud that went hand in hand with exotic drugs. In claiming "powerful and artful substances" as their area of expertise, physicians indicated that wonder drugs could be acceptable to learned medicine, if legitimized by the proper authorities.

Yet by attaching themselves to marvelous cure-alls, physicians made it harder to distinguish themselves from empirics. Chapter 6, "A Universal Cure," examines two alchemical empirics closely connected to antidote-cure-alls. The first, Andreas Berthold (d. 1610), became known as the peddler of Silesian terra sigillata, and he collected testimonial letters from respected princes attesting to poison trials on animals and condemned criminals. This evidence helped cement the Silesian terra sigillata's reputation as an effective new antidote and plague cure. In the midst of Berthold's efforts, a lawyer named Georg am Wald (d. 1615/16) advertised a poison and plague remedy called Terra Sigillata Amwaldina, which he touted as a "universal medicine" that was also good for dozens of diseases. A decade later, he claimed to have created a true cure-all called the Panacea Amwaldina. Am Wald referred derisively to the practice of using poison on "dumb animals." Instead, he provided reams of testimonial letters from patients praising the effects of his antidotes. Learned physicians eventually rejected am Wald's cures and his use of evidence, but their initial response was one of considerable interest. The cases of Berthold and am Wald highlight the difficulties physicians faced in figuring out what kinds of testimony counted as valid proof.

The book's conclusion, "Testing and Testimony," demonstrates that questions of testimony and proof did not disappear in the seventeenth century. Poison trials continued to be regarded as valid evidence in the process of obtaining special drug licenses, a common procedure run by both city and imperial authorities through the nineteenth century. Many of these licensing procedures involved the drug orviétan, a derivative of theriac. The problem of how to treat charlatans and their marketplace tests continued, and physicians continued to accept poison trials as valid evidence in drug licensing processes. Meanwhile, seventeenth-century physicians and other experimenters intensified tests on animals, using many of the descriptive reporting techniques developed in the sixteenth century. As the nascent field of toxicology shifted the focus from antidotes to poison itself, poison trials remained an important form of evidence.

A Note on Gender and Poison

Any book on this topic has to take into account the complex gender dynamics of poison in medieval and Renaissance Europe. Poisoning was frequently depicted as a female crime, and there was a long tradition of casting women as poisoners in Europe, lasting from Roman antiquity through the modern period.[48] In his *Discoverie of Witchcraft* (1584), Reginald Scot cited various Greek, Roman, and early Christian authorities in claiming that "women were the first inventers and practisers of the art of poisoning," while Italian physician Giovanni Battista Codronchi stated in 1595 that for one male poisoner (*veneficus*), one finds fifty females (*veneficae*).[49] Women, poison, and witchcraft often were seen as going together, as shown especially clearly in a series of plague-spreading conspiracy theories in Geneva in the sixteenth and seventeenth centuries.[50] As Garthine Walker notes, poisoning was typically a domestic crime perpetrated on those closely connected with the alleged criminal, and it belonged to the female household domain. The ambiguous boundaries between medicine, food, and poison, and the prevalence of arsenic-based substances as household pesticides, made poisoning an obvious murder weapon for women.[51] Those connections also left women open to accusations following sudden deaths of people in their households.[52]

Women were not alone in being tarnished with the crime of poisoning. Jews, Muslims, lepers, and foreigners were also portrayed as typical poisoners, leading Franck Collard to term poisoning "a crime of the other."[53] Moreover, the trope of women as poisoners did not mean that they actually *were* poisoners. Walker's careful study of poisoning cases in the city of Cheshire suggests that although poisoning was by far the most common female homicidal crime, absolute numbers of poisoning cases show little gender difference. (Men simply committed more homicides, especially more violent homicides.)[54] Accusations of poison magic in the heyday of the Bavarian witch trials tended to fall especially on tavern keepers, butchers, bakers, and apothecaries—that is, professionals with access to food, drink, and medicine.[55] Despite the continual fear of poison, poisoning appears to have been a fairly infrequent crime in comparison with others. Randall Martin notes the relative rarity of poisoning cases in his study of the sensationalist English pamphlet literature on murders, although those that did appear were described with special drama.[56] While both the incidence of poisoning cases and the representation of women as poisoners may have been overstated, the *idea* of women as quintessential poisoners was widespread across early modern Europe.

In contrast, there was a public perception of poison antidotes as part of a male domain. While there were hundreds of different kinds of poison antidotes in early modern Europe, the most sought-after tended to be exotic materials or ancient remedies. Among the most popular antidotes were animal objects such as unicorn (narwhal) horn and bezoar stone; precious gems, especially emeralds; clays such as terra sigillata and Armenian bole; and the electuaries theriac and mithridatium. The first three categories were all exotic objects and prized commodities.[57] The last were complex compounds that in theory needed to be purchased from an apothecary's shop. All of them involved the intervention of vendors—merchants, traders, drug sellers, pharmacists—who would have been almost exclusively male.[58] Sixteenth-century Europe also saw a rise in proprietary poison antidotes, that is, antidotes attached to a specific person's name. Nearly all of the antidotes that circulated widely and appeared in printed works were tied to men, including empirics, charlatans, and a few physicians.[59] Numerous prominent princes became known for their antidotes, which they used in the context of early modern courtly patronage and gift exchange. The tradition of prince-as-antidote-creator reached all the way back to antiquity and the famous example of Mithridates, the "poison king," and it was revived with fervor in the sixteenth century.[60] Dozens of German princes had "poison powders" (*Giftpulver*) that circulated under their names, the most prominent from Elector August of Saxony (1526–86), Archduke Ferdinand II of Tyrol (1529–95), and Emperor Maximilian II (1527–76). A version of Caravita's oil, the Oleum Clementis, was tied to Pope Clement VII, and the Medici grand dukes of Tuscany developed their own antidote oil.[61] In public depictions, the healing of poison appears to have been gendered male, and only male condemned criminals were used to test antidotes.[62] Throughout this book, the main actors will be prisoners, princes, physicians, and empirics, all of them men.

As is always the case, however, women had a strong role in the background, and they appear throughout this book. In a letter to the Marchesa Margherita of Mantua (1510–66), Cardinal Benedetto Accolti (d. 1549) described an antidote test he had conducted on a prisoner (chap. 2).[63] The Spanish physician Nicolás Monardes (1493–1588), who wrote a glowing account of bezoar stone, credited the duchess of Bejár with alerting him to its potential efficacy (chap. 5). Women were frequently tied to poison antidotes in handwritten collections of medicinal recipes, and their documents led me to this topic. A German recipe collection compiled by Countess Palatinate Elisabeth of Saxony (1552–90) contained an account of dog trials conducted by Wilhelm IV of Hesse-Kassel (1532–92, chap. 3). Elisabeth also compiled a manuscript dedicated to poison antidotes and plague cures, with contributions from both women and men.[64] Another German noblewoman, Countess Anna of Hohenlohe (1520–94), acquired a faddish new antidote and suggested the possibility of a poison trial to her son, Count Wolfgang II (chap. 4). While Anna appeared as a driving force in the archival records, she was written out of the printed version.

Overt gender differentiations provide an important backdrop to this book. The events we will encounter played out in a male world of professional and princely disputes, in physicians' writings and print culture and prison cells, none of which were domains friendly to women. Important scholarship from Steven Shapin, Simon Schaffer, Mario Biagioli, and others has described the important role of gentlemanly culture in constructing the image of seventeenth-century experimental natural philosophy.[65] *The Poison Trials* demonstrates the direct role that (mostly male) battles over status and authority played in developing a sixteenth-century variety of medical experiment.

Poison in a Pandemic

When I began this project, I had no idea how relevant its key questions would seem in the spring of 2020. As this book goes to press in the midst of the COVID-19 pandemic, the fear of hidden disease agents has become a part of everyday life. So too has the search for a cure. We no longer link a virus like SARS-CoV-2 directly to poison (although "virus" originally denoted a toxin), but the anxiety provoked by coronavirus's stealthy transmission calls to mind earlier views of poison and poisonous diseases. While we now hope for an effective vaccine rather than a cure for all disease, the drive to find wonder drugs and the search for an elusive panacea suddenly feels very familiar. Tied up in that pursuit are debates about how to test promising new therapies, including the question of whether some established ethical guidelines should be modified owing to the severity of the crisis. Similarly, the delicate balance between political power, medical authority, and popular anecdote confronts us daily. By the time you read this book, the coronavirus crisis may have diminished (one can hope!), but like any public health catastrophe, it has revealed the instability of approaches to drug testing, ethics, and authority even in twenty-first-century medicine.

We live in a very different medical world than the sixteenth-century individuals at the heart of this book, and comparisons between then and now go only so far. Nevertheless, the familiar elements can help us imagine how much more bewildering the situation was before biomedicine. At a time when disease had no obvious cures, when physicians shared the medical landscape with a panoply of other healers, and when rules for medical experiment had not yet been established, the lines between legitimate and illegitimate medicine were virtually indistinguishable. Poison antidotes and panaceas represented only one strand of the many different therapies available in sixteenth-century Europe, but they were an especially important catalyst for discussions around testing, proof, and evidence and for disputes over professional boundaries. This book demonstrates how poison and its cures became a focal point for questions and anxieties about wonder drugs, experiment, and medical authority.

PART ONE

Authorities

ONE

Poison on Trial: Theriac

The tale is legendary. In order to develop immunity to poisoning, King Mithridates VI of Pontus tested a wide variety of antidotes. For many of these tests, he used criminals condemned to death—an ancient precursor to Pope Clement VII. He also tested both poisons and antidotes on himself. Contemporaries whispered that he took a daily dose of poison along with various antidotes to accustom his body to poison and protect it from harm. This careful conditioning worked a little too well in the end: when he wished to end his life after being vanquished by Roman forces, he found himself unable to poison himself and had to beg his faithful servant to stab him with his sword instead.[1] Or so the legend goes.

As with most aspects of Mithridates' life, this dramatic account is highly unreliable. It was transmitted and colored by later Roman authors, who had a vested interest in making the "poison king" appear especially clever (and thus the Romans' conquest of him even cleverer). Whatever the truth of the tale, Mithridates almost certainly tested antidotes on condemned criminals, a practice that appears to have been widespread in the early Hellenistic world. One of Mithridates' predecessors, King Attalus III of Pergamum (r. 138–133 BCE), was rumored to have used condemned criminals for the same purpose, and the Greek poet Nicander dedicated two poems on poison antidotes to him. Many other Persian, Egyptian, and Macedonian rulers behaved similarly.[2] In one particular instance, a physician named Zophyros of Alexandria sent Mithridates an antidote to test, along with a condemned man on whom to test it. The casual nature with which the doctor referred to the transfer of the criminal suggests that it was not unusual.[3] Through this combination of self-testing and testing on criminals, Mithridates developed a potent drug that would bear his name: mithridatium, a sweet, honey-based mixture called an electuary, which contained as many as fifty-four ingredients, including numerous herbs and spices, some animal products, and opium. A poison antidote and cure-all, mithridatium would remain a popular drug through the eighteenth century.

Three hundred years after these ancient poison trials, the Greco-Roman physician Galen praised the wonderful effects of theriac, a derivative of mithridatium thought to be even more effective against poison and many ailments. By that point, using humans as test subjects was considered taboo, at least according to an influential text attributed to Galen. As a substitute, the author described a test on wild roosters as evidence of theriac's marvelous effects against animal venom.[4] This poison trial became an iconic example that traveled first into Arabic texts and then medieval Latin medical literature. Through a millennium of medical writing, there was an unbroken—if limited—intellectual tradition that described testing theriac using contrived trials on poisoned animals.

Alongside this textual tradition, practical tests on poison took place in both the Islamic Empire and the medieval west. These tests focused not on antidotes but on methods to uncover the presence of poison. Princes and other aristocrats had always been preoccupied with the fear of being poisoned, a crucial impetus behind the testing on criminals in Mithridates' era. Later caliphs, kings, and popes presided over elaborate ceremonies to unmask the presence of poison in food. Some Persian rulers used peacocks; many medieval Arab and European aristocrats preferred to rely on precious gemstones and other exotica that were said to alter their appearance in the presence of poison. The medieval criminal courts, as well, occasionally tested substances suspected to be poison. The assumption behind all of the varied forms of assaying was simple: poison lent itself particularly well to testing.

In the later Middle Ages, the overall medical focus on poison intensified. Poison became an explanation for far more than just poisoning. As the Black Death swept through Europe, leaving incomprehensible waves of death in its wake, it dumbfounded physicians beholden to the individualistic humoral model of disease. Some sort of poisonous origin seemed a plausible cause for the widespread mortality. Poison antidotes thus became important plague cures, adding a new level of significance. To understand the Renaissance trend of testing poison antidotes on condemned criminals, one has to understand this backstory. The ancient tradition of poison trials and the ongoing fear of poisoning merged with a developing idea that poison was the source of deadly epidemic disease. Throughout these twists and turns, there was a major consensus that poison was a substance one could easily test. This chapter argues that poison's special status as uniquely testable and uniquely deadly made it an obvious focus for experimentation over the longue durée.

But what was poison? What made it so particularly testable? Why was it seen as a cause for epidemic disease? To answer these questions, we have to step back to the time of the Poison King.

Poison in the Ancient World

The notion of poison in the ancient world was not a straightforward concept by any means. The Greek word *pharmakon* could indicate either a drug or a poison; it had a neutral meaning that referred to a spectrum of substances and did not imply either healing or harm. One substance, however, was understood to be generally harmful. Animal venom, always a concern in regions infested by poisonous serpents and spiders, was viewed as uniquely injurious. Even so, the focus tended to be put on the effects of venom, or poisoning, rather than on the venom itself. As Frederick Gibbs has argued, poisoning as a concept existed, but poison as a substance did not.[5] Following from this idea, there were two main classes of substances that we might call antidotes: *alexipharmaca*, which had general prophylactic properties thought to prevent or cure poisoning, and *theriaca*, which were seen as specifically useful against animal venom injected into the body.

These two categories were fairly clearly defined. Nicander's two poems to King Attalus III were titled "Theriaca" (on cures for animal venom) and "Alexipharmaca" (on cures for oral poisons). Nevertheless, some special *theriaca*, such as mithridatium, were seen to have alexipharmic properties as well. Because poisoning (both malicious and accidental) was seen as a major threat, physicians and rulers throughout the ancient world had an interest in seeking new and more effective *alexipharmaca* and *theriaca*. This drive to find specific ways to combat poisoning appears to have encouraged a spate of testing. The reports of poison trials under Mithridates, Attalus, and their peers, along with the great interest in antidotes among physicians, hint at a much larger role for poison trials that has been lost to historical records.[6]

Long after Mithridates' death, his celebrated antidote engendered an even more famous derivative. The physician Andromachus, who served the Roman emperor Nero (r. 54–68 CE), created a new drug by adding additional ingredients to mithridatium, most importantly the flesh of venomous snakes. This concoction became known as theriac, placing it among the *theriaca,* although it had alexipharmic properties. The term "theriac" remained confusing for centuries, as it could refer to the category of drugs (*theriaca*) useful against animal venom or to a substance that was both an antidote and a cure-all. There were countless different recipes for theriac, and even simple plants like garlic could be referred to as a theriac, but the complex theriac of Andromachus was hailed as the best.[7] Its reputation took many years to cement, however, and it owed its fame largely to the efforts of Galen. These efforts engendered an influential pair of poison trials.

The Galenic Poison Trials

Galen was such an enthusiastic proponent of Andromachus's theriac that his writings are full of references to it. He described the emperor Marcus Aurelius's daily use of the drug, and his canon includes two separate treatises in support of it, although it is possible that he wrote neither of them. One of the treatises, *On the Use of Theriac, to Pamphilianus,* has long been seen as spurious. The authorship of the other treatise, *On Theriac to Piso,* has been a hotly debated subject since the sixteenth century. Until a few years ago it was generally thought to be an authentic work of Galen, but a meticulous recent study by Robert Leigh suggests that a follower of Galen penned the work, rather than the master himself.[8] Be that as it may, scholars in ancient Rome, the Arabic world, the Latin Middle Ages, and the European Renaissance all assumed Galen to be the author of both treatises, and the texts helped cement the theriac of Andromachus as a crucial antidote and cure-all, largely based on Galen's authority.

The second chapter of *Theriac to Piso* was intended to establish the "infallibility" of theriac and "the power of its effect." To do so, it presented two poison trials as evidence. These trials were not on condemned criminals, however. By Galen's time, conventions had changed. The author noted he could test the effects of poison only on animals, and he explicitly cited a shift from earlier days. In former times, he wrote, rulers had used men "already condemned to death" to test antidotes, an obvious reference to Mithridates, Attalus, and others. However, physicians in second-century Rome were "unable to test it on men," so he had to use animals instead. He chose wild roosters, which he noted had a drier complexion than domestic roosters (and were thus, presumably, closer to humans by nature).[9] As we will see throughout this chapter, birds often had the unhappy role of test subjects, likely because of their wide availability and expendability. In this case, the author described how an unspecified number of roosters were divided into two groups, one of which was given theriac to drink and one of which was not. He then let "poisonous beasts" bite all the roosters, after which "those who have not drunk the theriac die on the spot, while those who have drunk it survive."[10] As in the much later trial conducted under Pope Clement VII, the test subjects (in this case roosters rather than humans) were divided into two groups, one of which received no antidote to show that the poison worked.[11] The message his trial conveyed was simple: theriac was a powerful antidote to deadly poison.

The taboo against physicians testing fatal drugs on humans has interesting parallels to the well-known prohibition against using human cadavers for anatomical study in ancient Greece and Rome. That taboo may have met with an exception in the third century BCE, when the Alexandrian physicians Herophilus and Erasistratus allegedly obtained special royal permission to perform both dissections and vivisections on condemned criminals. There is no solid evidence of these tests, however—if such practices occurred, they were almost certainly restricted and brief.[12] By Galen's time, human dissection was not allowed. The solution offered in *Theriac to Piso*—animals—was the same one Galen used for his anatomies. Animals represented an imperfect alternative, as Galen himself realized, but it was the best he could do.[13]

Despite the ban on using human subjects for dangerous tests, *Theriac to Piso* made a clear case for why it was necessary to test theriac. The author noted that it was difficult to find an account of someone who just happened to take theriac immediately after being bitten by a poisonous animal, "nor of anyone who drinks it in advance and then not long afterwards is bitten and gets a strong enough dose to kill." In other words, it was nearly impossible to determine the efficacy of poison antidotes through the normal course of clinical experience, because the chances of happening upon a case of poisoning, theriac in hand, were simply too low. Poison trials on roosters offered a way out of this problem. A second proposed trial in *Theriac to Piso* did suggest using humans for a much less dangerous test. To probe the authenticity of a given batch of theriac, the author suggested giving a person a purgative drug along with the theriac. If the person did not feel the effects of the purgative, one could assume the theriac was good, "since it prevented purgation."[14] In this case, a nonfatal purgative stood in for poison, with the advantage that it could be used on a human. Notably, the author presented these tests as definitive: when the quality of the theriac was in doubt, a trial could provide a decisive result.

At the same time, *Theriac to Piso* emphasized that theriac had been created not by blindly observing these sorts of tests, but through physicians' careful reasoning.[15] This assertion was a blatant and explicit dig against the rival Empiricist healing cult, who Galen (unfairly) depicted as clueless followers of misguided experience and superstition.[16] In contrast, Andromachus and his predecessors had used their senses and their great knowledge to assess the qualities of each ingredient and figured out logically which ingredients would mix together in the perfect manner. The poison trials, in this telling, had no real role in *developing* theriac as a drug, but instead merely reinforced the conclusions that physicians had arrived at through reasoning. Unlike Mithridates, who had used multiple tests to help create his antidotes, Andromachus had used his intellect. Nevertheless, the author of *Theriac to Piso* placed the poison trials in the second chapter of his text, immediately following his strong claims about the "infallibility" and "power" of theriac, and he clearly intended them as evidence of the drug's marvelous properties. He followed these trials with two cases in which theriac had cured grave illnesses, including one in which it helped Piso's own son.[17] Even if he claimed reason as the main guide for creating theriac, poison trials provided a road map to its success.

The poison trials in *Theriac to Piso* set several important precedents that had the force of Galen's perceived authority behind them. First, the text presented deliberate, contrived trials on both humans and animals, with only the animal trials involving deadly poison. The rooster trial broke the animals into two groups—one that received the antidote, and one that did not. The human trial involved only one subject, and it was proposed exclusively to test the authenticity or shelf life of theriac, using a purgative that would not cause grievous harm. Most important, however, *Theriac to Piso* used poison trials to underscore the wondrous effects of theriac as an antidote and cure-all. Along with Galen's other writings on theriac, it helped transform Andromachus's concoction into the most central drug in the physician's repertoire (and the apothecary's shop). This interest in testing theriac would long outlast the Roman Empire itself.

Poison Trials in Islamic Medicine

After the fall of Rome in the fifth century CE, the center of Greek philosophy moved slowly eastward, eventually finding a home in the vibrant intellectual culture of the Islamic Empire. As Greek medical knowledge was translated into Arabic and read with great interest, the Galenic poison trials found their way into the writings of several Arabic physicians.[18] Most prominent and influential among these authors was the philosopher-physician Avicenna (Ibn Sina, 980–1037), whose five-book *Canon of Medicine* became the standard teaching text in medieval universities. The *Canon* is most famous for its succinct and clear summary of Galenic medical theory in book 1. Less well known are two practically oriented portions that pertained to drugs, book 2 on simples and book 5 on compound drugs (mixtures).[19] Unsurprisingly, the most prominent compound drug was theriac, the subject of the very first chapter of book 5.

Avicenna drew explicitly on *Theriac to Piso* in this chapter. He cited the text by name, listed Galen as the author, and included descriptions of the two Galenic poison trials. He led into these trials by warning the reader that theriac lost its potency after thirty to sixty years. It was thus important to test a batch of the drug to make sure it was still fresh and strong. One of the "usual" trials for theriac, he wrote, involved giving a man a purgative and then, once he began vomiting, giving him the theriac. If the man's body stopped purging, "the theriac is good and fresh; if the case is the converse, it is bad." He also noted that Galen had described another trial on a wild rooster, "which is to say a male pheasant," who should be exposed to "creeping poisonous animals" until he was bitten. The pheasant should be given theriac directly thereafter: "if he lives, the theriac is good."[20] Although he cited Galen directly, this version was slightly different from the account in *Theriac to Piso*. Avicenna suggested using only one animal rather than two groups, and the theriac was to be administered afterward rather than ahead of the animal venom.[21] Nevertheless, the influence of *Theriac to Piso* for both poison trials was obvious, and Avicenna presented the two tests as Galen's. Notably, both of these trials served the same purpose—to test whether a batch of theriac was still good.

Although Galen and the Greeks represented one influence on Arabic writings on poison, many authors mingled Greek ideas with other scholarly traditions, particularly Indian sources. In the early ninth century, the caliph al-Ma'mun (r. 813–33) invited both Greek and Indian intellectuals to his court in Baghdad, and Sanskrit texts were available to early Arabic writers even before that date.[22] The Indian tradition may have been the basis for the many Arabic treatises devoted exclusively to poison and its cures. Persian natural philosopher Geber (Jabir ibn Hayyan, c. 721–815) became one of the first Arabic authors to write a treatise exclusively devoted to poison and its antidotes, likely based on both Greek and Indian sources.[23] Although his treatise contained virtually no references to practical experience, it influenced the extensive *Book on Poisons* by ninth-century Persian physician Ibn Wahshiyya, which described the looming danger of poison and gave recommendations for antidotes. While these Arabic poison treatises did not suggest testing the antidotes, they did address another aspect of early toxicological testing: the attempt to detect the presence of poison in food or drink. Ibn Wahshiyya, for example, noted that the caliphs kept live peacocks at their tables not because of their beauty, but because the birds were particularly sensitive to the smell of poison and would alert their masters to any hazardous content.[24]

Later Islamic physicians appear to have conducted their own, unspecified tests of poison antidotes—and not just theriac. The Sevillian physician Avenzoar (Ibn Zuhr, 1094–1162) was a particularly influential authority. His work on food and drink, the *Kitab al-aghdhiya,* contained a list of special remedies (*khawass*), which included several remedies for poison such as emerald, terra sigillata, balsam, and bezoar. His masterpiece, *Taysir,* praised the marvelous virtues of theriac in its opening pages, and it was one of the first texts to give a firsthand account of the wondrous qualities of bezoar stone (the antidote at the center of chapter 5).[25] In addition to this evident interest in poison antidotes, he appears to have gained a reputation among his peers for testing drugs, including antidotes. His younger contemporary, the Jewish physician and philosopher Moses Maimonides (c. 1138–1204), proclaimed emerald the best remedy for poison, on the evidence that it had been "verified" by Avenzoar, "the greatest among men in testing drugs and one who devoted himself to this more than any other."[26] He did not explain the exact nature of these tests, however.

Maimonides appears to have followed in Avenzoar's footsteps. Although he is best known for his philosophical works, Maimonides wrote one of the earliest freestanding texts on poison antidotes, his *Treatise on Poisons and Their Antidotes*. In addition to his praise of Avenzoar's drug tests, he included information from his own personal experience. In one case, he noted that two types of bezoar—mineral and animal—were widely seen as useful against poison, but he claimed that only the animal variety had held up under testing:

> I have tried all the types of this mineral stone that can be found in our land against scorpion bites, and they were not beneficial at all. I gave it to many [bite victims], but to no avail. But the animal bezoar has been proven by experience and confirmed empirically.[27]

Maimonides seems to have tested antidotes by giving them to patients who had been (accidentally) bitten by poisonous animals. If you wanted to cure venomous bites, he claimed, you should grind up the bezoar and either give it to the bite victim orally or turn it into a plaster to put on the wound.[28] These directives underscore the very real fear of poisonous animal bites, an ever-present danger in the medieval Mediterranean and a crucial motivation for testing antidotes. Despite his clear interest in testing, however, Maimonides did not describe contrived trials of the sort mentioned in *Theriac to Piso*. His tests took place in the normal course of his clinical experience.

Aside from the brief references to Galen in Avicenna and other Arabic scholars, there is little evidence of poison trials on either humans or animals in Arabic sources, even in practical texts like the *Taysir*. The ninth-century physician Rhazes (al-Razi, c. 854–925) included a chapter on poison antidotes in his highly influential treatise *Almansor*, but he made no mention of testing the various cures he listed.[29] Avenzoar himself included no suggestion of any testing, even though he gave first-person accounts of cures in many other places. His contemporary Averroes (Ibn Rushd, 1126–1198) wrote a highly influential, scholarly treatise on theriac that did not include any reference to the Galenic-Avicennan poison trials.[30] Many Arabic authors had an avid interest in poison and how to cure it, but deliberate tests on antidotes were but a whisper in the background.

Antidotes in Medieval Europe

The Arabic authors proved immensely influential in the European Middle Ages, as Arabic and Greek sources filtered into Western Europe during several waves of translation from the eleventh through the fourteenth century. Awed by the wealth of medical knowledge that came before, medieval scholars put heavy weight on the voice of authority. They also continued—indeed, amplified—the Arabic interest in poison. The late thirteenth and early fourteenth centuries saw a significant boom in writings on poison, especially in southern Europe. Sometime around 1290, Juan Gil of Zamora (c. 1241–1318), a Castilian physician and Franciscan friar, composed a treatise on various poisons and their remedies. His little book presented a thorough summary of existing sources, especially information drawn from Avicenna's *Canon*. However, Gil did not cite any information from book 5 of the *Canon*, which included Avicenna's discussion of theriac and the Galenic poison trials.[31] *Theriac to Piso* itself did not become available to the Latin West until the fifteenth century.[32]

Theriac did become a major topic of discussion for other medieval authors. Around the turn of the fourteenth century, at least two professors in the French university town of Montpellier wrote treatises on theriac and poison.[33] The most prolific of these scholars was the erudite Spanish physician Arnald of Villanova (1240–1311), who wrote an influential treatise called *Letter on the Dosage of Theriac* sometime around 1295. This work was notable for developing a consistent theory of poison and for including many antidotes, not just theriac, in the discussion.[34] It also gave a vague nod toward the necessity of testing. Arnald claimed that theriac, more than most medicines, required experience to determine the dosage. For the most part, he skirted around the question of how exactly one should figure out that dose. The *Letter* focused mainly on theory rather than practice. Yet he did include a few specific cases of empirical testing in his later masterpiece *Mirror of Medicine* (1308), a comprehensive summary of Galenic medical theory. In one instance, Arnald cited a "recent case" of testing theriac—not on any living being, but on cheese:

> [When] fresh cheese . . . is poisoned and turns black . . . theriac applied to one end of the cheese sensibly repels the toxin, so that whiteness little by little is seen to creep back from the theriac-end to the other end; and thereafter the cheese stays edible, even though the toxin previously rendered it dangerous.[35]

Importantly, Arnald used this case as an example of materials that attract or repel other substances by their very natures, not as a statement on antidotes or theriac per se. He followed this instance with an example of a magnet's properties of attraction and repulsion. Nevertheless, his toxic cheese underscored the status of poison—and its antidotes—as substances that lent themselves to testing.

Two other treatises on poison connected to Arnald placed more emphasis on testing. A brief discourse *On the Art of Recognizing Poisons* gave advice on how to avoid, detect, and cure poison. Arnald appears to have composed it near the end of his life, possibly in the context of transfer to the papal court of Clement V in Avignon.[36] It acted as a very cursory overview, and Arnald included few references to empirical verification of any of the advice he presented, but he did recommend methods to determine the presence of poison.[37] A more lengthy treatise that circulated under Arnald's name (likely written by a student), *On Poison*, included far more references to practice.[38] It also introduced a wider variety of antidotes—theriac, of course, but also bezoar stone, terra sigillata, emeralds, and musk. In several places, the author made reference to empirical testing, including a rather distorted account of the Galenic poison trials. According to Galen, the author explained, Andromachus's theriac was tested "on the condemned" in order to determine the proper recipe. These trials led to the crucial addition of viper's flesh.[39] Pseudo-Arnald thus tied criminal testing explicitly to the development of Andromachus's theriac. This portrayal likely would have horrified the author of *Theriac to Piso,* who had insisted so vehemently that theriac had been developed logically by reason, not by experiment, and that all dangerous testing used animals.[40]

Arnald's colleague and rival Bernard de Gordon (fl. 1270–1330) echoed the idea that antidotes should be vetted empirically. Indeed, he proposed that *all* new drugs should be tested, particularly theriac. In his short treatise *On Theriac,* Bernard noted "as an aside" that the properties of theriac could be discovered only "by experiment," as Avicenna and Galen had already established.[41] He thus gave ancient precedent (indeed necessity!) for testing the antidote. Bernard emphasized the importance of testing theriac before using it on humans, in order to make sure it worked properly, and he gave two recommendations for how to do so. First, one could take a laxative, and if the theriac stopped the ensuing flux, it worked. Second, one could take two pheasants (which, he noted, had the driest complexion of all birds) and cut off their crests, apply poison to the wounds, and wait until the birds began to stagger. Once the poison had taken effect, one should put theriac on the wound and in the water of one pheasant, but not the other. Bernard assured his readers that "if this one lives and the other dies, the theriac is good."[42] Michael McVaugh has cited the pheasant test as an example of "experience-based medicine" in the Middle Ages, although it is unclear whether Bernard ever carried out this trial or whether he merely recommended it as a possibility.[43]

Bernard's pheasant test reads like a hybrid of Galen's and Avicenna's poison trials. It used what we might now call a "control animal," as in Galen's case, but it tested the pheasants' reactions to the antidote given after the poison, as in Avicenna's. It also instructed the reader to wait until the animals had begun to stagger—a new directive, one that waited for signs that the poison had taken effect. It is likely that Bernard took this trial from the *Canon* and modified it, as he cites Avicenna frequently in his text (and *Theriac to Piso* was not available to him). Yet the fact that he suggested using two groups of birds raises questions. Had he seen a similar recommendation in another text? Or was Bernard's trial based on actual practices he had either seen or heard of? It is difficult to know. As in Avicenna's *Canon*, the poison trial played only a small role in Bernard's overall information on theriac, as a means to verify a given batch of the drug.[44] Yet unlike Arnald or Gil, Bernard emphasized the importance of continuing to test theriac to make sure a specific batch worked, not just to determine the proper dose.

Around the same time as Arnald's and Bernard's scholarship in Montpellier, the Italian physician Pietro d'Abano (1257–1316) developed new theories of poison in Padua and Paris. In his influential treatise *On Poisons and Their Remedies*, Pietro combined a practical focus on avoiding poison and administering antidotes with a sophisticated discourse on the nature of poison itself. He argued that poison sometimes worked through the Galenic theory of humors and complexions, but that it could also operate through its total substance (or specific form). The latter explanation posited that some occult (hidden) quality in poison attacked the body in its entirety, rather than provoking a humoral imbalance. The idea that poison's mode of operation differed from that of most substances had been vaguely present in Galen and expanded in Avicenna and Arnald, but Pietro put a greater emphasis on it.[45] Specific form helped explain why poison could be so harmful to everyone, even in small quantities. In his extensive study of poison theories in medieval and early modern Europe, historian Frederick Gibbs has pointed to specific form as the key to understanding how medieval and early modern people viewed poison.[46] We will keep returning to the idea throughout this book.

This idea that poison could harm universally meant that everyone could follow similar steps to prevent poisoning. Pietro suggested two approaches. The first was "defense with caution"—that is, keeping things on hand that would warn you of poison. For example, the prassius stone, related to emerald, would lose its luster in the presence of poison. Observing this change was a proven method that had preserved many kings from poison, he claimed.[47] Serpent's tongue (shark teeth), similarly, would sweat in the presence of certain poisons, including napellus, the tyrus snake, and leopard's gall. These substances acted as a sort of test. Like ibn Washiyya's peacocks, they probed whether poison was present. Pietro saw them as working through hidden, occult forces, a sign of his general interest in astrological powers embedded in minerals and other substances and consistent with his reliance on specific form to explain poison.[48]

Pietro's second recommendation involved actively driving out poisons. One method to do so was to ingest food or drink that expelled poison as part of a regular diet. Chestnuts taken with dried figs, or rue with hazelnuts, or pulverized diptame root would all prevent poison, as would theriac or terra sigillata taken daily in wine. Other medicines, such as bezoar stone, would destroy the very substance of the poison, an indication that the total substance of the poison could be combatted with the total substance of an antidote. Pietro also included some vague markers of his own testing. The emerald, for example, would expel all poison if pulverized and put in a drink, a remedy that "I tested many times."[49] He gave no specifics on how exactly he had tested the jewel, however.

All of these medieval authors writing on poison had several ideas in common. They all pointed to poison as a dangerous substance that was especially concerning to powerful individuals (who were in grave danger of being poisoned). They all recommended specific antidotes, especially theriac. Much in the spirit of Maimonides, pseudo-Arnald and Pietro gave long lists of animal, vegetable, and mineral cures for poison. Most important, they all pointed to both poison and antidotes as substances that were inherently testable, even if they left the specifics of actual tests vague.

Testing Drugs

Poison was not the only substance that inspired interest in testing. The same Montpellier physicians who wrote on theriac also devised rules for drug testing in general. On this topic, as in so much else, they followed Avicenna, who had included seven rules for testing drugs "by experiment" (*per experimentum*) in the second book of his *Canon*:

1. All "accidental qualities" should be removed from the drug.
2. The drug should be tested on a simple disease, not a compound one.
3. The drug should be tested on two diseases with opposite complexions.
4. The strength of the drug should match the strength of the disease.
5. The length of time it takes the drug to operate should be noted.
6. Effects had to be observed in many cases, not just one.
7. The drug should be tested on a man, not a horse or a lion.[50]

All of these rules had one common goal: to produce meaningful results that would not be tainted by confusing or irrelevant information. For example, it would be far easier to see results on a simple disease (one that had the properties hot or cold) than a compound disease (one that had more than one quality). Similarly, testing a drug on an animal would make it difficult to say with certainty whether it worked on a person. With these rules, Avicenna was not advocating early randomized clinical trials, nor did he recommend experiments on patients as the only means of drug testing. The words for "experience" and "experiment" were interchangeable at the time, and Avicenna likely meant that physicians should follow his guidelines when using drugs on patients in the regular course of their clinical practice. He followed his rules for testing by "experiment" with a much longer list of rules for testing drugs by "reason," which meant using the senses, especially taste and smell, to analyze the drug. The goal of both kinds of testing was not to figure out whether a given drug "worked," but rather to place its properties on the spectrum of hot, cold, wet, and dry.[51]

Both Bernard de Gordon and Arnald of Villanova adapted Avicenna's drug-testing rules, the topic of a recent study by McVaugh.[52] Bernard argued that drugs ideally should be tried on patients with neutral complexions, or, if such a patient was impossible to find, on a patient with the opposite complexion as the drug. He also implied that a drug needed to be tested multiple times and suggested a careful hierarchy of test subjects when introducing a new drug. One should first test a new substance on birds, then on "dumb animals," then on hospital patients, then on mendicant friars, "and so on in turn, because it could kill if it turned out to be one of the poisons."[53] Fear of poison loomed in the background of this hierarchy, which also gives a sense that the lives of some patients were valued more than others.

Other physicians dispensed with the idea of using sick patients altogether. McVaugh has shown that Arnald of Villanova proposed a particularly coherent system of testing drugs using healthy and temperate human subjects with average bodies (a man who was not too tall, short, fat, or thin). Like Avicenna, Arnald insisted that it was important to test on humans, as animals did not have the same physiology. Unlike his predecessors, however, Arnald proposed testing drugs not on sick patients, but on healthy people.[54] This strategy aimed at determining the complexion of the drug in a neutral environment. Once the properties were established, the doctor could use the drug to treat sick people. The varying guidelines of Avicenna, Bernard, and Arnald all underscore the difficult problem of testing most drugs in the individualized Galenic system. Because poison circumvented that system via the concept of specific form, testing antidotes appeared far simpler.

Guido da Vigevano's Worm Theriac

Did these medieval physicians ever follow their own guidelines? Did they actually test the quality of theriac or the safety of a new drug?

We have no indication from the writings of either Bernard or Arnald: if there was a dialogue between their drug-testing rules and their empirical observations, it has gone unrecorded.[55] Another doctor working in France left far clearer evidence of his own testing. In the early 1330s, the Italian physician Guido da Vigevano (c. 1280–1349) composed a remarkable treatise for King Philip VI of France. The text, called *Texaurus regis Francie*, had four main purposes: to help the king recapture lands lost to the Saracens during the Crusades; to preserve the monarch's health; to prolong his life; and to protect him from poison. The *Texaurus* is best known today for its fantastical sketches of technological devices, especially military technologies, leading some to herald Guido as a precursor to Leonardo da Vinci. His section on poison has received scant attention, but it reveals a similar level of imagination.[56]

Guido began his poison discourse in a conventional fashion, by detailing previous authors' advice in countering poison—particularly Galen, Avicenna, and the Arabic physician al-Razi (Rhazes). At the same time, he criticized his peers for relying too heavily on past authorities. Physicians in his day remained ignorant of the complete science of poison, he complained, because they had paid too little attention to finding new cures through experience. To emphasize the importance of experience, he turned to the ultimate authority: Galen. When the Greek master wanted to "have certainty" about the complexion of vinegar in order to treat an abscess, Guido stated, he devised the following trial: he put rose oil on one part of his legs, on another part an unguent, and on another part a vinegar wash. In the end, Galen determined that vinegar's complexion was by far the coldest of the three remedies and thus the best to treat the abscess.[57]

With that justification out of the way, Guido felt free to relate his own experience with testing. He told a tale in which "I, Guido" had attempted to find a remedy against the aconite called napellus, which he described as a deadly poison commonly used by "bad men" and particularly feared because theriac was ineffectual against it. According to Avicenna, he explained, a certain kind of mouse nourished itself from the roots of napellus and could be made into an antidote against it, presumably on the same theory as the use of viper flesh in theriac.[58] However, the "masters" who made poison locally claimed that Avicenna was wrong on this point. In order to gain some "certainty" on this disagreement, Guido decided to investigate the matter. He set off one day in the month of May to search for the little rodents. He dug up napellus plants and saw no mice or even evidence of nibbling on the roots. He did, however, observe some worms on the plant, which the poison masters deemed similar to a type of worm that lived on rape plants. Guido gathered some of these wormy aconites "with my own hands" and took them home with him. After letting the worms feed for several days, he mashed some of them up and mixed them into a theriac.[59]

Guido then tested this concoction. He began by trying it on "beasts" and found that it worked "perfectly" against napellus, "this terrible poison, which is the most harmful poison in all the world." Once he was sure that the worm remedy worked, he wanted to try it on himself, "just as Galen tested the coldness of vinegar on his legs." (The modern reader might question his equivalence between ingesting poison and putting vinegar on one's legs.) He took some napellus and waited until he felt its effects, which took about a quarter hour. He then took the worm theriac, felt "a great disturbance in the stomach," and vomited. He continued to feel ill, took more worm theriac, and vomited again. He went through this cycle a third time, after which he did not vomit again and was certain that the poison was all gone. He then presented the recipe for this antidote to the king:

RECIPE.

Terra sigillata; laurel berries—each in the quantity of 3 drachms

24 napellus worms

Mix together with oil and honey. Take in a dose the size of two hazelnuts.[60]

Whether or not the king made use of this unusual prescription, Guido's worm theriac provides crucial, albeit rare, evidence of actual drug testing in the Middle Ages. His emphasis on the importance of "experience" and his dramatic recounting of his own self-testing demonstrate clearly that medieval doctors were not merely tied to their texts. If Arnald's and Bernard's testing guidelines and theriac treatises provide tantalizing evidence of interest in testing poison antidotes, Guido's case shows us that this curiosity was not abstract. It also reinforces the notion of poison as a substance that was particularly testable. Guido sought "certainty" in the question of Avicenna's napellus-eating mice. He believed he had found it in his worm theriac.

Added to the examples of Arnald of Villanova, Bernard de Gordon, and Pietro d'Abano, Guido's case underscores the extent to which medieval physicians were fascinated by poison. So too were their readers. In a surviving manuscript of Guido's text currently held in the Bibliothèque Nationale de France, one reader marked the section describing Guido's worm theriac adventure with a crudely drawn manicule, or pointing hand, a common device highlighting points of interest in medieval and early modern texts.[61] It is the only place that the manuscript is so marked. Poison was a subject of broad concern, and that anxiety spread well beyond the confines of the medieval university.

Poison in the Courts of Law

Thus far, all of the poison trials we have seen derived from a medical problem: how to cure deadly poisons. While this question was at the forefront of numerous medical texts in this period, actual instances of testing appeared relatively infrequently in medical writings, as we have seen. Far more common was a form of testing that attempted to answer a different question: whether a given substance was poisonous. Throughout the Middle Ages, trials on animals (and occasionally humans) helped settle cases of suspected poisoning in both legal and political contexts. The courts of law took accusations of poisoning very seriously, as it was considered a particularly heinous crime. Unlike a victim murdered by hand-to-hand combat, a poisoning victim could not be sure who the perpetrator was; it was a devious—and thus highly dishonorable—crime.[62]

Courts often drew on physicians' expertise to determine whether or not a poisoning had taken place. While criminal trial records often did not specify these forms of analysis, sometimes court magistrates would ask physicians to examine the victim. If the victim had died or a physician was not immediately present, the court could also request that a doctor assess a set of written symptoms. Starting around 1300, physicians occasionally conducted autopsies of poison victims at the behest of the courts of law. This practice appears to have begun in Italy, but it spread to France by the early fifteenth century. Autopsies in suspected poisoning cases usually examined the stomach, intestines, heart, lungs, liver, and spleen, but rarely was an autopsy definitive in a poisoning trial.[63]

When an alleged poison was recovered during investigations, courts frequently attempted to determine its toxicity. Sometimes they asked apothecaries or physicians to assess whether a substance was poisonous, although in most cases court records left the exact nature of this analysis ambiguous. However, there is enough evidence to suggest that courts occasionally tested the toxicity of a substance by giving it to a test subject. Medieval English courts, for example, occasionally used animals to test poison.[64] In a spectacular case of suspected poisoning against the French knight Geoffroy de Salagny, some of the powder found on the suspect, a servant, was first fed to a dog and then to the servant himself.[65]

These practices may remind some readers of the medieval tradition of trial by ordeal, in which a suspected criminal was subjected to a potentially fatal test to determine guilt or innocence. Most commonly, the accused was submerged in hot or cold water, burned with a hot iron, or set to a duel, on the theory that God would intervene on behalf of an innocent person. Ordeals were officially outlawed by the Fourth Lateran Council, convened by Pope Innocent III in 1215, and they gradually fell out of use thereafter.[66] Even when they were in use, however, poison never figured among the standard tests. The medieval cases of feeding a suspected poison to test subjects attempted to assess whether that substance was toxic, not whether the subject was innocent or guilty.[67]

In the legal context, a poison test was often termed a "proof" (*proba* in Latin), a very different meaning from today's idea of proof as incontrovertible evidence. The medieval *proba* was a test or trial—a means of gathering evidence rather than a verification derived from it.[68] This terminology differed from the wording used by medieval physicians, who tended instead to use the terms *experimentum* (experiment) or *probatio* (approbation) to describe medical tests. By the sixteenth century, physicians had begun to use *proba* (or *épreuve* or *tryall*) to describe antidote tests as well.

Trees of Proof

The judicial courts were not the only place where a kind of poison "proof" took place. A different variety of proof was also common in princely circles. The tradition of having food tasters on hand in aristocratic households to test for poison is well known, and it was often accompanied by a complex assaying ceremony that demonstrated the food's safety (see fig. 1.1). In Italy, the sideboard used for this ceremony of "credence" eventually took on its name—*credenza*.[69] Many wealthy households also owned objects that allegedly would alter their appearance in the presence of poison: serpent's tongue (shark teeth), the prassius stone, pieces of unicorn horn, or assorted other gems and minerals. In France, many noble houses prominently displayed a "tree of proof" (*arbre d'épreuve*), usually a tree made out of silver (or, later, coral), with shark teeth, emerald, and other poison-testing materials hung from its branches (see fig. 1.2).[70] These trees were listed among the treasures of popes and wealthy aristocrats, as were prized poison antidotes such as bezoar stone and unicorn horn.[71] Aristocrats appear to have followed Pietro d'Abano's "defense with caution" recommendations to the letter.

1.1 Poison-testing ceremony at Charlemagne's court. Miniature, c. 1500. Bibliothèque Royale de Belgique, Brussels. Wikimedia Commons.

Indeed, a poison treatise written for Pope Urban V in the 1360s specifically cited Pietro in recommending the pope use serpent's tongue, emerald, and other "proven" items at the table.[72]

The courtly interest in Pietro's treatise should come as no surprise. From antiquity through the Renaissance, nearly all treatises on poison were created in the context of aristocratic court culture and patronage. The Roman poet Nicander dedicated his *Theriaca* and *Alexipharmaca* to King Attalus III, who was known to have conducted poison trials on criminals. The ninth-century Persian physician ibn Wahshiyya listed an array of kings and princes who had sought knowledge of poisons, and he described his *Book on Poisons* as "a precious book possessing many uses and advantages."[73] Maimonides' *Treatise on Poisons and Their Antidotes* was created at the behest of his lord, the Egyptian vizier Qadi al-Fadil, who was counselor and secretary to Saladin. Similarly, Arnald of Villanova enjoyed patronage from both the court of Savoy and the papal court, while Guido da Vigevano dedicated his treatise to King Philip VI of France. Of the major writers on poison in the Middle Ages, only Bernard de Gordon appears to have had no ties to the aristocracy.[74]

The ever-present fear of poison hung like a cloud over the power brokers of medieval and Renaissance Europe. Rumors of attempted and successful poisonings were rampant in its warring houses. Poison figured

1.2 Tree of Proof (*Natternzungenkredenz*) with fossilized shark's teeth, 15th or 16th century. Treasury of the German Order, Vienna. Photo by Walter Schabel. Creative Commons.

prominently in struggles between the Capetians and the Carolingians, the English and the Normans, the Armagnacs and the Burgundians, the Ghibellines and the Guelphs, as well as in dynastic struggles at all major courts of Europe, including the papal court. Warring parties often used accusations of attempted poisoning to discredit their enemies. Throughout the 1240s and 1250s, partisans of Pope Urban IV and Emperor Frederick II hurled accusations of poisoning at each other, a pattern repeated around Europe.[75] This use of poison to taint political rivals underscores the dishonor attached to poisoning, a topic we will explore more deeply in chapter 4.

Poison fears were especially pronounced at the papal court. Contemporary documents suggest that popes felt themselves acutely vulnerable to being poisoned. The supreme pontiff sat above dozens of underlings who hoped for his job, especially ambitious cardinals. The continual cycle of power shifts in the papacy helps explain why "trees of proof" appear in particularly large number in papal inventories.[76] Popes also avidly sought advice on how to avoid poisoning. Arnald of Villanova appears to have composed his practical treatise *On the Art of Recognizing Poison* as he was preparing to move to the papal court in Avignon, shortly before his death. Arnald's nephew, Armengaud Blaise, later dedicated a translation of Maimonides' practical *Treatise on Poisons* to Pope Clement V, one of three different translations of Maimonides' poison treatise created for popes. A translation of Avenzoar's discourse on the emerald was dedicated to Pope Boniface VIII, and Guglielmo de Marra warned in the early 1360s that popes faced special peril from poisoning attempts.[77] Suspicions of poisoning surrounded the death of nearly every pope who expired in unexplained circumstances, as we will see in the next chapter.

The various methods of poison "proof" echoed the assumption in medical texts that poison was a substance one could test. Indeed, poison testing in the Middle Ages appears to have been focused on methods for uncovering poison rather than methods for curing it. The use of "trees of proofs" and other assaying methods was a regular occurrence at aristocratic courts, while attempts to assess toxicity were common in poisoning cases at the courts of law. Tests of poison antidotes, in contrast, appear not to have been widespread. Nevertheless, the Montpellier treatises and Guido da Vigevano's manuscript suggest that antidote trials had the potential to make a greater impact. The bold tone in which "I, Guido" recounted his discovery of the worm theriac shows that he expected the tale to reflect well on him. In the middle of the fourteenth century, a deadly new threat would strengthen both poison fears and the interest in finding effective antidotes.

Poison and Plague

On September 15, 1348, by the picturesque castle of Chillon on the shores of Lake Geneva, a Jewish surgeon named Balavigny was arrested on suspicion of poisoning Christians. After being tortured on the rack, he confessed to multiple counts of poisoning wells and springs in his hometown of Thonon, about thirty miles away. Prompted by the executioner, he told of an international conspiracy of Jews intent on doing harm to Christians, led by a Master Jacob from Toledo, who had sent him a letter containing the poison and instructed him to put it into the wells and other drinking sources most frequented by Christians. Failure to do so would mean excommunication from the Jewish faith. Balavigny also confessed that he was not the only recipient of such a letter. On the command of their rabbis, Jews across the region had received the order to poison Christians. One Jew named Mussus had told Balavigny that he had poisoned wells in the nearby town of Villeneuve and put poison under the stones of a well in the Chillon tollhouse.[78]

On the basis of this confession, the tollhouse well was searched, a poison allegedly found. The substance was then "given to a certain Jew, as a trial, who died from this poison." With this casual reference, a mere footnote to the overall story, Balavigny's confession hinted at a practice of testing suspected poisons on suspect humans, as in the case of the servant accused of poisoning Geoffroy de Salagny. After the presence of poison was (allegedly) established, the court officials brought Balavigny across the lake to the town of Clarens and instructed him to show them a well he had (allegedly) poisoned there. A small linen cloth was found in a pipe inside the well, and Balavigny confessed, again under duress, that it had once contained his poison. He said the poison was of two colors, red and black, and he speculated that it must come from a basilisk. He also claimed that once in a person's body, this special poison could be passed on to other people and spread.[79]

The sinister scenes described in these documents, and others like them, took place in the first years of the Black Death—the epidemic of bubonic plague that swept from southern Italy north to England and Scandinavia in deadly waves from 1347 to 1350.[80] While many authorities assumed that a new disease lay at the heart of all the dying, large swaths of the population in parts of present-day Germany, Switzerland, and France blamed it on poison, especially Jewish poison, and demanded retribution. Balavigny's confession was one of several forced confessions from Chillon that detailed alleged cases of Jews poisoning wells, which were all sent to the city of Strasbourg in answer to a request for evidence of the Jewish threat. Many other towns came up with similar "proof" of a massive Jewish poison plot. On the basis of accusations like these, towns executed thousands of Jews in grisly pogroms. Strasbourg, which initially had sought to protect its Jewish residents, eventually burned two thousand Jewish men, women, and children in one of the most horrific incidents. Balavigny, as well, met his death at the stake.[81] Historians have rightly pointed to this brutal hysteria as evidence of the long-simmering animosity toward the Jews in Christian Europe. Court documents like the Chillon cases suggest a desire to provide seemingly sound legal backing for religious persecution.[82]

The confession of Balavigny also demonstrates how quickly early modern individuals began to tie plague to poison. Jews—along with lepers—

had been persecuted for allegedly poisoning wells even before the advent of plague, but the Black Death killed people at an unprecedented rate.[83] At least a quarter of the population of Europe perished as a result of the first epidemic, and some cities lost over half of their populations. Plague moved quickly and ruthlessly: people who got sick either died within a few days or recovered. In that sense, it seemed to operate much like a deadly poison.[84] The accusations against Jews took that resemblance to an extreme. The crime of poisoning had long been attached to marginalized communities, lepers and old women alongside Jews, and the Black Death intensified this connection with a malicious Jewish poison plot. Other observers, however, noted that Jews died from the new ailment at the same rates as Christians and that burning Jews did not increase a town's prospects in plague epidemics.[85]

Even for individuals who did not blame Jews, poison could provide a useful explanation for the public health crisis. While most physicians who tried to make sense of plague attempted to explain it using the conventional theory of the four humors, the disease did not fit neatly into existing categories. It killed people of all humoral complexions. Most physicians explained this phenomenon through some kind of catastrophic event (a severe earthquake, or a calamitous astrological conjunction, or both) that caused corrupt vapors, known as miasma, to mingle with the air.[86] Some physicians went even further and described this miasma as poisonous, and a few early writers specifically designated the disease as a poison. Italian physician Gentile da Foligno explained that corruption from the air caused a "poisonous matter" to be generated in the heart and lungs, which was the root of the disease. He called plague "the most venomous of all poisons."[87] Soon after writing his treatise, Gentile became one of the Black Death's early victims, dying of plague on June 18, 1348.

Not all medical writers agreed with Gentile. The Neapolitan physician Giovanni della Penna explicitly criticized Gentile's characterization of plague as a poison and described the disease as an imbalance of the humors instead, specifically an overheating of choler. A well-known report by the Paris medical faculty from 1348 reported a dangerous "corruption" that could quickly penetrate the heart, but it did not use the term "poison" or "poisonous" in its description of the disease's etiology.[88] Yet the tie between plague and poison was a common early notion. The Catalan physician Jacme d'Agramont stated that plague acted "like a poison" in his treatise written in April 1348, and the Paris medical faculty report did link plague to poison in its section on remedies. Even Penna argued that the choleric matter had a poisonous nature that corrupted the blood.[89] While physicians disagreed on the specific mechanism by which the disease operated, plague and poison became inextricably linked.

Antidotes and Disease

This twinning of plague and poison had major implications for poison antidotes, particularly theriac, which suddenly became a go-to remedy for the most deadly of diseases. The use of theriac to treat plague has sometimes been taken as a sign of medieval physicians' conservative response to the epidemic: they developed few new theories or medicines but instead relied on a familiar old drug.[90] Yet that reading does not give the whole picture. The use of poison as an explanation for plague expanded the importance of antidotes significantly. Theriac had long been seen as a kind of cure-all, good for a variety of ailments and (according to some physicians) general disease prevention, in addition to its primary function as an antidote to poison.[91] With the advent of plague, it also became a prominent remedy for acute epidemic disease.

Theriac was not the only antidote given new billing as a plague cure. Gentile da Foligno recommended terra sigillata, mithridatium, emerald powder, and amethyst set into a gold ring "to preserve you against all poison and pestilential air."[92] The Paris medical faculty praised terra sigillata especially for its "miraculous" properties against all poison, "whether [taken] after the poison or before." It also recommended theriac and mithridatium, Armenian bole, and emerald, the last described as "a remarkable medicine against all poison."[93] One of the most influential early plague writings, the Paris medical report spread in both Latin and vernacular manuscripts.[94] A rhymed German-language plague regimen from the second half of the fourteenth century echoed its recommendations:

And theriac in the size of a pea
Mixed with wine is as good as can be.
It strengthens the heart so very fast,
The poison will not stay and last.
Armenian bole's good with vinegar before food,
Terra sigillata brings joy to the blood.[95]

Poison antidotes were not the only suggested remedies for plague. Treatises also focused on the usual physicians' mainstays of diet, environmental considerations, and bloodletting, in addition to various methods to treat the characteristic buboes that erupted on lymph glands. However, most plague treatises contained some recommendations for medicaments, and poison remedies figured prominently among them.

If poison antidotes were regularly recommended as plague treatments, plague remedies also began to find their way into writings on poison. Guglielmo de Marra's *Papal Garland on Poison* appears to have been written for Pope Urban V shortly after his election in 1362, in the midst of a raging plague epidemic. Unsurprisingly, he included a chapter on remedies for "venomous and pestilential air." For both poison and plague, he mentioned a number of commonly cited antidotes, including theriac, mithridatium, emerald, unicorn horn, and bezoar stone. However, he also added several new antidote-plague cures. He praised a powder made from the herb tormentil, created by an herb seller named Benedicto of Mantua, as an excellent cure for pestilence and animal bites—and he noted that it had even saved a man accidentally poisoned with arsenic.[96] Against plague he especially recommended "miraculous" pills shown to him by his mentor, Master Balthassar of Padua, which would cure a patient "within twelve hours" and also worked as a general preservative and against a number of other ailments, especially epilepsy, apoplexy, and tremors.[97] These remedies all blurred the boundaries between curing disease and curing poison, and many of them, like theriac, could be used as a near cure-all.

Authors of treatises on plague and poison also tended to include empirical information, in order to demonstrate the effectiveness of new cures. Gentile da Foligno gave recipes for pills and other medicines that he had found effective against venomous humors "from experience, not from the authority of authors."[98] Guglielmo de Marra noted that Benedicto of Mantua had used his tormentil powder to great effect in "notable experiments" at the court of Padua, which Guglielmo had observed himself. He also could vouch for Master Balthassar's plague pills personally, as "I saw [them] tested on many."[99] Unfortunately, the authors left no indications as to the exact nature of these tests, but they used their own eyewitness experience to substantiate a cure's efficacy.

As plague epidemics continued to threaten European cities every decade or two, the connection between plague remedies and poison cures only strengthened. The Italian physician Antonio Guaineri (d. after 1455) wrote a two-part treatise on poison sometime before 1440, the first part on plague and the second part on poison itself.[100] In contrast to the practical bent of Gentile or Guglielmo, however, Guaineri's main interest was a theoretical discussion on the action of plague/poison and its cures.[101] He did relate experimental incidents from past authors, including prominent medieval physicians. In discussing the merits of theriac against poison, he recounted Guido da Vigevano's experimentation with the worm theriac in fairly exact detail and even included Guido's recipe, although he did not mention him by name.[102] Similarly, he used Arnald of Villanova's poisoned cheese as evidence for theriac's ability to expel poison, including the poison of plague.[103] Guaineri included a few cases drawn from his own incidental experience, such as his nasty case of food poisoning from a piece of pork, or his observation of the laxative effects of certain pills, but his practical examples were generally in service of his theoretical aims.[104]

A near-contemporary of Guaineri, Sante Arduino, similarly privileged theory in his enormous compendium *On Poison*, written in the 1420s. This work, an encyclopedic compilation from nearly forty different medical authors, presented a comprehensive overview of knowledge on poison and its antidotes. Although Arduino mostly drew on the scholarship of others, he presented the material in a way that made a theoretical argument about the action of poison.[105] Yet comments on experience threaded throughout the volume. In many entries, he stated whether an author had (or had not) found a certain remedy to be successful in practice, frequently with versions of the ubiquitous term *probatum est* (it has been tried). In a section on Galen near the beginning of the book, he cited excerpts from *Theriac to Piso*, including the "approbation" (*probatio*) to test whether a batch of theriac was good by seeing whether it would halt purgation in a human subject.[106] The rooster experiment using poison appeared later, in a discussion of Avicenna's *Canon*.[107] Importantly, he also mentioned the legend of Mithridates, king of Pontus, in several places, although he did not specifically cite Mithridates' tests on condemned criminals. Arduino's book gathered together the most important works on poisons from the Greek, Arabic, and Latin traditions preceding him, and he included empirical validation as part of that story.

Despite the empirical bent to parts of both Guaineri's and Sante Arduino's works, both authors were far more interested in explaining poison's action in theory than in validating antidotes they considered effective. Nevertheless, their theoretical ideas, which focused on poison's action by its specific form, helped cement the notion that poison was a universal harming agent that could manifest as disease—making antidotes potentially useful for far more than just poisoning.[108] As we have seen, Guaineri discussed antidotes as remedies for plague, and Arduino described them as useful for disease in numerous places. While theriac and mithridatium had always been seen as cure-alls, the potential uses of antidotes like bezoar, terra sigillata, and emerald began to broaden as well. Yet there appears to have been little interest in testing these remedies directly in the first half of the fifteenth century, at least among academic physicians.

Ancient Precedents, Renaissance Trials

On a Saturday morning in 1480, a small group met at the city hall in Tours on the command of King Louis XI of France. The gathering included the city's mayor and four of its aldermen, two members of the king's entourage, two men from the household of the governor of the Touraine, an apothecary named Simon Moureau, the city clerics, and a dog. The motley assembly had gathered to test a "certain poison" on the dog. Moureau gave the animal a "strong dose" of the poison mixed into some fried mutton and an omelet. The dog gobbled up the feast and promptly died. The city's barber-surgeons then dissected the unlucky animal. The circumstances that led to this test are unclear: the town scribes merely described it as a test of a poison. It may well have been an attempt to solve an alleged case of attempted poisoning, perhaps of the king himself. King Louis had vocally denounced several thwarted poisoning attempts in the 1470s.[109]

This incident resembled tests we have already encountered in this chapter. It aimed to ascertain whether the "certain poison" would indeed cause illness or death. There were, however, two major differences. First, this test was a spectacle witnessed by both aristocrats and city officials, an event worthy of note in a chronicle. Its scale, in terms of the number of people in the room as observers, was far larger than that of any test of poison recorded previously. Second, and particularly crucial: the dog was dissected after the spectacle, indicating an interest in what the poison did to the interior of the animal's body, in addition to apparently killing it. As we will see in the next chapter, the trial in Tours presaged a larger shift, one that put a greater emphasis on testing in practice.

By the late fifteenth century, the special status of poison had been cemented—a status that wove through medical theory, aristocratic court culture, and the courts of law. Not only was poison seen as a universal harming agent and an ever-present threat to the powerful, it also was implicated in the deadly plague that continued to claim lives throughout Europe. All of these trends put a new focus on poison antidotes. Even before the Black Death, antidotes had seemed ripe for investigation, as we saw in the case of Guido da Vigevano's worm theriac. Poison, as we have seen, appeared to lend itself especially well to testing because it caused immediate harm. But despite a textual tradition of testing antidotes that dated back over a millennium, actual tests appear to have been few and far between. In the sixteenth century, the interest in contrived tests of poison began to increase, and the focus widened to encompass its cures.

TWO

Condemned Bodies: Oleum Clementis

A half century after King Louis's dog trial in Tours, in 1524, Pope Clement VII ordered a test of the antidote oil created by surgeon Gregorio Caravita, described in the first pages of this book. The bare outlines of this event were reminiscent of the Tours trial: on the command of a powerful ruler, a test subject took a dose of poison in front of a small crowd of invited personages. In both cases, apothecaries provided the poison—Simon Moureau in Tours, Tomasso Bigliotti in Rome. Yet both the purpose and the methods of the two tests differed drastically. The Tours trial aimed to uncover the presence of poison, likely in the service of solving a crime—part of the long medieval tradition of testing for poison that we saw in chapter 1.[1] Pope Clement's trial of Caravita's oil had a completely different goal. Rather than test a poisonous substance, it probed the value of an antidote, in the vein of Galen's poison trials on roosters. Although a pharmacist was present, the papal physician, Paolo Giovio, oversaw the trial. Most significantly, the test subjects were not roosters or dogs. Pope Clement commanded that the antidote be tried on "condemned bodies, faithfully and diligently, for the benefit of the public."[2] Not long after the trial of Caravita's oil, Pope Clement himself became known for an antidote oil called the "Oleum Clementis," which one of his cardinals tested on criminals. While we saw glimpses of antidote testing in the Middle Ages, as well as tests using poison on humans suspected of poisoning, the use of condemned criminals as medical test subjects was new. Something had changed in the intervening years.

What prompted this shift? Why did antidotes suddenly become a focus of testing, when they had not been fifty years earlier? And how did convicts come to be seen as appropriate test subjects? There is no smoking gun for this mystery. Pope Clement and his physicians never penned an explanation of their reasoning. Nevertheless, there are a number of clues. The long history of poison trials provides one hint, as Renaissance humanists poured over ancient texts and put great weight on the ideas of Galen and other Greek and Roman authorities. The ancient example of King Mithridates VI of Pontus, who had also used condemned criminals for poison trials, certainly had an influence. At the same time, fears of malicious poisoning at Renaissance courts reached a fevered pitch in the fifteenth and sixteenth centuries. Poison was suspected in the death of nearly every Renaissance pope, and reports of attempted—and successful—poisonings flooded the letters of Renaissance aristocrats. The letters of the Medici family in Florence suggest that many of these fears were justified.[3] From both medical and political perspectives, antidotes became ever more important, and poison trials had instant appeal.

The shift from dissecting dead criminals to poisoning living ones was nevertheless quite a leap, and as this chapter will argue, the actors involved with the poison trials knew it. We are lucky to have a firsthand account of the test of Caravita's oil from the point of view of the testers: the pamphlet put out in Pope Clement's name and written by Giovio, Bigliotti, and the senator Pietro Borghese. This document, which they called a *Testimonium* (testimony) of the "true and admirable virtues" of the oil, allowed the testers to shape the narrative as they saw fit.[4] While we should not take any of it at face value, it is enormously helpful in suggesting what the authorities in charge of the poison trials viewed as concerning and, conversely, what they viewed as unproblematic.

The *Testimonium* provides us with a variety of clues about the context into which the testers fit the trial of Caravita's oil. The supreme pontiff, it states, commanded that the antidote be tried on "condemned bodies"—a curious and very telling phrase. It presents the condemned prisoner as a body, not as a human being, and thereby evokes another kind of testing on criminals: the use of their cadavers for anatomical study. The first human dissections in Europe began in Italy by the early fourteenth century, but the late fifteenth century saw an uptick in the number of dissections and scholarly interest in them. By the 1520s, the criminal body was well established as an object of medical study.[5] Emerging conventions and conversations surrounding anatomy almost certainly

2.1 Paolo Giovio. Woodcut by Tobias Stimmer, 1589. Wellcome Collection.

provided the central impetus for the decision to use poison on condemned criminals. Similarly, the description of Pope Clement's trial as being for the "benefit of the public" reflects a growing legal trend that subordinated individual rights to the civic good.

At the center of the trial of Caravita's oil, two people in particular were pulling the strings: the physician Paolo Giovio and Pope Clement VII himself (see figs. 2.1 and 2.2). Both men had motivations to conduct this sort of test. Today Giovio is far better known as a historian than as a physician, but he had trained under an ambitious anatomist and had

2.2 Sebastiano di Piombo, portrait of Pope Clement VII, c. 1531. Wikimedia Commons.

almost certainly experimented on (deceased) criminals in the course of his medical education. Pope Clement, for his part, had been a strong advocate of anatomical investigation even before he became pope. Under his watch, Rome became known for a relatively steady supply of cadavers for dissection.[6] Clement also came to power amid grave concerns about both poison and plague. He had seen his cousin and friend, Pope Leo X, die of presumed poisoning in 1522 and had lived through the deadly wave of plague that hit Rome in 1523. The plague had been a central problem for his immediate predecessor, Pope Adrian VI, a fact that likely did not escape Clement's attention—nor the fact that Adrian, too, was rumored to have been poisoned. With this threat of poison looming in the background and the existing use of "condemned bodies" for other forms of testing, the move to use criminals to test poison appears an unsurprising progression.

In fact, as this chapter will show, Giovio and his co-authors did their best to portray the test of Caravita's oil as a conventional event that did not disrupt existing medical, religious, or legal expectations. The outcome proved the efficacy of a marvelous new antidote, but the procedure, in their telling, closely followed existing traditions. This test was not a brash demonstration by innovative medical outsiders; it was an orderly event conducted by respected professionals. Poison trials were entirely consistent with other developments in late fifteenth-century medicine, especially the growing humanist interest in anatomical investigation, and indeed, it is possible that Pope Clement's trial was *not* the first of its kind. We have a murky view of the transfer and use of bodies for dissection at the time, and tests of poison antidotes may have occurred in some limited way that went unrecorded (or have been lost to the record—or sit in an archive waiting to be found). Be that as it may, the trial of Caravita's oil was the first such test to make waves in printed texts and private letters, and it spawned many copycat poison trials in the following decades. Whether or not poison trials had occurred previously, conditions in sixteenth-century Italy—and Rome in particular—made it possible for them to gain widespread recognition. The testers' overriding aim was to make the outcome of their poison trial appear valid and laudable. To do so, they appealed not to claims of novelty but to business as usual. The closest analogy came from the anatomical study of "condemned bodies."

The Anatomy of Poison Trials

Anatomy had long been important to cases of poisoning, as we saw in the previous chapter. From at least the early fourteenth century, courts occasionally ordered autopsies in cases of suspected poisoning. The dissection of the dog in Tours was an outgrowth of that forensic interest, which would blossom in the seventeenth and eighteenth centuries.[7] Autopsies were not, however, a major focus of medical attention, at least in medieval texts. Although physicians often acted as medical experts for autopsies and occasionally noted this role, they recorded little about them. Pietro d'Abano, writing in the early 1300s, only briefly mentioned a pharmacist who was autopsied after he accidentally swallowed some sublimated mercury.[8] As Katharine Park has argued, the status of autopsy began to change in the later fifteenth century, as Italian patricians increasingly requested autopsies for their deceased loved ones.[9] Physicians, in turn, placed more scholarly emphasis on autopsies as a means to study the body, and they began to write about them in more detail.[10]

As the interior of the body became an ever-greater source of fascination, the frequency of public dissections grew in leaps and bounds alongside that of autopsy. In tandem with this growth, the criminal body gained a central role in medical study. Although human dissection had been a teaching tool since the early 1300s, the medical curriculum expanded its focus on anatomy in the last two decades of the fifteenth century. Autopsy represented only one influence on this burgeoning interest. Another strong impulse came from the intellectual trend called humanism. In its medical iteration, humanism focused on efforts to recover original texts from Galen and other ancient Greek and Roman medical authors. In the process of that recovery, Galen's own focus on anatomy became clear. Although the newly discovered texts made it clear that Galen had dissected only animals, Renaissance physicians applied his methodology, with increasing vigor, to the existing practice of human dissection.[11] As many historians have noted, the bodies used for this investigative push were nearly always cadavers of executed criminals.[12] Because the first human poison trials on prisoners occurred in the midst of this continuing anatomical boom, the dissection of executed criminals provides crucial context for understanding this new impetus to test poison antidotes on human subjects.

In anatomical investigations, the use of criminal bodies helped diffuse the significant cultural and religious anxieties surrounding public dissection, as Andrea Carlino and Katharine Park have shown. Although the myth that the Church banned human dissection still persists today, Park long ago demonstrated that it never issued a ban or even criticism.[13] Although dissection was not taboo from an institutional perspective, the practice nevertheless provoked widespread cultural apprehension in a way that private autopsies did not. Dissection evoked public dismemberment, a disgraceful punishment for the worst crimes. It was viewed as depriving the deceased of the dignity they deserved after death, and it interrupted funerary rituals that involved the full body. Even more alarmingly, it elicited concerns about whether the body could be fully reassembled at the Last Judgment.[14] As Carlino put it, dissection "bordered on the sacrilegious" in public opinion, if not in the laws of the Church.[15] Using deceased foreign criminals of lowly birth for public dissections represented a compromise that allowed for the opening of the body while softening public disgust.

These complexities were reflected in physicians' writings on dissection. In 1502, Alessandro Benedetti (d. 1512), a humanist physician in Verona, published the first major Renaissance treatise on anatomy. It began with a full-throated defense of human dissection, especially for its potential to "help the living."[16] Benedetti went to some lengths to portray public dissection as an acceptable practice, both religiously and culturally. He noted first that "pontifical regulations" had long permitted the dissection of cadavers: "otherwise it would be regarded as most execrable and abominable or irreligious."[17] Benedetti thus emphasized the Church's endorsement of human dissection. At the same time, he clearly realized the practice was not as unproblematic as he claimed, as he also noted that "ritual purifications of the physicians' souls take place" following dissections, in which they "propitiate their offense with prayers."[18] After alleviating all potential religious concerns, Benedetti explained why cultural discomfort was also unnecessary. He assured the reader that physicians used only cadavers of "unknown and ignoble" criminals "from distant regions" who had no family nearby. Their corpses thus could be used "without injury to neighbors and relatives."[19] With these careful measures in place, he implied, dissection was entirely unproblematic.

Despite Benedetti's assurances, the fact that the executed criminals were foreign did not completely soothe anxieties. As the demand for bodies grew, discomfort about public dissection only increased, especially suspicions that the rules were not always followed as stated. Cities and universities took additional measures to ensure that dissections proceeded in as calm and orderly a fashion as possible. Public anatomies often took place during Carnival, a time when "life turned inside out" and blasphemies were more accepted.[20] The university of Pisa appointed two senior students specifically to make sure that "there should be no cause for tumult."[21] In Rome, bodies were chosen for dissection secretly and taken away literally in the dead of night, in order to avoid conflicts.[22] Officials were aware of the potential for public unrest surrounding dissections, and they made every effort to head off such concerns. As we will see, the physicians and other officials who conducted poison trials followed a similar process.

Using "condemned bodies" for university dissections meant that the body of the executed criminal was considered representative of human anatomy in general, for all bodies. Despite all of the anxieties surrounding dissection, there was surprisingly little discussion about this point. Authors occasionally attributed particular features to criminality, and physicians made occasional reference to differences between upper-class and lower-class bodies in general, especially as pertained to diet.[23] Yet there was virtually no expressed concern that a criminal's body might have fundamental differences that could hinder its usefulness to teaching and research. Bodies slated for public dissection were usually chosen to be as *representative* as possible—young, of strong build, reasonably healthy, not too fat and not too thin.[24] From the sixteenth to the nineteenth century, there was a steady increase in the use of lower-status populations—hospital patients, soldiers, prisoners, and eventually slaves—that provided a convenient supply of experimental subjects.[25] Even in the eighteenth century, when ideas of bodily difference had expanded to include clear racial hierarchies, physicians in the Americas used enslaved peoples as subjects for both anatomical investigation and drug testing. Historian Londa Schiebinger has pointed out the dissonance between the rhetoric of Africans as inferior and the use of their bodies as typical.[26] Physicians were willing to overlook difference to take advantage of vulnerable populations as medical subjects. In short, the Renaissance anxieties about anatomical dissection arose from the rituals and cultural ideas surrounding death, not from the worry that the investigation of criminals' cadavers would yield insufficient or tainted information. That same lack of concern about the criminal body as representative would also characterize poison trials.

The Living and the Dead

If using executed criminals for anatomical investigation came with such heavy anxieties, what about the use of living criminals for medical experiments? There is no uniform answer to this question. There was deep concern about the practice of using living criminals for one kind of experimentation—vivisection, or cutting into the body while still alive.[27] An explicit discussion of vivisection appeared in a rediscovered ancient text, the first-century Roman physician Celsus's *On Medicine* (*De medicina*, c. 50 CE), brought to renewed prominence in the early fifteenth century.[28] In his proemium to the work, Celsus discussed clashing ideas about dissection and vivisection between two Roman sects of physicians, the Empiricists and the Dogmatists. While the Empiricists found both dissection and vivisection to be vicious and unnecessary, the Dogmatists did not find it "cruel that in the execution of criminals, and but a few of them, we should seek remedies for innocent people of all future ages."[29] The argument, in this case, was that the knowledge gained would be helpful enough for the broader populace to be worth the sacrifice of a few criminals.

Although he supported the practice of dissection, Celsus disagreed with the few-for-the-many argument in the case of vivisection. He argued that "to lay open the bodies of men whilst still alive is as cruel as it is needless." Rather than cutting open living bodies, he recommended a different way to investigate them. He asserted that "actual practice" in treating the wounded would yield a knowledge of the interior of the body "in a somewhat slower yet much milder way."[30] With this statement, Celsus effectively laid out the difference between setting up a contrived trial and gaining knowledge through the course of normal clinical experience. Although vivisection would yield valuable information about the body, it was far too objectionable for its advantages to outweigh the disadvantages of simply observing wounded patients. This view held true in the Renaissance, with some potential exceptions among particularly brazen physicians.[31]

The Renaissance use of living criminals for poison trials, in contrast, does not appear to have caused the same revulsion as vivisection, at least in the Italian cases presented in this chapter. As we shall see, the *Testimonium* describing the trial of Caravita's oil addressed a number of perceived concerns, but none were associated with the basic act of handing a prisoner a deadly drug. However unpleasant the effects, giving poison to a criminal already condemned to die was very different from cutting into the living body. There was, moreover, an example drawn from the ultimate ancient authority: Galen. In his book on antidotes (*De antidotis I*), Galen specifically mentioned the use of condemned criminals for testing poison under both Attalus III of Pergamum and Mithridates VI of Pontus, discussed in the previous chapter.[32] As humanists rediscovered Galen, they also rediscovered Mithridates' tests.

Benedetti left his readers with the sense that in the eyes of both medical and religious authorities, the handling of condemned criminals before death was less fraught than anatomizing the dead body. Criminals facing execution, he claimed, "have sometimes asked to be handed over to the colleges of physicians rather than to be killed by the hand of the public executioner," with the understanding that they would be dissected afterward. He added, "cadavers of this kind cannot be obtained except by papal consent."[33] He made no other comment on the subject, leaving the impression that this unusual situation was not particularly problematic for the physician involved, as long as proper consent was obtained. Written a quarter-century before Pope Clement's trial, Benedetti's statement raises two tantalizing possibilities: first, that physicians sometimes had a direct hand in the death of criminals slated for dissection, and second, that popes would have known about—indeed, approved—these instances.

Whether or not such executions ever occurred is unclear, although there is a hint from later decades. Sometime in the 1540s or 1550s, the esteemed anatomist Gabriele Falloppio (1523–62) described a test he had conducted on a condemned criminal granted to him by Grand Duke Cosimo I de' Medici. Duke Cosimo gave him the convict with the understanding that Falloppio could kill the man as he wished and dissect the body afterward. Falloppio decided to experiment with fatal doses of opium, a substance often described as a poison in the scholarly literature. The prisoner survived the first test and asked for the dose to be doubled on the second attempt, on the assumption that he would go free if he survived, which he did not.[34] Falloppio made no mention of papal sanction, but he also gave no indication that the event was unusual or untoward. This kind of private execution was almost certainly a rare practice, and this particular instance occurred long after the trial of Caravita's oil—but if executing a prisoner via private experimentation was possible, the leap to experimenting with antidotes was not far.

Even before the encounter under the dissector's scalpel, then, contact between physician and prisoner was possible. Falloppio's opium experiments and Benedetti's assertion about papal consent suggest that in certain rare circumstances, physicians could usurp the role of executioner, and that they sometimes used this access for their own experimental investigations. Yet in these particular special cases, the ultimate aim was dissection. Falloppio's opium experiments were simply a means to that end. The poison trials presented quite a different situation, as the experiment itself was the goal. Dissection was not even mentioned in the trial of Caravita's oil, which took place at the wrong time of year for a public anatomy, in mid-August. The antidote test represented the main event, not the prelude.

The pope's role also went beyond mere consent. According to the *Testimonium*, the trial took place "on the order of" Clement VII.[35] These words were surely not accidental. The pamphlet emphasized the pope's power over life and death—and his power to intervene in the regular process of executing a condemned criminal. Execution was a highly ritualized act in all areas of early modern Europe. It simultaneously punished the perpetrator and provided a warning (and, it was hoped, a deterrent) to others. As French philosopher Michel Foucault put it, public execution "made the body of the condemned man the place where the vengeance of the sovereign was applied, the anchoring point for a manifestation of power."[36] In most cities in Italy, the number of executions tended to rise in times of difficulty, such as famines, and fall during prosperous periods; public executions provided a way for authorities to show action in troubled times. In many cases, the specific manner and place in which the prisoner was executed came as a direct response to his crimes.[37]

The symbolic role of execution had a special meaning in Rome, owing to the constant push and pull between papal and civic authorities. At any time the pope could—and did—assert his position as the ultimate judicial authority and either demand an execution or intervene to prevent one. Pope Clement VII himself used execution as a highly symbolic act in other well-documented cases.[38] The *Testimonium* put out under his "command," then, must be seen as a document imbued with deeper meaning. To uncover that meaning, we need to dive into the specific context of Renaissance Rome in the early years of Clement's reign.

Anatomical Rome

Rome initially lagged behind northern Italian cities in the study of human dissection, but it began to catch up in the early sixteenth century. By the time of Clement VII's papacy, the Roman medical college was conducting an ever-increasing number of dissections, with full papal sanction.[39] Clement made some significant contributions to these anatomical studies. Although he is best known today for his vacillating leadership and for his powerlessness during the brutal Sack of Rome in 1527, at the time of his election he was fêted as a patron of humanists.[40] Many of his pet projects involved literature or the arts, but he also lent his support to medical humanism. Even before he became pope, while he was still Cardinal Giulio de' Medici, he urged the physician Euphrosynus Boninus to translate Galen's anatomical works from Greek into Latin. Boninus eventually published a collection of three Galenic anatomy treatises in 1524, and he mentioned Clement's encouragement in the book's dedication.[41] Clement later supported an even more ambitious humanist undertaking, Ferdinand Balmio's translation of Galen's treatise on bones. Although the work did not appear in print until 1535, the year after Clement's death, Balmio credited the pontiff's support as crucial for the success of the project.[42]

In the later years of his papacy, Clement took an active role in establishing formal procedures for anatomical dissection in Rome. He issued statutes for the Roman College of Physicians in 1531, which were updated by subsequent popes. As we will see, these careful guidelines—likely practiced long before they were codified—were directly relevant to the trial of Caravita's oil in 1524. According to the statutes, all dissections first required a license from the cardinal vicar (the vicar general for Rome, proxy for the pope). Once that permission was granted, the cadaver was to be obtained from the senator or governor, the authorities in charge of criminal law.[43] As in other places in Italy, the Roman statutes established clear steps that would help assuage the discomfort with public dissection—in this case, by requiring the oversight of the city's two legal authorities, papal and civic. Any dissection needed both sacred and secular sanction.

Clement's patronage also benefited another prominent humanist: Paolo Giovio, the physician who oversaw the trial of Caravita's oil. Giovio became widely known as the papal historian, and most scholars have characterized his commitment to medicine as halfhearted. Nevertheless, in his early career he relied on his medical credentials to gain patronage.[44] He moved to Rome in 1512 as personal physician to Cardinal Bandinello Sauli (d. 1518), drawn by the vibrant group of humanists that had poured into the ancient city. There he caught the eye of Pope Leo X, who appointed him to teaching positions at the university and introduced him to his cousin, Cardinal Giulio de' Medici. Giovio became a favorite of the cardinal and moved over to his court in 1517, when his former patron suffered a spectacular fall from grace. Six years later, Cardinal de' Medici became Pope Clement VII, with Giovio part of his intimate household. He remained a close companion to the pontiff until Clement's death in 1534.[45]

As Clement had several other physicians, Giovio's medical role at the papal court has generally been viewed as minimal. It is thus highly significant that he appears as a central figure in the trial of Caravita's oil. His involvement in the test was almost certainly a marker of Clement's trust in him, but it may also signify a genuine interest on Giovio's part. According to the great Italian literary critic Girolamo Tiraboschi (1731–94), Giovio printed an assessment of an antidote oil in 1523, which he signed with his name and the initials D.P.P., signifying his role as papal physician. This document no longer exists, and Tiraboschi gave no additional details, but if there was a publication prior to the *Testimonium*, it suggests that Giovio was particularly attentive to poison antidotes.[46]

Giovio also likely had ample experience in the use of criminal bodies for anatomical investigation. He had trained in Pavia under the tutelage of a physician named Marcantonio della Torre, an ambitious young anatomist who is best known today for his collaborations with Leo-

2.3 and 2.4 Historiated initials in Andreas Vesalius's *Fabrica* (1543) depicting putti performing medical student tasks. Pictured here: removing a cadaver's head to dissect the brain (D) and gathering bones from a graveyard (P). Courtesy of the Harvard Medical Library in the Francis A. Countway Library of Medicine. Photos by Jessica Murphy.

nardo da Vinci. During Giovio's time in Pavia, in the winter of 1510–11, Marcantonio and Leonardo dissected around thirty cadavers together, memorialized in Leonardo's private notebooks. Their collaborative work ended when Marcantonio died suddenly of plague in the spring of 1511.[47] Giovio never specifically addressed his own role in Marcantonio's dissections, although he wrote glowingly of Leonardo's anatomical illustrations and of Marcantonio's talents as a physician. In a later eulogy, he specifically praised his mentor's anatomies on "the bodies of the condemned."[48] Medical students generally were expected to help their professors in the more tedious anatomical tasks, reflected in the whimsical block initials of Andreas Vesalius's famous *Fabrica* (1543) (see figs. 2.3 and 2.4). Whatever his own involvement in Marcantonio's work, Giovio made it clear that he viewed his mentor's experiments on the "condemned" as a praiseworthy enterprise.

The Pope's Body

Anatomical practices also helped alert Pope Clement to the dire threat of poison. From the early sixteenth century, autopsies were conducted regularly on deceased popes, especially those who expired in circumstances that smacked of poisoning.[49] One such case occurred at the death of Pope Leo X, who died suddenly, following a brief illness, on December 1, 1521. As reported by the pontiff's master of ceremonies, Paride de Grassi, "the pope's body was opened, and his heart and spleen were found to be corroded and the spleen similarly partially cauterized, which the surgeons and physicians viewed with amazement." The assembled medical experts declared "for certain" that the pope had been poisoned. The pontiff's cupbearer was arrested and examined but found innocent, and the case ended inconclusively.[50] This event took place only two years before Clement assumed the pontificate, and Leo was Clement's first cousin and close friend, both from the powerful Medici family of Florence. Partisans of Leo's immediate successor, the Dutch pope Adrian VI, similarly raised concerns about poison when he died only a year into his controversy-ridden papacy.[51] Adrian's body showed no suspicious signs, but the concerns about poison certainly would have reached Clement's ears.[52]

The first sixteenth-century poison treatise also appeared in Rome in the 1520s: Ferdinand Ponzetti's *Book on Poison* (*Libellus de venenis*), published in 1521. Ponzetti (d. 1527) was a cardinal of the Church, elevated by Leo X in 1517 in the same ceremony in which Clement himself became the Cardinal de' Medici. He participated in the bitterly contested conclave that elected Pope Clement, on the Medici side.[53] For the most part, Ponzetti's book was a compilation of previous authors' notions of poison, although it privileged ancient texts over Arabic and medieval works. He discussed poison mainly on a philosophical rather than a practical level. Nevertheless, he included a long chapter on an itinerant group known as the "house of St. Paul" or *pauliani*, who conducted marketplace shows with poisonous snakes. The men, according to Ponzetti, would ingest various antidotes that would make them immune to the bites of poisonous beasts. They also sold an antidote "of which they promise so many miracles," made of terra sigillata, Armenian bole, and juniper berries mixed with goat's blood. This section is remarkable for the credibility that Ponzetti gave to the antidotes and the extent to which he based his testimony on his own discussions with members of the *pauliani*.[54] As Katharine Park has shown, Ponzetti was among the first authors to connect the *pauliani* with poison demonstrations, an association that soon became common.[55] These dramatic marketplace tests of poison antidotes would later become a foil for learned poison trials, as we will see in the next chapter. In the 1520s, however, the practice seems to have been novel enough that Ponzetti took it more or less at face value.

Poison was in the air in 1520s Rome, both figuratively and literally. Plague, the disease most connected to poison, struck the city in 1522, in the first summer of Pope Adrian VI's pontificate. In a letter written August 11, the nobleman Baldassare Castiglione (1478–1529) claimed that four thousand people had already fled.[56] By early September, plague was raging in all quarters of Rome, and Pope Adrian's curia were counseling him to leave the city. The pope refused. Mortality rates reached one hundred per day in early October, and most of the city's cardinals fled (including the future Clement VII). One cardinal died, but the pope continued to stay in the city. Castiglione reported in late October that "only a few men have survived."[57] By December, finally, the brunt of the epidemic had passed and the city slowly began to recover. In August of 1523, Paolo Giovio wrote with relief that "Rome triumphs without plague," although he had not been in the city during the main epidemic.[58] Nevertheless, plague continued to trouble Rome during Clement's pontificate. Castiglione noted in July 1524 that the plague was "much diminished, but not extinct."[59] If Giovio's population estimates from 1516–17 are correct, the total population of Rome decreased by over 30,000 in a decade, from approximately 85,000 to 53,897.[60]

In August 1524, then, Pope Clement VII had every motivation to view a poison trial as a measure that made both medical and political sense. The twin threats of malicious poisoning and plague made the search for effective antidotes a matter of utmost urgency. Ponzetti's recounting of the *pauliani* marketplace tests suggests a genuine interest in new, tested antidotes. The timely example of Caravita's oil—used as a plague antidote and promised as a cure for poison—likely offered too good an opportunity to pass up. In various ways, condemned criminals would have seemed the obvious test subjects. In addition to the use of criminal bodies for anatomical dissection and the classical examples of kings who used condemned criminals to test poison antidotes, legal developments in Renaissance Europe more broadly may have made such a test easier to justify. Increasingly, individuals were seen as "important resources of the state to be managed for the common good, as defined by the prince," as John Brackett noted, a trend that was accompanied by a consolidation of legal power under the Renaissance prince (including the pope).[61] For Clement, a poison trial, if successful, could help prove his power to defend himself from enemies and to protect his population from the plague.

Although this course of action made sense contextually, it nevertheless lay outside of normal procedure. Despite the rampant use of deceased condemned criminals for anatomical investigation, there was no established practice of using living criminals for tests of poison. The key, then, was to portray the trial of Caravita's oil in a manner that both underscored its deference to convention and trumpeted its results. The *Testimonium* published under Clement's name and signed by three "trustworthy men" did just that. It highlighted Clement's power vis-à-vis his predecessor, his attention to the public good, the orderly progression of events following both legal and religious customs, and the unassailable medical success. Although the text was written as a straightforward narrative, a careful examination reveals a deliberate attempt to present a highly unusual event as legitimate and incontestable.

Caravita's Oil Revisited

Published on "the ides of August," only a few days after the poison trials took place, the *Testimonium* was a slim pamphlet of just four pages. Although brief, it delivered a large amount of information with a strong core message, as signified by its opening lines. The pamphlet's title described the trial of Caravita's oil as an experiment conducted by "eminent men" at "the pope's command," and the document opened with a greeting "to all good mortals" from the three signatories: the Roman senator Pietro Borghese, the physician Paolo Giovio, and the pharmacist Tomasso Bigliotti. The involvement of these three men emphasized the trial's legitimacy; they explicitly presented the pamphlet as "the testimony of trustworthy individuals" (that is, themselves).[62] Giovio was a highly regarded intellectual, Bigliotti from a respected family of pharmacists, Borghese the city senator. Borghese's presence as the top signatory was imbued with particular meaning. As we have seen, the 1531 statutes delineating the procedure for obtaining corpses for university anatomical dissection required permission from both a papal representative and the Roman governor or senator. Although this policy was not official in 1524, it already may have been common practice. The exact procedure was followed in this instance, suggesting an attempt to conform to custom. The *Testimonium* emphasized both papal sanction (the pope's "command") and the written permission of the senator.

Although both the senator and the governor oversaw criminal courts in the complex Roman justice system, the senator was the head of the old civic court of Rome.[63] Pietro Borghese, member of a prominent Sienese family, had been senator under Pope Leo X in 1515–16; he is generally not cited as returning to that role in 1524, but documentation from the period is spotty, and it was previously unclear who occupied the senatorial role in the first part of 1524.[64] The *Testimonium* provides a partial answer. Borghese's signature on the document implied secular legal sanction on behalf of the Roman population and emphasized that regular procedure had been followed regarding the use of the criminals' bodies.

At the same time, a number of departures made this case very different from the process in university anatomical dissections. Borghese had

2.5 The Campidoglio in Rome after Michelangelo's redesign, 1536–46, in the *Speculum Romanae Magnificentiae.* Etching by Étienne Dupérac, 1569. Metropolitan Museum of Art.

a *physical* presence at the poison trial, which would not have been the case for a dissection, and the event took place at the Campidoglio, site of the senator's notorious prison (see fig. 2.5). Borghese's signature not only confirmed his permission but also emphasized his role as a witness to the event: "this is, and is testified by, Pietro Borghese, City Senator."[65] This uncustomary role for the senator suggests that the poison trial was quite unusual. It was not a conventional use of condemned criminals' bodies and thus required extra evidence of its legitimacy.

That unconventionality shone through the opening lines of the *Testimonium,* which explicitly addressed both the validity of the poison trial and the dire need for it. The authors asserted that "it was reasonable and useful to everyone [for us] to have tested the virtues of the marvelous oil that Master Gregorio Caravita of Bologna created." They also noted that Caravita had used the oil in "that time under Pope Adrian VI when Rome was devastated by a most terrible pestilence" and had "cured all sorts of men infected with pestilential disease in the hospital of S. Giovanni Laterano," one of Rome's smaller hospitals.[66] By putting plague in the background, the authors both reminded their readers of the city's misery under Clement's predecessor and created a public justification for testing whether the antidote worked against poison. Pope Clement, indeed, specifically commanded that the test be conducted on "condemned bodies, for the public good." Whatever Clement's own fears of poison, the test of Caravita's oil was depicted as a public benefit.

The two criminals chosen for the initial test were Corsican mercenaries named Gianfrancesco and Ambrogio. Corsicans, a minority population in Rome, were widely considered to be particularly violent and scurrilous, and mercenary soldiers doubly so.[67] The two men had been convicted and sentenced in the court of the magistrates, although their specific crimes were not listed, and had been handed over to the executioner to be beheaded with an axe. The prisoner used for the second test, a Mantuan named Antonio, had been sentenced to death for murder.[68] None of these test subjects was likely to cause an uproar: all of them came from outside of Rome, the regular procedure of trial and conviction in the criminal courts had been followed, and all appeared deserving of punishment. These men would have been excellent candidates for a public dissection, had their death sentences occurred in winter.

The authors of the *Testimonium* similarly indicated that they hewed scrupulously to religious conventions in the course of the trial. Accompanying the terrified convicts, they noted, was "a crowd of pious men," who "by custom urged them to ask for forgiveness of their sins and said numerous prayers on their behalf."[69] These men were members of the Confraternity of San Giovanni Decollato (St. John the Decapitated), an organization that took charge of guiding condemned prisoners in Rome to a good, pious death. If they died in the right way with God, accepting their sentence and asking forgiveness for their crimes, executed criminals had the possibility of getting time off from purgatory—or even skipping purgatory altogether and going directly to heaven.[70] Their presence was customary at all executions and once again signified that the proper procedure was being followed.

Not only did the criminals receive the expected religious ministrations, the pamphlet suggests, they also experienced a just punishment. Giovio and his colleagues instructed Caravita to anoint only one of the criminals (Gianfrancesco) with the marvelous oil, while the other (Ambrogio) was left to die "without any remedy." They depicted Gianfrancesco as pious and repentant, ready to face his fate. Ambrogio, in contrast, was portrayed as sullen and rude, and he refused to cooperate with the testers. This unsubtle distinction between the "pious" and the "nefarious" criminal calls to mind the story of the two thieves crucified alongside Jesus, one penitent, the other unrepentant. The good and bad thieves were a familiar motif in Renaissance literature and art—including the images held up to condemned criminals by confraternities like San Giovanni Decollato, for whom the good thief represented the certainty of redemption.[71] The contrast between Gianfrancesco and Ambrogio thus contained an unsubtle message: Gianfrancesco was worthy of being saved—spiritually, but also, in this case, physically—while Ambrogio deserved his agonizing death. Yet Gianfrancesco did not completely escape retribution for his misdeeds after he survived the poison trial. As punishment for his crimes, Pope Clement relegated him to the slave galleys. Antonio, the condemned criminal who survived the second poison trial, received a similar punishment: "the pope saved his life, yet banished him to perpetual rowing as a penalty for murder."[72]

This last point highlights the complex considerations involved in using condemned criminals as medical test subjects. The two criminals who survived the poison had a special status that made Pope Clement hesitant to condemn them to death a second time. Yet they still had to atone for their crimes. The galleys provided a solution to the balancing act that weighed the need to set an example against a desire to recognize the remarkable survival of the two men. In later poison trials, surviving criminals tended to be set free, as we will see in the coming chapters—but in this first instance, Pope Clement regarded the galleys as reward enough. The death of Ambrogio, the "bad" criminal who received no antidote, passed with little enough mention that it does not appear to have been a matter of concern; after all, it was the expected outcome of the originally scheduled execution. The pope's bigger problem was how to handle the criminals who survived.

Significantly, survival was attributed to medical rather than divine intervention. There were obvious religious undertones of the entire scene—in addition to the points already discussed, Caravita was a former monk, and the antidote was applied by anointing, with clear analogies to holy oil. Nevertheless, the *Testimonium* gave credit for the prisoners' survival to the oil itself, rather than to God's intervention on behalf of the criminals. In that way, it was very different from the old practice of trial by ordeal discussed in chapter 1. While the prisoner in the second test, Antonio, described his participation in the trial as "a miracle," the authors of the *Testimonium* defined the oil by its "true and admirable virtues." Even in an era of dramatic religious upheavals, the pope's physician presented the trial as a medical event, not a miracle.

In fact, Giovio and his co-authors went to some lengths to emphasize that the trial of Caravita's oil produced a valid and meaningful medical outcome. They described exactly how they sourced the poisonous aconite from the Apennine mountains, reduced it to a powder, and made it into sweet marzipan cakes. When the villainous Ambrogio refused to eat his marzipan, Caravita gave him an additional dose of the napellus mixed into an egg combined with sugar, "in order to subject both [criminals] to the same experiment." Ambrogio was thus explicitly used as what we might now call a control subject, echoing the use of roosters in the Galenic trials. The authors also carefully detailed the effects of the poison—Gianfrancesco and Ambrogio were stricken by pains of the heart, paleness, nausea, and vomiting, while Antonio experienced great pain and became numb and cold. This gruesome detailing of symptoms provided evidence that the poison had worked, amplifying the recovery. Even more remarkably, the authors took rather drastic measures to allay potential suspicions that Caravita might have tricked them, "after the custom of itinerant surgeons." They gave Antonio both the poison and the antidote "with our own hands."[73] In their efforts to prove the validity of the trial, the testers assumed the role of poisoners without any apparent anxiety about this position, despite the long history of poison as a particularly nefarious crime.[74]

The purported suspicion of Caravita should not be taken at face value. Originally from Bologna, Caravita had been a respected surgeon in the city of Perugia. The ambitious young Pietro Andrea Mattioli had moved from Siena to Perugia to study surgery with him and then followed him to Rome, a testament to Caravita's reputation.[75] He was a foreigner in Rome and thus technically itinerant, but so were Giovio and Pope Clement VII, for that matter. Caravita certainly was no marketplace empiric, and Giovio and his fellow authors knew that. Indeed, Caravita's positive reputation induced them to bring his remedy in front of Pope Clement in the first place. Their emphasis on trickery was more likely an attempt to acknowledge and assuage potential doubts about the test.

The *Testimonium* is significant not only for the extensive detail it provides about the first known Renaissance poison trial, but also because it allows us to pinpoint the specific concerns it tried to dispel. The basic act of giving poison to living criminals does not appear to have been a matter of any great anxiety. Instead, the reassurances focused on legal and religious conventions (following established procedure) as well as public order (relegating the surviving criminals to the galleys to atone for their crimes) and professional identity (showing that Caravita was not tricking them). By implying that the test followed all norms and simply represented business as usual, the authors opened a space in which this very unusual medical event could occur.

The Oleum Clementis

If tests of poison antidotes on condemned criminals had taken place before the trial of Caravita's oil, they did not make waves. Caravita's oil stood out because it appeared to work. This success gave the pope a valuable weapon against attempted poisoning—or at least the appearance of one, perhaps the main impetus behind the publication of the *Testimonium*. The explicit mention of a generous payment to Caravita may also have acted as an invitation to other vendors of marvelous antidotes. With the combined forces of Pope Clement, Senator Borghese, and Paolo Giovio, the pamphlet signaled strongly that the results should be taken very seriously.

They were. As mentioned in the introduction to this book, the Mantuan agent L'Abbatino reported the marvelous event to Marquis Federico II Gonzaga with palpable excitement.[76] He named Giovio as his contact, and Federico appears to have followed up on this information. In September 1524, Giovio wrote a letter to Federico that indicated the marquis now had a sample of the oil but did not know how to use it. After the usual epistolary pleasantries, Giovio noted: "I send herein in print the directions for using the poison oil and what the Testimonial of its trial [*prova*] testifies," an almost certain reference to the printed *Testimonium*.[77] It is unclear whether Federico ever received the *Testimonium*, however, as the ambassador Francesco Gonzaga was still seeking it in November. "I have met with Dr. Giovio about obtaining a copy of the form and method of operation of this miraculous oil," he wrote to Federico. "I was unable to find one in the Roman booksellers, but [Giovio] promised to let me have at least two copies."[78] It is possible that Federico never received Giovio's letter, which is held in Houghton Library at Harvard today, rather than in the Gonzaga archive. It is also possible that he simply wanted additional copies. Either way, Caravita's oil grabbed his attention.

What happened to Caravita's oil after that point? While there is scant evidence in the historical record, we can piece together a story from a few clues. A similar poison trial appears in a popular book published in 1561, *The Secrets of Lady Isabella Cortese*, a collection of alchemical recipes. The very first recipe in the book relates an oil "against poison and plague," proven in a trial conducted by Pope Clement VII on two criminals—a nearly identical description to the trial of Caravita's oil. In this case, however, the purveyor of the antidote oil was Brother Gregorio Mezzo Capo, the Roman senator was listed as Simone Tornabuoni, and the pope's chamberlain was also in attendance. It is possible that this anecdote is merely a garbled retelling of the trial of Caravita's oil. However, Tornabuoni took up his role as Roman senator in October 1524, so it is more likely that this brief description relates a separate trial of a different antidote oil.[79]

In any case, Pope Clement's interest in poison continued. He commissioned a mounted unicorn horn in 1533, described by the artisan Benvenuto Cellini, who entered (and lost) the competition for the mount.[80] Matteo Corti, who served as Clement's personal physician from 1531 to 1534, wrote two unpublished treatises on poison and its cures during his time in Rome.[81] The connection between Clement and antidote oils also appears to have remained strong. Years later, Falloppio penned a fascinating description of a marvelous antidote oil called the Oleum Clementis, "of which Pope Clement was the inventor." He claimed that the oil, also known as "scorpion oil," had been created by Clement in 1527 during a bout of plague that had stricken all of Italy and that "everyone who used this oil survived." The method of applying the oil, he noted, was by rubbing it into the pulses, spine, head, and the area around the heart.[82]

The Oleum Clementis bears such a strong similarity to Caravita's oil that it is hard to imagine they were not closely related, if not one and the same. The circumstances Falloppio described warrant further examination. The year 1527 was indeed a plague year in Rome—it swept through the city in the summer, following on the heels of the brutal Sack of Rome on May 6 and compounding the misery of the city. Clement did not emerge a hero from the sack (quite the opposite), and he spent much of the year imprisoned in the Castel Sant'Angelo. It is rather difficult to picture him compounding remedies in that setting.[83] Yet one can imagine an attempt to distribute Caravita's oil (or similar) under Clement's name in order to improve the pontiff's low standing in the city. Falloppio's account certainly should not be taken at face value; it was written with historical distance and without a firsthand knowledge of Clement's role. Such a blithely positive take on the 1527 plague in Rome is certainly out of step with the stark historical reality. At the very least, however, it suggests that Caravita's oil had not lost its aura of success.

Falloppio drew a direct line between the Oleum Clementis and a later poison trial conducted in Ferrara by Benedetto Accolti, cardinal of Ravenna. In this case, Duke Ercole II d'Este of Ferrara (d. 1559) granted Accolti a "miserable man condemned to death" in order to test the oil. Falloppio personally witnessed the trial ("I saw it"), as did his mentor, the respected Paduan professor Antonio Musa Brasavola (d. 1555). According to Falloppio's report, the convict took approximately one ounce of realgar (arsenic sulfide) in a potion, vomited, and collapsed half dead several hours later. He was then salved with the oil. After half an hour, he felt quite refreshed, and he lived for five days with the continual reapplication of the antidote, "although he always had a hiccup." Falloppio claimed that "all present who previously derided the oil, Brasavola among them, started to believe." The assembled physicians speculated that the antidote had the power to repel the corrosive poison by vomiting. Falloppio added that he himself had had great success with the Oleum Clementis against vermin.[84]

Falloppio's offhand testimony, merely a side note in his text, provides two crucial points of evidence. First, Caravita's oil and Pope Clement's role in the proof of its success had not been forgotten. Even several decades later, the Oleum Clementis was worth mentioning. Second, the poison trials did not end with Pope Clement and Paolo Giovio. In the years after the trial in Rome, similar tests began to take place elsewhere. Yet there is an intriguing twist to Falloppio's testimony. He was not the only viewer to record a poison trial conducted by Cardinal Accolti. A second account appeared in an unsigned letter to Federico II Gonzaga, who apparently remained avidly interested in antidote oils. The letter described a very similar trial in more detail, with a much different portrayal of the outcome.[85]

In January 1539, according to the report, Cardinal Accolti wished to try "a certain oil" that allegedly had the power to counteract poisons. The duke of Ferrara granted him a condemned criminal as a test subject. In the dungeon of Ferrara's castle, the Castello Estense, the prisoner swallowed a dose of "sublimated mercury" (mercury chloride). Immediately thereafter, Accolti anointed the man with the oil, spreading it on his pulses, forehead, nostrils, and heart. After three quarters of an hour, the prisoner began to suffer multiple evacuations "from above and from below." The cardinal gave him an egg mixed into a glass of Malvasian wine and repeated the administration of the oil, but the vomiting did not cease. For two days, the poor prisoner suffered bouts of vomiting and diarrhea despite multiple attempts with the oil. Eventually the testers brought him to the hospital of Sant'Anna, a medieval institution that is still a working hospital today. Five days later, he died. The physicians who had been caring for him then decided to dissect his body to see the effects of the corrosive poison. They found his stomach completely burned, his intestines twisted and torn.[86]

In contrast to Falloppio's report of apparent success, this view of the trial depicts it as an abject failure—the criminal died after a week of suffering and was dissected. Was Falloppio's test the same trial, spun in a positive light? The evidence is murky. On one hand, there are a striking number of similarities between the two trials. However, other details leave open the possibility that this trial may have been a separate event. The prisoner described by Falloppio lived for five days, while the second prisoner survived for seven. The two convicts took different poisons (arsenic sulfide and mercury chloride), and the suffering of the first prisoner seemed far less dramatic than the fate that befell the prisoner described in the second document. These differences could be put down to faulty memory on the part of Falloppio, writing years later, but it could also suggest that a series of tests using mineral poisons took place in Ferrara. That strategy would follow the pattern set in Rome, in which a test using arsenic followed the test on napellus. Whatever the case, only Falloppio's positive account found its way into print, many years later.

Another failed trial took place in Bologna the following year, although the details are even patchier. A brief line in the archival record of executions in Bologna, dated February 27, 1540, noted only that a convicted thief, a porter named Bastiano di Gianfrancesco, had been taken to the "usual place of the Conferteria inside the Palazzo del Podestà," that is, the chapel run by the confraternity of Santa Maria della Morte. He was to be hanged the next morning, but instead died in the night "as the result of poison, which he was given in prison to make an experience, but without his knowledge." Bastiano's body was then hung from the gallows.[87] This brief account left no sense of how the "experience" with poison came to occur or who (besides the prisoner) was involved. The notary recording the prisoner's death focused instead on demonstrating that usual procedures had been followed—the visit with the confraternity and the eventual hanging of the corpse from the gallows.

Despite the missing details, conventions from the other poison trials around the same time can help us reconstruct a plausible version of the Bolognese "experience." As in other cases, it almost certainly took place with official permission. In this case, sanction likely would have come from Bologna's papal legate, who meted out justice in criminal matters.[88] At least one physician was present at all other poison trials to that point, and the phrase "to make an experience" hints that the same might have been true in Bologna. If so, the most likely candidate would have been Matteo Corti, the physician who had served Pope Clement VII and written two treatises on poison. By 1540, Corti was an eminent anatomy professor in Bologna and was just coming off a now-legendary tussle with the young anatomist Andreas Vesalius, who had given a spectacular series of lectures at the University of Bologna in

January 1540. In the course of his anatomical demonstrations on the cadaver of an executed criminal, Vesalius humiliated Corti by pointing out numerous places in which Corti's lectures, based on Galenic ideas of anatomy, were wrong.[89] It is tantalizing to imagine Corti attempting to redeem his tarnished reputation by overseeing a poison trial (which failed), although that is mere speculation. Whether or not Corti was present for the Bolognese poison trial, the event, coming on the heels of the anatomical showdown, evinced a common interest in investigating the condemned body. Had the trial in Bologna been successful, it may have appeared in more than just a brief notarial remark.

Condemned Bodies and Defiled Men

The northern Italian poison trials in 1539–40 demonstrate that the test of Caravita's oil was not a singular event. Fifteen years on, princes remained interested in testing antidotes on condemned criminals, and highly respected physicians participated with intellectual engagement—Brasavola and Falloppio for certain, and possibly Corti.[90] None of this testing produced the same astounding results as Pope Clement's trial, which is likely why they did not make it into printed works (save for Falloppio's account many years later). In fact, it is very likely that additional tests took place. The venerated Brasavola explicitly endorsed a broader use of condemned criminals for drug testing, writing in 1552 that unknown or dangerous medicines should be tried only on "defiled men" who had been condemned to death. He indicated that he had conducted tests on many convicts himself.[91] The few poison trials in the historical records likely hint at additional (mostly unsuccessful) cases that have gone unrecorded.

Brasavola's discussion of using condemned criminals for medical tests also touted the practice as a benefit to "all people," a justification related to Pope Clement's claims about the public good. We will return to this topic in chapter 4, at a time and place in which concerns with using criminals for fatal tests came into focus. In mid-sixteenth-century Italy, however, there was no sense that poison trials as such were a problem, as long as they fit within existing social and religious customs. Pope Clement's command that his surviving prisoners be relegated to the slave galleys and the Bolognese authorities' decision to hang the corpse of Bastiano di Gianfrancesco from the gallows both indicate concern with what came *after* a poison trial. But the interest with which physicians tested antidotes—and their openness about doing so—suggests that poison trials were viewed within the context of already accepted forms of empirical investigation on criminal bodies, living or dead. Norms of testing had traveled far from the dog trial in Tours.

Physicians' role in poison trials would only increase in the following years, expanding beyond Italy and beyond the focus on Caravita's oil and its derivatives. While the failed tests of Caravita's oil remained hidden in archival documents, physicians soon began to discuss their failures along with their successes openly in printed texts. As we will see in the next chapter, a far better documented set of poison trials took place between 1560 and 1570, giving an important forum to new ideas about experimentation and medical communication.

PART TWO

Experiments

THREE

Experimenting with Drugs: Mattioli's Scorpion Oil

A small group of physicians gathered in Prague Castle in the dark days of December 1561. They had come to watch a soldier, condemned to be hanged for thievery, take a dose of poison and then an antidote. The test aimed to see whether a special "powder against poison" created by Archduke Ferdinand II of Tyrol could overcome the deadly aconite known as napellus.[1] Previously, the archduke's physicians had found the princely poison powder to work wonderfully on a condemned criminal given arsenic, and the archduke's father, the Holy Roman Emperor Ferdinand I (1503–64), was intrigued. He asked the physicians to "make an experiment" of the powder using the aconite, thought to be especially deadly to humans. The famous Italian physician Pietro Andrea Mattioli led the proceedings. As the doctors looked on, the prisoner ate a dram (about 4g) of powdered napellus root mixed with rose sugar.[2]

The trial went awry from the start. Nothing happened. After an hour and a half sitting in a warm room, the prisoner still felt completely fine. The physicians worried that the colder climate made Bohemian aconite weaker than the Mediterranean plant described by Dioscorides, and they decided it would be wise to give the young man an added dose of the leaves and flowers. After two hours, he still felt nothing and was sent back to his prison cell. Most of the physicians left, with Mattioli, who lived nearby, left to keep tabs on the progress. After another hour passed, a prison guard notified Mattioli that the prisoner had become ill, and the doctor hurried to the cell. He found the man suffering from "ulcerous fatigue, a great imbecility, and grave complaints in his heart." Although the prisoner asserted boldly that he felt fine, the cold sweat on his brow and his fading pulse indicated that the poison was finally working.[3] He then drank a potion of the archduke's powder mixed with wine. Immediately thereafter, his eyes rolled back in his head gruesomely, he thrashed about, grabbed his throat, and would have fallen to the floor had the prison guard not grabbed him. Mattioli instructed the guard to splash wine on the man's face and pull his hair, at which point he revived. They let him lie down on the straw in a corner of his cell, as he complained of being very cold, and he soon vomited out "a putrid matter, part livid and part choleric." After that evacuation, he claimed to feel a bit better and turned on his side as if he wanted to sleep, but then "died suddenly."[4] The antidote had failed to overcome the poison.

This dramatic eyewitness report came from Mattioli himself, who began to add this poison trial to revised editions of his best-selling *Commentaries on Dioscorides* from 1563 onward.[5] Also present at the Prague trial was Mattioli's friend and colleague Georg Handsch (1529–78?), a Bohemian physician, who translated Mattioli into German in 1563 and included a slightly different version of this anecdote.[6] Together, the two men committed to print a poison trial that proved fatal. Ironically, both Mattioli and Handsch had an obvious self-interest in the trial's success. Archduke Ferdinand II, the creator of the poison powder, was their patron. The two men served together as the archduke's personal physicians. Why, then, did they include this detailed account of the archducal antidote's failure in a book destined to be read by thousands?

The answer, this chapter argues, lies in physicians' deliberate efforts to portray poison trials as learned medical experiments. As in the case of Paolo Giovio's poison trial under Pope Clement VII, Mattioli laid out the exact circumstances of the test in excruciating detail. He described the effects of the poison in ways that conformed with accepted medical theory—it first affected the heart, then moved to the stomach and intestines. The failure, too, merely confirmed existing scholarly opinion. According to Mattioli, the inability of the archducal poison powder to prevent the prisoner's death had nothing to do with the powder itself (which, he speculated, might have been an old batch). Instead, it reinforced the toxicity of the poison, proving the correctness of Avicenna's ideas about aconite.[7] Poison trials, in this telling, were a scholarly enterprise conducted in the footsteps of learned medical tradition. In a medical culture that increasingly focused on firsthand observation as evidence, they represented *con-*

trived experiences—events that could be controlled and manipulated—that yielded meaningful scholarly conclusions.[8]

There was a foil for this depiction. In the sixth book of his *Commentaries*, which dealt entirely with poison, Mattioli contrasted learned physicians' poison trials with another, less legitimate kind of testing. He warned his readers to guard themselves against "counterfeit theriac" sold in the piazzas by charlatans (*ceretani di banca*).[9] In contrast to the proper testing described by Galen, these "dishonest deceivers" used all sorts of "deceptive tricks" to appear to test their pseudo-theriac in public and hoodwink many "vulgar ignorants."[10] While Mattioli was known for his vitriolic attacks against perceived enemies, he described the charlatans' tricks in unusually extensive detail. His bluster calls to mind the concern about potential trickery in the trial of Caravita's oil, with its derogatory reference to "itinerant surgeons."[11] On the following page, however, Mattioli revealed that he, too, had an interest in promoting novel antidotes based on demonstrative trials. He advertised his own "oil of scorpions," which, he noted, was the same oil tested by Caravita in a trial he had witnessed personally.[12] Together, these passages point to a major problem for physicians in their nascent interest in testing poison: how to distinguish their poison trials from the tests conducted by lower-class, empirical practitioners, known varyingly as snake charmers, mountebanks, theriac peddlers, empirics, charlatans, and other, more derogatory names.

The period between 1560 and 1580 saw the heyday of the poison trials, when over a half dozen tests on condemned criminals took place in France, Italy, and the Holy Roman Empire. Mattioli was central to this expansion. At the imperial court in Prague, he personally oversaw several poison trials and described them in his widely read *Commentaries*. By proudly promoting poison trials, he opened the door for other physicians to do the same (and they did). At the same time, a different kind of poison trial lurked in the background. In the vibrant marketplaces of Renaissance Europe, itinerant healers demonstrated the efficacy of their antidotes in public spectacles intended as entertainment and advertisement, by testing them on poisoned animals—or sometimes, even more dramatically, by swallowing poison themselves.[13] It was standard practice for physicians to criticize these healers. It was something else entirely for them to conduct tests that, to the neutral observer, might appear vaguely similar.

To set themselves apart from lower-class empirics, physicians developed a model for a learned poison trial. This model involved both methods of testing and strategies of communication that presented a direct contrast to mountebanks' marketplace shows. In so doing, they explicitly cast experiment as a scholarly endeavor. A rich body of historical scholarship has demonstrated that the meaning of the term "experiment" before the later seventeenth century was quite different from our understanding today.[14] The words "experience" and "experiment" were relatively interchangeable, and many other words were also used to signify knowledge drawn from empirical evidence. Poison trials were no exception: they could be described as a "proof" (*proba*), an "experiment"(*experimentum* or *periculum*), an "experience" (*experientia*), a "history" (*historia*), or any combination of these terms. The vocabulary around the phenomenon was very murky.

Yet poison trials *were* experiments in the sixteenth-century conception of the term: that is, they were singular instances drawn from hands-on practice.[15] In the Middle Ages, this kind of experiential knowledge held a relatively low status, as physicians emphasized their learning in order to separate themselves from empirics. An experiment, while perhaps interesting, had no broader bearing on overall theory.[16] By the late fifteenth century, however, scholars began to place more weight on experiential knowledge, and physicians led the way. Historian Paula Findlen has highlighted the extent to which sixteenth-century physicians and apothecaries engaged in experimental practices, and Evan Ragland has demonstrated that sixteenth-century medical writings were chock full of references to "making an experiment" (*periculum facere*).[17]

Poison trials are particularly interesting in this context because they intersected directly with the actions of empirical healers. In order to claim poison trials as a scholarly enterprise, I argue in this chapter, physicians came up with new techniques of conducting and reporting experiments that vastly expanded the purview of sixteenth-century experimental thinking. The process was not pretty. Meticulously and unsympathetically, they recorded the terrible effects of poison on the human body, and often the symptoms of poisoning received more focus than the test's outcome. For physicians, these (to our minds) horrific details, combined with a recounting of classical examples, helped reclaim poison trials as a scholarly endeavor. A host of scholarship on scientific experiment by sociologists of science has demonstrated the deliberate effort that went into constructing a specific narrative of knowledge-making in science and medicine, from the early modern period until today.[18] This chapter demonstrates the role that poison trials played in that process.

Theriac Peddlers

The contrast between learned and popular poison trials grew out of a centuries-long attempt to distinguish learned physicians from other healers, particularly unlicensed "empirics" who sold drugs. In the early decades of the sixteenth century, those efforts focused particularly on theriac, which remained extraordinarily popular. Itinerant theriac sellers were common across Europe from at least the fourteenth century, and they became more prevalent in the fifteenth and early sixteenth centuries. They peddled their wares in German marketplaces, Italian piazzas, and French street spectacles. While the specific details varied from place to place, there were common characteristics among these healers: they were generally lower-class outsiders who traveled from town to town and market to market, and their sales technique usually involved a poison-laced public spectacle (see fig. 3.1). In Germany, "theriac hawkers" would give their theriac to frogs and toads (considered poisonous) or, in some cases, would put on dramatic shows of self-poisoning.[19] Mattioli similarly complained of charlatans who appeared to poison themselves.[20] The common complaint in nearly every physician's account was deception.

Not all itinerant drug peddlers hawked theriac. As we saw in chapter 2, the *pauliani* had begun to sell a version of terra sigillata in Italian marketplaces by the late 1400s. French physicians, meanwhile, complained of empirics' powders and herbs.[21] However, theriac particularly rubbed physicians the wrong way since, in their view, it was a drug with the full force of learned tradition behind it. In 1532, the German physician and poet Euricus Cordus (d. 1535) published a short treatise that vilified the widespread sale of inferior and fraudulent theriac. Part humanist study, part polemic, the slim book poured invective on "the loose itinerants and disloyal apothecaries" who were ruining the drug.[22] A respected humanist with a medical degree from Ferrara, Cordus chose to write this book in German in order to reach the large number of Germans who (in his telling) were buying medicaments from dubious sources. He seconded Pliny's observation that "every person from the lesser folk who claims to be a doctor is readily believed," and he complained about people's willingness to rely on "faithless Jews, sorcerers, crystal-gazers . . . and whatever other loose folk" for their remedies. Things had gotten so bad that every "rumored whore or runaway nun" claimed to be a doctor. True, many of their remedies were very popular, but "I would rather reach for the proper arts and hidden truths and be scorned . . . than follow popular error and be honored."[23]

3.1 Charlatan as portrayed by Giuseppe Maria Mitelli, 1660. Wellcome Collection. CC BY.

After building up this dichotomy between "proper arts" and "popular error," Cordus turned to theriac, the "greatest and highest medicine," as an example of the riches that the ancients had to offer if properly made.[24] The majority of his treatise was taken up with a long recipe for the "true" theriac of Andromachus and Galen, which included dozens of ingredients (many of them expensive imports) and required making three different kugel doughs or "little breads." Although the recipe was absurdly complicated, Cordus tried to make it easier for his readers to understand by explaining unfamiliar concepts and terms.[25] He later provided detailed advice on when to obtain the best vipers (early summer), how to catch them without being bitten (grab them by the neck), and how to keep them alive until needed (feed them).

The theriac recipe directly followed a chapter that introduced a different topic: how one should test theriac to see whether it was "good or bad, strong or powerless." Cordus suggested three different trials. The first two replicated the Galenic trials that we saw in chapter 1: giving a purgative drug to a person to see whether the theriac inhibited purgation, or testing poison on two hens, only one of which would receive the antidote. But Cordus did not stop there. In addition to these trials, he continued, one also could test theriac's ability to fend off certain poisons in human subjects, although he declined to name the specific poisons in order to keep the knowledge away from "evil people." He cautioned, "the trial is dangerous and should only be attempted on condemned people."[26] While Cordus gave no specific account of a poison trial, he made it clear that one could be done (with care). To my knowledge, this brief statement was the first reference to a poison trial on condemned criminals outside of Italy.

Like Mattioli, Cordus followed his suggestions for testing theriac with an attack against other, inferior kinds of poison trials. After suggesting that one could test poison on condemned criminals, he immediately pivoted back to his complaints about empirical practitioners. He warned the "common folk" to beware of "itinerant country frauds" who tested their theriac in marketplace shows by using "cunning and deceptive words" while pretending to drink poison. These flashy performances were nothing but "vain deceit" and their remedies an "evil thing" that caused physical harm. If, however, you could have true theriac, properly made, "you would have a precious treasure that would see you through all sorts of danger."[27] Cordus thus used poison trials to set up one of his many contrasts between proper learned medicine and shady empiricism. Just as charlatans' wares were false, so too were their fraudulent trials of poison a far cry from authentic, physician-recommended methods, which included tests using poison on condemned criminals.

Where did Cordus get these ideas? His invective against country mountebanks likely came from a marketplace context easily recognizable to his readers. There is, however, no evidence that German princes or physicians were conducting poison trials on condemned criminals in the 1530s. That reference almost certainly came from elsewhere. It is possible that Cordus drew the idea from Galen, who, as we have seen, mentioned the ancient practice of testing drugs on condemned criminals.

It is also possible, however, that he had heard about the trial of Caravita's oil from colleagues in Italy. Cordus had received his medical degree at the University of Ferrara in 1521, a student of the eminent professor Niccolò Leoniceno (d. 1524), and he remained in contact with (and respected by) physicians south of the Alps after his mentor's death.[28] Given the casual manner in which he mentioned the poison trial, it seems likely that he was referring to a known practice.

Whatever the case, Cordus's small treatise provides evidence that testing antidotes on criminals had piqued the interest of physicians in northern Europe in the decade after the trial of Caravita's oil. It also shows that this practice immediately called to mind a popular arena in which poison antidote testing already existed. For Cordus, the differing approach to poison trials was just one point in his overall contrast between properly compounded theriac and empirics' sloppy imitations. Physicians would continue to hammer the distinction between true theriac and its useless knockoffs for the next two centuries. Nevertheless, the problem of how to distinguish learned poison trials from empirics' marketplace shows remained, especially for new drugs that, unlike theriac, did not have millennia of learned medicine solidly behind them.

Scorpion Oil

Cordus's theriac treatise reached a limited audience: it appears to have gone through only one printing and remains a very rare work today. Similarly, only a few copies of Pope Clement VII's *Testimonium,* describing the trial of Caravita's oil, survive today—and even back in 1524, the marquis of Mantua had a hard time getting a copy. Both works were slim pamphlets, and both addressed local concerns (the recent Roman plague in Clement's case; German drug-selling practices in Cordus's). Even though a handful of poison trials took place in the 1530s, they made waves only very locally and did not appear in printed documents. While clearly intriguing, poison trials had a limited currency in their initial introduction.

That state of affairs changed in 1544, when Mattioli published an Italian translation of and commentary on an ancient herbal, *On Materia Medica,* by the first-century Greek author Dioscorides (c. 40–90 BCE). A commentary on a Greek medical tome may not sound particularly groundbreaking, but humanist botanists in the early sixteenth century had settled on Dioscorides as the expert on medicinal materials writ large.[29] In his initial publication, Mattioli aimed to provide Italian readers with a complete edition of Dioscorides' five books. He seamlessly

3.2 Moretto da Brescia, Portrait of Mattioli as a Young Doctor, 1533. Musei di Strada Nuova. Wikimedia Commons.

combined the original text with information drawn from other Greek and Arabic medical authorities and his own experience.[30] The approach was an instant success, far more than Mattioli himself ever expected. In 1548, he added illustrations and a sixth book on poisons, and in 1554 he translated the entire work into Latin. He continued to update the *Commentaries*, publishing new Italian and Latin editions throughout his life (and thereafter). The book was translated into Czech (1562), German (1563), and French (1566). In 1568, Mattioli boasted that he had already sold over thirty thousand copies. Eventually, the work became known simply as the "Mattioli," and Dioscorides faded from view (see fig. 3.2).[31]

Mattioli's astounding success helped transform poison trials from little-known local tests into a broader phenomenon across Renaissance Europe. Beginning with the first edition of the *Commentaries*, Mattioli included a brief description of Pope Clement's poison trial in his chapter on napellus, widely seen as especially deadly to humans. Mattioli could confirm its poisonous nature firsthand, he stated, because "I saw its effects in Rome, in the time of Pope Clement VII," when the pontiff had requested "an experiment of a certain oil against all poisons." Mattioli had a very personal tie to this test. He noted that the oil was the creation of "Brother Gregorio Caravita the Bolognese, my tutor in surgery." He then related an abbreviated account of the trial of Caravita's oil on the two Corsican prisoners. The prisoner who ate a larger dose of napellus received the antidote and recovered. The prisoner who ate less "was left to die without any remedy." The outcome, Mattioli noted dispassionately, served to reinforce Avicenna's opinion on the particularly deadly nature of the aconite.[32]

With this anecdote, Mattioli tied himself not only to the trial of Caravita's oil, but to Caravita himself. Although he eventually became one of the best-known physicians in Europe, little in Mattioli's early career would have hinted at later fame. After finishing a medical degree in Padua, he moved to Perugia to study surgery with Caravita and then followed him to Rome. There he worked with sufferers of the French disease (syphilis), about which he published a short treatise in 1530.[33] In 1527 he fled the Sack of Rome and escaped to Trent, where he worked as the city physician.[34] The success of the *Commentaries* offered the ambitious Mattioli a way out of small-town doctoring. Yet even though he eventually became a prominent defender of learned medicine, he remained interested in practical pharmacy and especially in new drugs. His letters are full of pharmaceutical recipes and queries about drugs, and in the 1560s he added a segment on distillation and alchemical medicine to all editions of his *Commentaries*.[35] His book on poison, added in 1548, was a testament to his enduring fascination with poisons and antidotes, continuing the centuries-long tradition of poison treatises discussed in chapter 1.

Alongside his fascination with novelty, Mattioli wielded the classical tradition to his advantage. The book on poisons was littered with classical references, often in service to his modern aims. In his dedication to the princely ruler of Trent, Cardinal Cristoforo Madruzzo, Mattioli described the grave threat that poison posed to Renaissance men and women of all walks of life. Evil people plotted murder with poison, poisonous animals lurked around every corner, and poisonous diseases could wipe out entire families.[36] Thankfully, there were ways to counter this ever-present danger: eminent men of the ancient world had created diverse antidotes to poison. In particular, "the great Mithridates, king of Pontus," had sought to discover "true knowledge of the substances that would valiantly overcome all poisons." Similarly, Attalus of Pergamum, "no less celebrated by Galen than Mithridates," had compounded "not merely one antidote, but many and diverse mixtures." There was also, of course, the shining example of Andromachus, physician to Emperor Nero, whose theriac had "preserved all of the Emperors and other grand personages" of ancient Rome against poison. Andromachus was not alone, however. Mattioli listed over twenty classical physicians who had created marvelous antidotes, including Galen himself. His new book of poisons, he stated, presented an attempt to recover some of this great knowledge "in our current most flourishing age"—a Renaissance paean to ancient wisdom.[37]

Mattioli later contrasted these marvelous ancient cures with villainous modern-day charlatans and their quack medicine. The very term he used to describe the healers, *ceretani di banca*, highlighted their deceit. As David Gentilcore has shown, *cerretano* was a common epithet for charlatans that emphasized fraud, as it evoked phony mendicants who falsely claimed to beg alms for the shrine of Cerreto in Umbria.[38] Mattioli thus linked marketplace theriac hawkers to sacrilegious deception, while also drawing parallels between himself and ancient physicians. Galen, he noted, had been similarly "fed up (as he said) with deceptive deceivers of infinite deception" who sold false theriac to make a profit and hoodwinked "vulgar ignorants deceived by the antidote's name." However, Mattioli asserted, one could see through these scammers by knowing their various tricks. As Dioscorides and Avicenna had both attested, poison was less harmful on a full stomach. The charlatans thus would eat "large quantities of lettuces" in a salad swimming in oil or gobble up fatty ox-tripe in their chambers, until their stomachs were "as taut as a drum." In this way, they could appear to swallow arsenic without harm and then vomit it up later, in secret.[39] Mattioli continued his invective for pages, with the consistent argument that these empirical practitioners were nothing but "ruffians" who "steal money from others' purses."[40] Their cures had no merit.

Mattioli was not, however, opposed to new medicines in principle. Quite the contrary. Immediately following his diatribe against the charlatans, he noted that he could not let the topic go without mentioning that he, Mattioli, had a few "very valorous" antidotes of his own. The first was a distilled water comprising over a hundred ingredients, including a smorgasbord of remedies already considered to have powers against poison: theriac, mithridatium, unicorn horn, terra sigillata, emerald, viper flesh. It also contained opium, in addition to dozens of herbs, spices, gums, oils, and roots. The preparation was complex, involving multiple steps and over a month of labor, including distillation. The effort was worth it, he claimed, for the miraculous effects against all kinds of poison.[41]

The second antidote was a "marvelous oil against poison." Mattioli explained that this was the same oil he had already mentioned in his entry on aconite, which I have called Caravita's oil, but Mattioli termed his "oil of scorpions." The title was no misnomer: the recipe required three hundred live scorpions "gathered in the heat of the summer." It also called for a one-hundred-year-old oil or, lacking that, "the oldest you can obtain," which was to be infused with herbs in stages in a bain marie (water bath) until it achieved a blood red color. The scorpions were then put in an alembic in hot ashes, until they began to sweat and become enraged, after which one was to transfer them to the warmed oil, immediately seal the alembic, and let it steep in a bain marie for three days, then squeeze out and discard the dead scorpions. A host of other ingredients were then added to the oil, including both theriac and mithridatium. This remedy, Mattioli explained, worked "miraculously" against all poison, but especially aconite, as shown in the experiment on the two Corsican prisoners.[42]

In his invective against charlatans, then, Mattioli created a narrative of useless, fraudulent medicines, sold with a sleight of hand. He used this characterization to set up a contrasting tale of his own effective, near-miraculous antidotes, which he presented openly, "for the benefit of everyone," rather than keeping them a secret "as most of today's physicians tend to do."[43] In this way, he turned a medicine that could have been viewed with skepticism into a valid, desirable remedy. His gambit worked. As Piero Gambaccini has noted, scorpion oil soon entered the Italian pharmacopoeia, where it stayed for centuries. It was almost too successful, as it was quickly imitated.[44] The Medici dukes created their own, highly prized scorpion oil based on a simplified version of Mattioli's recipe, while charlatans fashioned cheaper scorpion oils.[45] Today the phrase "scorpion oil" calls to mind a quack medicine. Mattioli introduced it as exactly the opposite.

Mattioli's Poison Trials

Mattioli wrote little about how, exactly, his antidotes had been developed, but the glimpses he did give pointed to poison trials as one crucial method. In the 1548 dedication to his book on poisons, he noted that Mithridates had conducted "trials of this or that" antidote, using "this or that poison" on "great numbers of men condemned to death."[46] Just before his discussion of charlatans' fraudulent methods, he also directly referenced the Galenic poison trials on wild roosters. Initially, he noted, Romans had not fully known how to make theriac. However, diligent physicians used their knowledge to compound antidotes and to test them "in the manner described by Galen, writing to Piso," and had thereby developed the true theriac.[47] Both of these references put a new focus on poison trials. As we saw in chapter 1, the legend of Mithridates and his tests on criminals appeared in Galen's writings, but the few late medieval authors who mentioned Mithridates tended to emphasize his self-poisoning rather than his use of criminal test subjects. Mattioli brought criminal testing back to the fore.

In subsequent editions, Mattioli's focus on poison trials only increased. In the expanded and overhauled 1568 Italian edition, he moved the legend of Mithridates to the very beginning of the *Commentaries*, adding it to the volume's dedicatory letter to Princess Johanna of Austria, daughter of Emperor Ferdinand I and wife of Francesco I de' Medici. He also added new details to the tale. Mithridates, he asserted, was highly learned and spoke twenty-two languages, on its own enough to gain him "eternal fame." But he did not stop there. Later in his life, he managed to "obtain greater fame and an immortal name" by learning the properties of all simple medicines, in order to find remedies for poisons. He conducted numerous trials (*prouva*) to test the effectiveness of specific simple medicines against specific poisons, using "wicked men who were condemned to death for their crimes." He thus learned "by experience" the best way to fend off poison, and he compiled all of these substances into his famous mithridatium. Similarly, Attalus of Pergamum compounded multiple antidotes using the same methods.[48] Once again, Mattioli catalogued a learned tradition of poison trials, this time in the opening pages of the *Commentaries*.

Mattioli then placed himself within this scholarly tradition and contrasted it with the fraudulent antidotes sold by "deceivers." Unfortunately, he continued, physicians of his day no longer had access to most of these marvelous ancient medicines, even though the danger from poison was even greater. To fill the void, he had decided to compound "new antidotes" that would "correspond to the ancients' in their worth." He had thus, with great effort and "long experience," created his own antidotes, which, he noted, were described further in book 6 of the *Commentaries*. These remedies were meant not to replace the great antidotes of the ancients, theriac and mithridatium, he assured his readers, but to add to them, especially given the high incidence of fraud in the sale of those prized drugs. He recommended that everyone buy theriac—and his oil of scorpions—from the trusted apothecary Francesco Calzolari (1522–1609) in Verona, who was widely admired for his authentic theriac.[49] Mattioli thus placed his oil of scorpions on a par with theriac, compiled and sold by a trustworthy apothecary. It was nothing like the fraudulent antidotes sold by deceitful, money-grubbing empirics.

From the mid-1560s, Mattioli also began to include three new poison trials in his entry on napellus. The most extensive of these accounts was the failed test of Archduke Ferdinand's poison powder at the Prague court in 1561, described at the beginning of this chapter, which he added to the *Commentaries* from 1563 onward.[50] As we have seen, he described the progression of the Prague trial in blow-by-blow detail. In the 1565 Latin edition, he also made it abundantly clear that he had played a central role in this trial as the main scholarly expert. When the Bohemian aconite initially failed to work, all of the other physicians went home, "leaving the cure to me alone." Against the prisoner's own assertions that he felt fine, Mattioli was able to read the signs (including pulse) suggesting the opposite.[51] He described the prisoner's vomit with words such as "putrid," "livid," and "choleric," all terms drawn from learned medical theory. Even in the description of the man's eventual death, Mattioli asserted his scholarly perspective, arguing that it occurred "partially because of the deadliness of the poison and partially because the antidote, owing to its age, was very much past its prime."[52] The trial did not make a definitive statement about Archduke Ferdinand's powder, but it made clear the toxicity of the napellus. It had an epistemological purpose.

As a caveat to this story, Mattioli explained that a different result had occurred in another test. The same dose of aconite had been given to a twenty-seven-year-old condemned criminal, in order to "make a trial of bezoar stone." This time he did not describe any direct involvement; he did not even indicate whether he had personally witnessed the trial or simply heard of it from others. Nevertheless, he was meticulous in relating every detail. After drinking the "deadly beverage," he wrote, the young man said he felt a burning in his throat, as if it were full of pepper. An hour later, after the prisoner began to vomit, he was given seven grains of powdered bezoar mixed in wine, after which "various and diverse symptoms began to occur." The man vomited a thick, green matter, and he said he felt as if he had some sort of ball in his navel that was ascending toward his stomach and sent a cold wind to his forehead and body. Soon thereafter a dullness came over him, "not dissimilar to paralysis," which stretched from his hands and legs to his "hind parts," so that it was difficult for him to move his fingers. Slowly, however, the paralysis eased, first in the left side of the body, then in the right, and he complained that all of his veins felt cold. He then experienced varying symptoms in his head, face, body, and digestive system, with a fading pulse and a disturbed mind. Finally, after seven hours, "the antidote conquered the poison," and all of the symptoms disappeared. The torments the man experienced, Mattioli once again noted, made it impossible to disagree with Avicenna's opinion on the aconite's toxic nature (and he took the opportunity to criticize the conflicting views of Leonhard Fuchs on the matter).[53]

Mattioli's extensive description of these two Prague tests reinforced his view of poison trials as useful medical "proofs." Whereas his mention of the poison trials conducted by Mithridates (on criminals) and Galen (on roosters) depicted poison trials as part of a venerated scholarly tradition, these symptom-by-symptom accounts of prisoners' painful battles with poison established them as a valid modern tool. This version of Mattioli's poison trials spread to nearly every corner of Europe, as more or less the same account appeared in all Latin, French, and Italian editions after 1565.

Georg Handsch's Poison Trials

In its initial publication, however, the portrayal of Mattioli's involvement in the Prague poison trials was far less dramatic. The test of Archduke Ferdinand's poison powder first appeared not in the Latin or Italian editions, but in the 1563 German translation by Georg Handsch, a Bohemian physician who has been the subject of recent studies by Michael Stolberg.[54] After receiving a medical degree from Ferrara, the young Handsch returned to Prague in 1555 and acted as a kind of apprentice to Mattioli. From the 1560s, he became a personal physician to Archduke Ferdinand in his own right, and he stayed with the archducal family for most of his career, eventually following Ferdinand to Innsbruck. He left behind over 4,000 pages of handwritten records on his medical practice, but his translation of Mattioli's *Commentaries* was one of his only publications (see fig. 3.3).[55]

Handsch's translation of Mattioli did not follow the original exactly, even though Mattioli praised his colleague for rendering his Latin edition "accurately and diligently" into German.[56] The book was titled *New*

3.3 Woodcut of Mattioli in Georg Handsch's German translation, 1563. Courtesy of Houghton Library, Harvard University.

Herbal (*New Kreütterbuch*), and accordingly, it focused on plants. It left out the animals that Mattioli described in book 2, and it omitted books 5 and 6 entirely. In many ways, it closely resembled the 1562 Czech edition, which also appeared in the context of Emperor Ferdinand I's court. As in the Czech version, it included large new woodcuts of the plants and a richly illustrated excerpt on distillation, which eventually became standard in Mattioli editions.[57] Yet unlike the abbreviated Czech edition, Handsch's German translation included the full text of Mattioli's entries on plants and added new details: the title page advertised "many fine new experiments" and "artful distilling ovens." Among these "new experiments" was the Prague poison trial. There was no published original text from which to translate this anecdote, and while the basic outlines corresponded with Mattioli's later account, many details differed—suggesting that Handsch himself had significant latitude in shaping the text.

In contrast to Mattioli's enthusiasm, Handsch's translation was more circumspect about poison trials. Handsch-as-Mattioli introduced the story as one that "I witnessed in the year 1561 at the Prague court." He cautioned that he was relating it "only so that anyone who comes across this herb will know to protect himself from it." His version included a far more extensive description of the previous, successful trial of Archduke Ferdinand's powder on a condemned prisoner: the man took arsenic, shook violently, and swelled up before he took a dose of the archduke's powder, quickly recovered, and was set free. Because the emperor was holding court in Prague, it was decided to see if the drug worked against napellus, as "this herb is more poisonous than all other plants." In contrast to Mattioli's physician-centered account, Handsch's also put more people in the room: "In the presence of His Imperial Majesty, all of the illustrious imperial physicians, and other high-born people, the executioner gave [the poison] to a strong young man who had squandered his life through thievery and was to be hanged on the scaffold the next morning."[58]

Handsch's translation thus subtly diffused the responsibility for the trial away from Mattioli. The trial was done only as a follow-up to a successful test, and only because the emperor was in town and had commanded it. There were many witnesses, including the emperor himself, and the executioner (not a physician) handed the poison to the prisoner. Most significantly, however, Handsch's account did not leave Mattioli alone at the scene during the long period where the poison did not seem to work. Like Mattioli, he noted the extended delay in the poison taking effect. However, he simply explained that two hours after taking a second dose, the prisoner began to complain that his "whole body was tired, the heart heavy and exhausted." During the entire dramatic scene that ensued after the prisoner took the antidote, he did not name the main actors, but wrote impersonally that "one touched his forehead and pulses and found a cool sweat on his forehead, and that the pulse had begun to diminish."[59] As far as the reader could tell, there were still multiple physicians in the room, and it was unclear who was in charge.

Whereas Mattioli continually emphasized his direct involvement in the trial and frequently used the pronoun "I," Handsch-as-Mattioli used the collective "we" and left the actors vague, except for the executioner.

These two versions did not contradict each other directly—in fact, they match almost exactly in many places. Handsch's version merely added background context while omitting any specific mention of Mattioli's role in the trial. Those elisions are especially significant because Handsch's private notebooks confirm Mattioli's leading role. "On December 11," Handsch wrote in one entry from 1561, "Dr. Mattioli showed, in the presence of other physicians, [the effects] of one dram of napellus root mixed into a ball of rose sugar, on a twenty-nine-year-old man condemned to death." In his small, sloppy handwriting, Handsch recorded the hour-and-a-half wait for the poison to work, and he too noted that the prisoner "was returned to prison," where Mattioli gave him an additional dose of the aconite's leaves and flowers. His description of the progression of symptoms matches Mattioli's. Yet Handsch's account also casts one aspect of his mentor's version into doubt. After Mattioli administered the archducal powder, "he called me," and "I arrived to find the miserable [man] in a worrisome state." Mattioli could not feel a pulse, and "neither could I," and the vomiting began soon thereafter.[60] In contrast to Mattioli's portrayal of the archducal trial as an affair that he managed "alone," Handsch's contemporary notes suggest he assisted his mentor for most of the prisoner's battle with the symptoms and for his eventual death.

Why did Handsch's published translation fail to mention Mattioli's central role in the poison trial? And why, conversely, did Mattioli apparently exaggerate his solo role in the final stages of the prisoner's ordeal? I think the signs point toward a concern on Handsch's part about poison trials, or at least about disclosing them to the public, that Mattioli did not share. In his private notebooks, Handsch recorded several poison trials and left no indication that he found them problematic in any way.[61] Yet he not only minimized Mattioli's (and, indeed, his own) involvement in the test of the archducal poison powder in his published translation, he also left out the trial of Caravita's oil and the bezoar trial entirely, even though he recorded the latter trial in his private notebooks (dating it January 1562).[62] His translation did not include the book of poisons, the story of Mithridates' poison trials, the theme of deceitful charlatans, or Mattioli's marvelous remedies. As far as the topic of poison was concerned, Handsch's translation was a very different work from the Latin, Italian, and French versions of the *Commentaries*. Whereas Mattioli depicted poison trials as helpful tools for physicians to use in developing new antidotes, the German translation avoided (or at least evaded) the topic.

This dissonance may point to cultural differences between Italy and lands north of the Alps on the perception of condemned criminals, a topic to which we will return in the next chapter. However, it is impossible to know whether the elisions in Handsch's version represent a deliberate choice coordinated with Mattioli or whether they were executive decisions on Handsch's part. I suspect that Handsch feared a more complete accounting could be harmful to the physicians' reputations, but there is no firm proof. In any case, his omissions also serve to highlight the parts of the Prague poison trial that *did* make it into his translation. Although the text remained hazy about who, exactly, was conducting the trial, Handsch's version did not spare any details. The German version reflected the other Mattioli editions nearly exactly in describing the excruciating effects of the poison. Handsch also added a few points missing from Mattioli's anecdote. The dose of aconite, he explained, was obtained "from the Bohemian mountains, which the Bohemians call Krkonoff, at the origin of the Elbe; it lies on the border of Bohemia and Silesia, two miles away from the little city called Hohenelb—there this deadly poisonous herb grows in great quantity." This observation reinforced Handsch's focus on plants throughout his translation.[63] The toxic napellus, not Mattioli, took center stage in Handsch's version of the test.

Handsch and Mattioli may or may not have been in complete agreement about the way poison trials should be depicted to a wider audience, but they both contributed to the notion of poison trials as learned medical experiments. Through their detailed descriptions, made available to the public, they set themselves apart from charlatans' poison trials. Mattioli and Handsch pulled back the curtain to reveal (more or less) what took place in the prison cell, and they did not hide from instances where an antidote failed. Indeed, the most striking aspect of their account was the dispassionate catalogue of intentionally inflicted suffering. Handsch, in his private notebooks, referred to the Prague trial as Mattioli's "demonstration on a poisoned prisoner."[64] Detailing that demonstration took precedence over the outcome.

"More certain of the trial"

Handsch and Mattioli were not the only medical men to write about the Prague poison trials. Around 1564, the imperial surgeon Claudius Richardus penned a letter to Nicolaus Olahus (d. 1568), archbishop of Gran in

Hungary, describing the success of bezoar stone in two poison trials on condemned criminals. The original letter is lost, but two physicians later published copies: Thomas Jordan included the Latin version in his 1576 book on plague, and Johann Wittich translated it into German as part of his 1589 book on wondrous antidotes (see chapter 5).[65] The letter, which reads more like a treatise than a personal missive, may have circulated at the imperial court before Jordan printed it.[66] Its overall purpose was to highlight the marvelous properties of bezoar stone, both in cases of illness and in cases of poisoning. Richardus used a similarly thorough strategy in describing the two poison trials, and, if anything, he recounted the prisoners' symptoms in even more grisly detail. If we piece together his account with Mattioli's, it is possible to draw ties between a series of poison trials at a single court.

Richardus himself is something of a mystery, although it is clear that he was Emperor Ferdinand's chief surgeon and part of his official courtly household (*Hofstaat*). Mattioli's *Commentaries* mentioned him once, in his description of the imperial physicians' dissection of a beaver, and Richardus was listed among the imperial medical personnel present for the coronation of Emperor Ferdinand's son, Maximilian II, as King of the Germans in 1562.[67] There is evidence that he remained at the imperial court to serve Maximilian after Ferdinand's death.[68] He appears to have been a native German speaker, but if so, he was rather unusual.[69] Surgeons in the Holy Roman Empire were generally lower-status practitioners with little or no university education, but Richardus wrote in Latin. He also described the phenomena he observed in much the same way as a learned physician might, using terminology consistent with scholarly humoral medicine. The coronation pamphlet listed him incongruously as "D. Claudius Richardus, surgeon," the "D." (for Doctor) an honorific not generally attached to surgeons (who were called *Magister* or Master). Absent additional information, we must simply assume that he somehow obtained a deeper education than was the norm.[70] Whatever his background, Richardus recorded Ferdinand's poison trials at two different courts (Prague and Vienna) with meticulous care.

Indeed, the first trial Richardus described appears to be a more detailed account of the final test that Mattioli related in his *Commentaries*, in which a criminal took aconite and bezoar, survived, and was set free. In fact, Richardus hinted at an attempt to avoid some of the mishaps that occurred during Mattioli's test of the archducal poison powder. Although he never mentioned that trial specifically, the bezoar test appeared to try to mitigate the failure of the Bohemian napellus to take effect. Richardus noted that the criminal was given nothing but bread and water for a few days before the trial and received no food or drink on the evening that he took the poison, "so that from this . . . meager diet his body was very accessible to the poison, for the veins were free, all passageways wide open, and the poison could easily be spread to the most external appendages."[71] The poor man's body was thus carefully "prepared" to better accept the poison. The condemned prisoner then took a quarter lot of the powdered root of napellus, the exact same dose that had been given initially in the trial Mattioli described. In contrast to the previous case, the poison worked quickly. Following the dose, "the poor man felt great anxiety about the heart and pains in the stomach, and he felt his eyes were quite dark and everything spun around him." His pulse slowed and became faint, and he exclaimed, "'my lords, if you do not come to my aid, I will surely perish!'"[72] In the eyes of all present, the poison had taken effect.

Following this plea, the doctors gave the prisoner five grains of bezoar dissolved in wine. As this warmed him, he felt better, but soon thereafter he vomited. This expulsion made him hope he was getting better, but "the poison was now gathering in the stomach." The prisoner soon complained of pain, his body swelled up as if he had dropsy, there was a hardness under his ribs and a pain in his kidneys, he could not urinate, one arm was lame, and he had a faint pulse. He vomited frequently, and when he had stools, he complained of a cold hardness, "as if a stone lay in his stomach." As the pains began to attack his eyes, "the poor man said he would rather be dead than live even a little longer and endure such pain." His symptoms, Richardus once again speculated dispassionately, were made worse by the fact that "he had eaten nothing nor drunk a drop of liquid for five hours before he took the poison, so that one could be more certain of the trial." Nevertheless, the prisoner slept very well during the night, and in the morning he woke up and felt fine. "And he once again became hale and healthy and has also remained healthy. And the Emperor gave him his life, set him free, and gave him a generous payment."[73] In the end, the trial had succeeded. Despite the prisoner's vehement reaction to the poison, he had lived to tell about it.

This strategy of creating a "more certain" test was reinforced in the second poison trial Richardus described, which took place in Vienna on December 10, 1562, and tested the power of bezoar to combat arsenic. The criminal in this case was "a young man who had forfeited his life and was not yet 20 years old." In the lead-up to this trial, the youth had "lost a lot of weight in prison, for one gave him little to eat." Even without this treatment, "he was already of a gaunt and slender build and also had an ill appearance and color." Richardus postulated that the prisoner would be particularly susceptible to poison given his "thin and scrawny" nature.[74] The Viennese criminal had also remained awake the entire night before the trial was conducted, although whether out of nerves or per the physicians' request is not clear. Either way, the testers (however cruel their methods) once again tried to ensure that the poison would work quickly.

It did. The prisoner took a dose of arsenic mixed with sugar and, within an hour, became dizzy and began to feel a burning in his stomach, "and his throat felt exactly as if he was full of fire, and he was very dizzy and his vision had become very dark." He then received ten grains of bezoar in borage water, after which he initially felt quite a bit better: his stomach and throat cooled off, and the cooling sensation spread to his whole body. Soon, however, his condition started to worsen:

> When asked if he had pains in the knees, he answered: "Yes, with a great stabbing," which I believe is the property of all poison. After this he began once again to have strong pains in the stomach with a burning in the throat, and he spat out salty saliva and then vomited copious phlegm, which was mixed with the poison. He felt somewhat better from this but not long thereafter complained of great pains in the body. He was liberated from the chains around his feet so that he could pass a stool, after which the pains settled a bit. Within a half hour, the stomach pains and nausea returned, with the spewing of bitter material, and his outward appendages felt cold, and he said he felt as if a ball of wire or flax lay in his stomach and was climbing up toward his gullet. After that he vomited out a thick and glassy phlegm, and following the vomit his vexing symptoms were somewhat alleviated. But he was once again taken with pains of the head, belly, and knees, and the area beneath his ribs was tense, and his stomach swelled. Another stool followed thereafter. After this happened, he once again felt pain in the body with great restlessness, and he then had the third bowel movement, the *materia* of which was quite black, thin, and burned, and a little vomit came as well.[75]

Richardus finally gave the prisoner some meat broth "after the clock struck 12" (the poison having been given "around 7 o'clock," or five hours previously), but the prisoner ate it with difficulty, "for he could not stand the smell of cooked meat." An hour later, he could barely bring himself to take a taste of bread softened in wine. Slowly, however, he began to feel better, and they began to hope for a recovery. By the evening the man was able to eat and sleep, and "in the morning when asked how he was feeling, he responded '*Gantz wol,*' which is to say, very well." His symptoms continued to fade, with no additional medicine.[76]

As in Mattioli's anecdotes, these two poison trials used careful observational details and emphasized the extreme level of suffering that the poison provoked. Richardus went one step further, however, by introducing the idea that manipulating the prisoner's body beforehand could increase the effects of the poison, thereby making the test run more smoothly. His claim that the first prisoner was given very little food "so that one could be more certain of the trial" suggested the potential for the outcome to make a meaningful statement about bezoar's efficacy, and his description of the second prisoner's thin, scrawny body and his lack of both food and sleep reinforced this point. In his telling, the food deprivation allowed the physicians to be certain that the poison had taken effect.

This attempt to create a "more certain" trial can help us draw two important conclusions about poison trials at Emperor Ferdinand's court. First, the testers appear to have responded directly to the problems with Mattioli's trial in their attempts to ensure a quicker response to the poison. In that sense, the poison trials can be seen as experiments in poisoning as well as attempts to test antidotes, and Richardus demonstrated no concern or pity for the criminal. Second, the very notion of a "more certain" trial hints at a willingness to see singular events (like poison trials) as carrying meaningful epistemological weight. Based on these "more certain" trials, Richardus could assert confidently that bezoar was an effective antidote. Although his aim may have been different from Mattioli's and Handsch's, his methods matched theirs very closely: he referenced "the property of poison" and described its passage through the body with phlegmatic vomit and choleric stools. He was thus able to send Archbishop Olahus a scholarly report of the antidote's success. The fact that it was printed—and translated—by two later physicians suggests that it was viewed as a document to take seriously.

"Herewith are noted the symptoms"

Emperor Ferdinand's court was not alone in conducting poison trials in the 1560s. Both Duke Cosimo I de' Medici of Tuscany and King Charles IX of France oversaw tests on condemned criminals in Florence (1566) and Clermont (late 1560s). Both of these cases solidified the notion of poison trials as definitive experiments and underscored Mattioli's strong influence on poison trials across Europe.

The French royal surgeon, Ambroise Paré (1510–90), published an account of King Charles IX's trial in 1579, five years after the king's death. He was explicit about Mattioli's influence. Paré did not read Latin, but Mattioli's *Commentaries on Dioscorides* was translated into French in 1561, and an updated edition appeared in 1572.[77] According to Paré, in the previous decade the king was holding court in Clermont when a Spanish man brought him a bezoar stone and claimed it was good against all poison. Paré disagreed with this assertion. To settle the matter, the king wanted to "conduct an experiment [*faire experience*] on some rascal condemned to hang." His head of household, a Monsieur de la Trousse, noted that a cook had recently been sentenced to hanging for stealing silver from his master. In the presence of de la Trousse, an apothecary's apprentice gave the man a dose of "a certain poison" (sublimated mercury, as Paré mentioned later) along with the bezoar. Immediately, the prisoner began vomiting and defecating, felt as if there was a fire in his body, and demanded a drink, which was refused. An hour after taking the poison, de la Trousse summoned Paré, accompanied by "three of [de la Trousse's] archers." Paré arrived to find the prisoner in a sorry state, vomiting and bleeding from all orifices, and he died "miserably" within seven hours. Paré then conducted an autopsy, in the presence of de la Trousse and four archers, and he found the man's stomach black and cauterized.[78]

There are some obvious differences between this poison trial and the previous examples we have seen. Writing with the benefit of hindsight, almost a decade after the failed trial took place, Paré portrayed himself as skeptical from the start. The main impetus came from the former king (now deceased), and de la Trousse oversaw the test, not a medical professional. Nevertheless, Paré's narrative followed the overall pattern established in Mattioli's texts, including the grisly recounting of the prisoner's symptoms. Certain parts matched up nearly exactly. Paré claimed, in particular, that the prisoner preferred to die of poison in prison rather than be hanged in public, a direct rewording of Mattioli's *Commentaries*.[79] In case the reader missed that reference, however, the printed marginalia noted that "Mattioli describes a similar history of Pope Clement VII, who wanted to test an antidote for the benefit of the public."[80] Given Paré's discomfort with this trial, the invocation of Mattioli may have acted as a measure of protection (with good reason, as we will see in the next chapter). In any case, Paré's anecdote demonstrates that medical practitioners with no known ties to Mattioli read his poison trials with interest and followed his model.

In Italy, Mattioli's influence was less explicit but even more forceful. The Medici dukes, in particular, took up his methods and his recommended remedies. As Jo Wheeler and Sheila Barker have shown, Duke Cosimo I de' Medici and his son, Duke Francesco I, created their own poison antidotes in the palace laboratory (*fondería*) and used criminals to test them. Their signature antidote was an "oil against all poisons" that closely resembled Mattioli's scorpion oil. Wheeler has published the recipe for the Medici oil, kept by the Florentine ducal apothecary Stefano Roselli (d. 1587), and he notes that it appears to be a simplified version of Mattioli's antidote.[81] The Medici dukes used this oil as a gift for political allies. In 1547–48, Cosimo twice sent the oil to Ferrante I Gonzaga, the duke of Milan, via Ferrante's secretary, Francesco Vinta. In both cases, he suggested that the duke might want to try it on condemned criminals before using it himself.[82]

In at least one case, the Medici dukes also tested a powder against poisons, reminiscent of Mattioli's trial using Archduke Ferdinand's poison powder. On September 26, 1566, a man named Jacopo Gimignano from the village of Monsagrato (near Lucca), held in the Bargello prison in Florence and about to be hanged for theft, was given two drams of arsenic "in order to make an experiment of His Illustrious Excellence's [i.e. Duke Cosimo's] antidote." Six physicians oversaw the trial, and the notary Paulus Mutius Bibenas carefully recorded it; a copy can be found in the archive of the Strozzi family today. The document begins with the inauspicious words, "In the name of God, Amen: herewith are noted the symptoms [*accidenti*]" that followed the ingestion of the poison, a sign of the torments that were to come.[83] Even more than in other cases we have seen, the physicians who oversaw Duke Cosimo's trial focused on two particular things: the specific timing of the symptoms and the prisoner's pulse.

The unfortunate Jacopo took the poison mixed with sugar, along with a communion host(!), at 12 noon.[84] After three quarters of an hour, he complained of bloating in his body, but the pain was not terrible. The physicians took his pulse and found it "very frequent." At 14 hours and a third, he felt a bitterness in his throat. The physicians found his pulse "more frequent but weaker," which, they speculated, came from the anticipation of worse symptoms. Around ten minutes later Jacopo felt a coldness in his stomach, and the physicians found his pulse to be lower and reported that his head had begun to sweat. They decided to administer the antidote, after which "he said he felt warmer than before." At 15 hours and a quarter, he felt like he had to vomit a little and had aches in his stomach and body. At 15 and two thirds, he once again complained of bloating in the stomach, and his pulse was slower and "troubled." At 15 and three quarters, the bloating and gurgling in his stomach and body worsened, but his pulse was the same. At 16 and a half hours, he reported basically the same status, but about a quarter hour later, he vomited, and his pulse was rather thin. After the vomit, he felt a lessening of the bloating and the pains in his head. He vomited again at 17 and a half hours, and he had a vomit full of phlegm and choler at 18 hours. At 18 and a quarter hours, he vomited again, his pains increased, and the physicians began to fear for his life. They administered another dose of the antidote, but Jacopo immediately threw it back up. He was breathless, sweating profusely, vomiting blood, the doctors said his pulse had slowed, and he had massive pains. The doctors prepared the antidote once again. This time, they gave it to Jacopo in two bites along with the communion host and a few sips of wine, although he threw up one bite around 19 hours.[85]

This gruesome recounting of symptoms, always with detailed markers of time and pulse, was just the beginning. It continued for a full twenty-five hours. Poor Jacopo continued to vomit, defecate, and experience a variety of pains, and the physicians continued to record the exact timing of the symptoms as well as his pulse. He received several additional doses of the antidote. In the dead of night he appeared to improve and was able to eat a bit, but the pains and vomiting returned. The last entry in the record notes that at 13 hours on the second day, he received a dose of the antidote given in the communion host. Curiously, the report stops abruptly after that, and there is no indication of his fate.[86] An attached letter from the ducal physician, Baccio Baldini, described the test as a "proof of His Excellence's powder," perhaps a sign that the prisoner survived.[87] As in the examples of Mattioli, Handsch, and Richardus, however, the physicians focused more on the symptoms of poisoning and the physiological response once the antidote was administered than on the outcome.

That detailed focus on symptoms has been a thread through all of the poison trials we have examined thus far. In some ways, it mimics physicians' case studies of patients, which became popular in two different kinds of scholarly print genres in the sixteenth century, *Curationes* and *Observationes*. The copious record-keeping also calls to mind the extensive note-taking about natural history that Brian Ogilvie has described, as well as the general interest in historical anecdotes, or *historia*, among physicians and naturalists.[88] Poison trials were different, because they involved an illness intentionally created by the testers. The symptoms were also particularly dramatic and drawn out (none of the poisoned criminals keeled over dead immediately), a fact that lent itself to the focus on timekeeping and details of symptoms. Nevertheless, the way in which physicians recorded and communicated these trials put them in line with related scholarly genres. By the 1570s, then, Mattioli had, in a sense, achieved his aim. His concern with distinguishing physicians' tests from those of charlatans did not appear as an explicit issue in later poison trials, largely owing to the strength of Mattioli's own reputation. By the mid-1560s, the precedent and road map for learned poison trials had been established.

Trials as Proof

Yet the comparison with empirical practitioners and their own poison trials never went away. In 1563, a man from Brescia named Maffeo Bertolani sought a license to sell his "miraculous remedy against the bites of all poisonous animals" in the city of Venice. He provided a recipe to the city's Magistrate of Health (a board of noblemen), and he asked to be allowed to "conduct an experience" in which he would allow himself to be bitten by venomous animals and then take the antidote. This scenario might have set off alarm bells for anyone who had read Mattioli; it sounded exactly like the charlatans' trials that he derided as fraudulent. Yet the superintendents on the Magistrate took Bertolani seriously.[89] They examined his recipe and determined that it was composed of materials "that all of the doctors who have written about these matters praise against the bites of poisonous animals." On July 9, the board voted unanimously to grant Bertolani a license contingent on the outcome of the test, which took place on August 18, 1563. After taking the remedy, Bertolani and another man, Aluigi Salvalagio (a partner?), subjected themselves to unspecified venomous creatures, and the Superintendents found the cure to be "good and proven." The men received a license.[90]

Why would the Magistrate have agreed to such a thing? Bertolani's case demonstrates the enduring appeal of empirics' remedies as well as the poison trial as a method of proof. The Magistrate of Health heard countless petitions for special licenses in the 1560s and 1570s, many of which they granted. David Gentilcore has noted that health magistrates in Venice—and across Italy—accepted charlatans' demonstrations surprisingly often and continued to do so through the eighteenth century.[91] Bertolani also appears to have argued his case cleverly. He pointed out that Venice sought "virtuous people" who had remedies "of universal benefit"; he provided a recipe for physicians to examine; and, most crucially, he offered to prove his case with a poison trial on his own body.[92] Whatever the skepticism about such marketplace shows, the poison trial carried official weight; the license would be approved only if the trial succeeded (although Bertolani presumably would have been dead or gravely ill if it had failed). Given the confusing array of new remedies that arrived by the day (see chapter 5), princes, town magistrates, and even physicians were willing to grasp at a seemingly solid method of proof, a *pruova*.

This notion of proof allowed poison trials to function on several different levels, as seen especially clearly in a final group of tests from the Holy Roman Empire. In 1580, a schoolteacher, mine foreman, and jack-of-all-trades named Andreas Berthold traveled across the western Holy Roman Empire selling small earthen medallions from Silesia that he touted as a marvelous new poison antidote and cure-all, which he called German terra sigillata. Unlike Bertolani, he did not personally conduct any tests. In the city of Jülich, a man named Crisant von Cronenberg tried the antidote "in the name of Mr. Andreas Berthold" on two dogs, in the presence of the town surgeon, Johannes Ottweiler. As in the Galenic poison trials, one dog received the antidote and the other did not. The dog receiving the antidote survived, and the mayor and town council of Jülich provided him with a certificate, signed by the notary Paulus von Herl, testifying to the experiment's impressive outcome.[93]

Fresh off this success, Berthold then traveled south to Hesse, which happened to be in the throes of a plague epidemic. The land's ruling prince, Landgrave Wilhelm IV of Hesse-Kassel, took an active role in directing the response to plague and in seeking effective remedies for the illness.[94] Unsurprisingly, he was intrigued by Berthold's new remedy, but he was not willing to "trust him on his bare words." He asked his physicians, Maritius Thaurus and Laurentius Hyperius, to "make a trial" of the drug before he purchased any of it.[95] This trial was, in a sense, a merging of Bertolani's test before the Venetian Magistrate and the courtly trials on condemned criminals. The purpose was similar to that of the Venetian case: if it succeeded, Landgrave Wilhelm would deem the drug worthy of purchase. As with the other courtly poison trials, however, the prince commanded it, and his physicians conducted it "in the upper room of the castle, in the presence of His Princely Grace."[96] Yet Wilhelm did not offer condemned criminals for this trial. As in Berthold's first test, the physicians used dogs.

Using animals rather than humans, the Hessian physicians (perhaps with input from their scholarly prince) devised a far more extensive test than any of the previous poison trials we have seen. Eight dogs of different breeds and colors were chosen to test four different poisons: mercury sublimate, aconite, nerium (oleander), and apocynum (dogbane). As in the Galenic trials on roosters, the physicians divided the animals into two groups. One group received the antidote, and the other received only poison. In pairs, two dogs with roughly similar external features were given a lethal dose of poison, and then one of the two received a scruple of terra sigillata.[97] Wilhelm's physicians carefully recorded the color, breed, and markings of each dog, the time of day of each instance of vomit, defecation, or cramping, and the ultimate outcome, in a manner reminiscent of the Medici test. In the case of the first pair, for example, "a yellow-gray male dog" took mercury sublimate and Silesian terra sigillata "between eight and nine hours in the morning." Within half an hour after the second hour he "vomited copiously." In the third hour, he "vomited bilious material, the same in the fourth hour, and he escaped [death] with no other serious harm." The second dog, "a yellow and white cur," received no antidote:

> Within half an hour he urinated and shortly thereafter defecated; after [another] half hour he fell to the ground with spasms; after they ceased he stood up panting and grinning for a while, and soon thereafter again was taken with a convulsion and swelling in the belly. At one o'clock he suffered a violent convulsion and lay as though dead for a quarter of an hour. Half an hour after two, he began to move, and for half an hour he stood with deathly still feet. Between four and five he once again convulsed violently, and he died in the night.[98]

The other cases progressed similarly, carefully recorded. All of the dogs who received poison and antidote survived. All of the dogs who received only poison died—save for one. A shaggy brown dog with a white tail given aconite and no antidote suffered for so long, with such terrible paroxysms, that the "most merciful" prince commanded it be given a half dose of terra sigillata. It too survived, building a seemingly overwhelming case for the efficacy of the antidote.[99] Landgrave Wilhelm promptly purchased a quantity of the drug.

The Hessian test represented a seamless merging of commercial "proof" and medical experimentation. The initial purpose may have been transactional, but Wilhelm's physicians used the opportunity to devise an extensive trial and observe and record the effects of the poison and the antidote. As in the Medici case, they produced a thorough report on the trial. Unlike the Medici case, however, their report, which they called an *observatio*, was written to be shared with the broader medical community.[100] Historian Gianna Pomata has described the *observatio* as a genre that was "fundamentally geared to the goal of exchanging and circulating information within communities" in the sixteenth century, and the Hessian *observatio* quickly circulated to other German courts.[101] One scribal fair copy can still be found in a recipe collection in Heidelberg, which belonged to Countess Palatinate Elisabeth of Saxony (1552–90).

Letters from contemporary princes suggest that they, too, had read the report.[102] On the basis of this *observatio*, another German prince, Count Wolfgang II of Hohenlohe, conducted a poison trial on a condemned criminal, as we will see in the next chapter. The medical opinion of the respected Hessian physicians was taken seriously.

Yet the landgrave and his physicians also provided Berthold with the *observatio*, along with a separate testimonial letter that summarized the "experiments" and certified them as "proven miraculously." Berthold used these documents to his advantage. In 1583, he published a book on the German terra sigillata. The title page of the book highlighted the "multiple experiences" testing the drug, and the heading on the first page read: "On the true and proven German terra sigillata." At the end of the book, he published "duplicates" of the testimonial letters from Jülich, Hesse, and Hohenlohe.[103] As much as the Hessian physicians may have been interested in the medical effects of poison and antidote, in the end their efforts functioned as "proof" of the drug's efficacy that validated Berthold's commodity. Significantly, he published his book in Latin, with one obvious audience: learned physicians and aristocrats. He emphasized that his validation came from princes, physicians, and other powerful individuals. In a sense, then, Berthold created a virtual marketplace show for Latinate readers.

Experimenting with Poison

Experiments were nothing radical in sixteenth-century Europe. By the 1580s, an extensive experimental culture had sprung up across Italy and the Holy Roman Empire, and physicians and apothecaries led the way.[104] Poison trials formed one branch of these experiments, and not all of them involved dramatic courtly trials on condemned criminals. As Findlen has noted, the eminent physician and naturalist Ulisse Aldrovandi (1522–1605) poisoned roosters in order to test theriac, as in the Galenic poison trials.[105] Richard Palmer has highlighted the extensive poison research on animals in Venice in the 1560s, "producing a steady toll of chickens, wolves, dogs, and pigs as the news spread."[106] It is for good reason that Jutta Schickore chose snake venom as the central uniting theme for her book *About Method*, which examines the development of experimental methodologies from the seventeenth through the twentieth centuries.[107] As I argued in chapter 1, poison had been viewed as a particularly testable substance since antiquity, and it should come as no surprise that sixteenth-century individuals used it as one focus of their expanded experimental purview. Schickore's study of method had a long prehistory.

What separated poison trials from other kinds of experimentation in early modern Europe (in anatomy, alchemy, natural history, and many other fields) was the explicit attempt to distinguish poison testing as a scholarly undertaking. Because the familiar genre of marketplace poison trials already existed across early modern Europe, physicians and other medical professionals had to work actively to achieve this separation. They thus emphasized that the trials took place at the command of a powerful ruler; they were run by physicians or other trusted medical practitioners; the outcomes were meticulously recorded and described in terms that other scholars could understand. By recording, sharing, and in some cases even publishing detailed accounts of the trials, physicians—led by Mattioli—created a new scholarly genre, the poison trial account. Peter Dear has pointed to the development of experimental writing as a crucial tool in cementing experimental practices in seventeenth-century mathematical sciences: "an account of an experiment," he argues, "is an essential part of its performance."[108] With the efforts of Mattioli and his medical colleagues, poison trials became firmly ensconced in scholarly medical literature in the period 1560–80.

Despite their best efforts, however, learned medical authors still mimicked empirical practitioners' marketplace tests. Every single poison trial described in this chapter began with the question of whether or not an antidote worked: the test was intended to settle that question. However much physicians and other courtly medical practitioners may have focused on specific symptoms, or timing, or pulse, or the humoral composition of bodily effluvia, their poison trials echoed charlatans' tests in the stated goal of proving or disproving an antidote. Proof was built into the very semantics of the tests, as the terms *pruova/prova* (Italian) or *proba* (Latin) were used interchangeably with various linguistic versions of "experiment" or "experience." Empirical practitioners like Bertolani or Berthold could thus use poison trials, with the approbation of physicians, to receive official recognition for their commercial products.

In practice, however, the issue of "proof" faded into the background in physicians' accounts of poison trials. Mattioli described his failed Prague trial not as evidence that the Archduke Ferdinand's antidote was worthless, but as a sign of aconite's toxicity. (Indeed, in another part of his *Commentaries*, he lauded the powder's "proven" powers against arsenic.) The Medici physicians did not even include the outcome in their notarized "proof" of Duke Cosimo's poison powder; instead they, too, focused on the poison. Even Richardus, who wrote expressly to demonstrate bezoar's marvelous powers, focused on the prisoners' dramatic symptoms far more than the outcome. The one thread connecting all of the poison trial accounts we have seen was an almost sensationalistic focus on symptoms, both the reaction to poisoning and the interaction between poison and antidote. While physicians (and surgeons) began with the question of *whether* a drug worked, they ended up focusing on *how* both poison and antidote affected the human body. For that reason, the use of condemned criminals as test subjects was centrally important.

Yet after their heyday in the 1560s, poison trials on condemned criminals fizzled out. While isolated trials continued in Italy through the end of the sixteenth century, there are no detailed records after the early 1580s. Despite their perceived possibilities, poison trials on criminals never took on a life of their own—they never became an "epistemic" medical genre, to use Pomata's concept.[109] Poison trials broke new experimental ground not only because they aimed to test poison on humans (an impulse that dated back to antiquity, as we have seen) but because they engendered an extensive cache of experimental records. That record-keeping was easier with animals, as Landgrave Wilhelm's dog trials make clear, and medical experiments in general began to focus more extensively on animals in the seventeenth century.[110] As the next three chapters demonstrate, the underlying religious, political, and ethical problems with poison trials, combined with alternate testing methods and a shifting focus from poison antidotes to wonder drugs, made the practice of using condemned criminals for poison trials less and less appealing. In the period between 1525 and 1565, however, they offered an enticing way for physicians to produce contrived trials in a manner befitting medical learning. In that light, Mattioli's failed test was a runaway scholarly success.

FOUR

To Cure a Thief: Silesian Terra Sigillata

On the night of December 21, 1580, in the remote village of Unterregenbach in southwestern Germany, a man named Wendel Tümler made a terrible mistake. He stole a thaler coin from a tavern patron and was caught red-handed. The victim's drinking companions immediately seized Tümler and marched him down the river to Langenburg Castle, where he was thrown into the dungeon (figs. 4.1 and 4.2). As his case unfolded, the stolen thaler turned out to be the least of his misdeeds. Originally an upstanding citizen from the nearby village of Schmalfelden, he had, in the course of six years, allegedly bankrupted his farm, abandoned his wife and children, and turned into a notorious horse thief—even, in one case, stealing from a poor widow. Tümler was accused of five counts of theft before he was tortured, and although he initially denied all charges, he eventually confessed to seventy-four under torture, many of them horse thefts. As befitted the volume and severity of his crimes, he was condemned to death by hanging.[1] In any other situation, his life would have been over, his family disgraced—but the thief had an unexpected stroke of luck. His conviction was perfectly timed.

Langenburg Castle happened to be the residence of the local prince, Count Wolfgang II of Hohenlohe, who was known for his avid interest in natural history and alchemical experimentation.[2] In late 1580, Wolfgang's mother, the Countess Anna, had acquired a sample of a new terra sigillata from Silesia—the same drug tested in the trials on

4.1 Langenburg Castle today. The castle was remodeled as a Baroque palace in 1616. Photo by the author.

Landgrave Wilhelm IV's dogs, discussed in the previous chapter. Indeed, Countess Anna had received a report on the successful dog experiments and read it with great interest. Now she and Count Wolfgang were interested in testing whether the antidote worked "as well on humans as it does on animals." Tümler's arrest occurred at the exact moment that Count Wolfgang and Countess Anna were searching for a test subject, and the convicted thief quickly agreed to take part in a poison trial, with the provision that he would be banished from Hohenlohe rather than executed if he survived. According to the official account of Tümler's case, the timing of his arrest and conviction was a happy coincidence.[3]

Or was it? A peek behind the scenes casts doubt on this simple story. In most cases, the theft of a lone thaler would not have led to a capital conviction. Petty crimes in Hohenlohe were usually handled fairly leniently, and the victim himself had no interest in pressing charges.[4] Yet from the very beginning, Count Wolfgang drove the case toward a death sentence and a poison trial. The day after the thief's arrest, Wolfgang wrote to Tümler's home province of Schrozberg, asking for the chief magistrate's opinion on the thief's character and on the likelihood that more crimes might emerge if he were tortured. When the magistrate, Jörg Heber, responded with a negative portrayal of Tümler, Wolfgang personally requested that the executioner question the accused thief.

Shortly after Tümler's confession, the count proposed using him to test the Silesian terra sigillata.[5]

At every stage, these actions met with considerable skepticism from Count Wolfgang's councilors, who sent constant advice by letter from the princely residence in Neuenstein. His trusted secretary and magistrate,

4.2 The pull-up door leading to the underground prison cell at Langenburg Castle, used through the sixteenth century. The cavern is now home to a colony of bats. Photo by the author.

Zacharius Hyso (d. 1591), argued that there was no particular reason to view Tümler's misdeed as anything but a minor incident. Hyso expressed misgivings at the initial call for the executioner and insisted that the questioning begin without torture, following the usual practice. Wolfgang's query about the poison trial also caused a measure of consternation. His councilors supported the experiment in theory, but they had a number of concerns. Steps had to be taken, Hyso cautioned, to avoid unrest among the "foolhardy" peasants and to prevent Tümler, if he survived, from claiming that he had been coerced into the test. While the advisers saw "no real problem" with Wolfgang's conducting the poison trial, they were nevertheless quite anxious about potential pitfalls.[6] Nearly all poison trials after 1560 took place with the understanding that the prisoner would go free if he survived, and the Hohenlohe councilors worked from this assumption as well. That possibility raised some serious complications. If Tümler survived, he could complain if he felt mistreated. Conversely, the general populace might be outraged at the notion of a convicted criminal returning to their midst. Whether Tümler lived or died, the locals might protest the lack of a public execution. Whatever the outcome, potential trouble lurked ahead. So grave were the councilors' concerns that they sent a follow-up document laying out a careful template for making the test acceptable to all parties.

Their solution to this complex knot of problems rested on two main strategies, which will be the central focus of this chapter. First, they stressed that Tümler had to *be seen to have* consented to take poison "out of his own free will, without any tricks or force, also without being convinced or told to do it." There had to be absolutely no perception of coercion on the part of Count Wolfgang. Second, the councilors suggested that the prince emphasize the medical benefit to "all of Christendom." Should the trial succeed, the populace would have a proven new remedy for poison and plague. If Tümler died, they would at least have some certainty about the cure. Either way, his act of taking poison would "come to the good and help of many people."[7] In other words, the councilors underscored the importance of the prisoner's consent and appealed to the idea of a greater good that justified the test.

These ideas may sound familiar. The experimental subject's "informed consent" and the benefit to the broader public are the twin concepts on which modern biomedical ethics have been built. The first two principles of the Nuremberg Code, written in the midst of criminal trials of Nazi doctors in 1947, state, "the voluntary consent of the human is absolutely essential" and "the experiment should be such as to yield fruitful results for the good of society."[8] Sixteenth-century notions of consent and societal good were, obviously, very different from today's, and I do not intend to impose modern ethical concepts on early modern actors. The Hohenlohe councilors were concerned with appearances, not human rights, and they were not particularly nervous about the possibility that Tümler might die. Rather, they worried about what would happen if he lived. At the same time, the solutions they recommended—consent and the public good—sat at the center of discussions about human experimentation that developed from the eighteenth century into the post-Nuremberg era.[9]

It is tempting to assume that there was no such thing as medical ethics in sixteenth-century Europe. For starters, the term did not exist yet—it was not coined until 1803, when the British physician Thomas Perceival published a code of professional conduct for physicians and surgeons called *Medical Ethics*.[10] The concepts of human dignity and autonomy, which underlie the modern emphasis on informed consent, had not yet become topics of political or philosophical discussion. Renaissance princes, moreover, had enormous power over the condemned, and as we saw in chapter 2, the needs of the individual were viewed as secondary to the needs of the state. Philosopher Michel Foucault placed the beginnings of careful thought about human subjects a full century later, in the late seventeenth century, and historians have tended to follow his timeline even when disagreeing with some of his arguments about later periods.[11] In an era of witch trials and despotic princes, it has been assumed, human life came cheap, especially for subjects as powerless as condemned criminals. In short, there appears to be very little reason why poison trials should have required any justification at all.

And yet they did. In every poison trial I have found, surviving criminals received some sort of reward—either freedom or a reduced punishment—for their efforts. Nearly all poison trial accounts referenced the benefit to "all Christendom" as a justification for undertaking the experiment. Although the early Italian poison trials discussed in chapter 2 appear to have been conducted without the prisoners' knowledge, later accounts emphasized that the criminal had consented to take poison. While these measures had little to do with modern medical ethics, they nevertheless belie the notion that human experimentation was a free-for-all before the late seventeenth century. The various attempts to make poison trials appear more acceptable, I argue, derive from two main problems: the interruption to the ingrained cultural and religious ritual of public execution, and the fear that princes and physicians might be viewed as poisoners. A poison trial threatened not only the prisoner's soul, but also, as in the Hohenlohe councilors' worst-case scenario, the societal order.

Conforming to religious and cultural norms had been an important emphasis in Renaissance poison trials from the very beginning. In the earlier Italian cases, authors focused on the conventions involved in execution and in obtaining bodies for anatomical dissections. The actual act of giving poison to a condemned criminal, however, did not appear to cause any great anxiety. The Hohenlohe councilors, in contrast, feared that the poison trial itself might be seen as objectionable. This difference may reflect a greater distaste for poison trials north of the Alps. As we saw in the previous chapter, the Italian physician Pietro Andrea Mattioli was far more willing to portray himself as an active agent in poison trials than his Bohemian translator Georg Handsch. Yet there are good reasons why Handsch's view might have been widespread. Avoiding harm to a patient was a central theme in the many guidelines for physicians' professional conduct penned from antiquity through the Renaissance.[12] Today's most famous example, the Hippocratic Oath, explicitly prohibited a physician from giving patients "lethal medicines." While the oath was not widely known before the Renaissance, it experienced a boom in the sixteenth century.[13] Even before that resurgence, however, it was widely held that a physician should do no harm. Princes, as well, were supposed to work for the benefit of their subjects, and poison was sometimes used as a smear between political rivals.[14] Of course, the condemned criminal occupied a very different position than patients or princely subjects, and the poison trials rested on the premise that their liminal status changed the rules. Nevertheless, anyone setting up a poison trial had to confront the negative connotations of poisoning.

Tümler's case provides an ideal forum for uncovering these underlying complexities. All of the trial accounts we have encountered thus far were written from the perspective of a physician or a surgeon, a prince or a pope. In all of them, the process of obtaining the condemned criminal for the test appears straightforward. The Langenburg case, in contrast, gives us a behind-the-scenes look at the trial as it unfolded. The archive in Hohenlohe contains full documentation of Tümler's criminal trial as well as the councilors' opinions on the wisdom of using the thief to test the Silesian terra sigillata. In this light, the process appears far more fraught. Where the printed works portray an uncomplicated case, the archival documents show hesitancy throughout Tümler's journey from petty thief to test subject. Read next to other poison trial accounts, I argue, this hesitancy was more the rule than the exception. At the heart of the matter lay the basic facts inherent in every poison trial. A prince and his physicians aimed to give a condemned criminal a deadly poison and then, using the trendiest antidote available, attempt to cure him.

The Marvelous Antidote

As in the case of Caravita's oil, the impetus for the Tümler poison trial came from a promising new local antidote. In this instance, the remedy was a German version of the ancient poison antidote terra sigillata, a special clay found on the island of Lemnos that Galen himself had touted as highly effective against poison. Virtually unknown in the Middle Ages, Lemnian terra sigillata became a popular antidote and cure-all in Renaissance Europe, possibly owing to active marketing by the Ottoman Turks.[15] Despite the widespread interest, true Lemnian terra sigillata remained scarce, and complaints about fraud abounded (about which we will hear more in the next chapter). The German terra sigillata found in Silesia promised similar efficacy without the concerns about Turkish duplicity.

The Silesian antidote was tied to two people, who appear to have been partners of a sort: a mine foreman named Andreas Berthold of Oschatz (Saxony), who we met in the previous chapter, and the Paracelsian physician Johannes Schulz, known as Johannes Montanus (d. 1604). Both men touted the cure as a Paracelsian alchemical remedy—it was greasy, it was yellow, and it was mineral, which prompted Berthold to give it the name *Axungia solis*, or "grease of the sun," a remedy cited by Paracelsus.[16] Montanus was credited with discovering the Silesian terra sigillata, and Berthold's role, at least initially, appears to have been as a salesman (although their relationship eventually soured, as we will see in chapter 6). In any case, it was Berthold, not Montanus, who traveled across Germany selling the antidote in 1580–81.[17] The circumstances of the various trials of Silesian terra sigillata were thus very much tied up in the entrepreneurial interests of Berthold, who adroitly capitalized on the widespread interest in finding new and viable poison antidotes, particularly among the ruling princes.

In part, the Silesian terra sigillata's notoriety derived from its alleged virtues. In his 1583 book on the drug, Berthold claimed that it closely resembled the Lemnian terra sigillata—and even exceeded it, given the difficulty of obtaining true terra sigillata from Lemnos. First and foremost, it was helpful against all poisons and venoms, "strong and gruesome" poisonous drinks, and love potions "from which one frequently gets out of his senses." He also proclaimed it to be a useful means against "the gruesome and terrible poison of plague" and a safeguard against an epidemic if taken regularly as a preservative; it would cure those already affected by plague, even if they were "half dead" already. The terra sigillata would also strengthen the heart and the vital forces and guard against heart pains and palpitations; cure fluxes, especially in the eye; staunch intense bleeding and heal wounds; and help against various other illnesses.[18] It thus mirrored many other poison antidotes that had a broader application, such as theriac and bezoar stone. The claims about plague certainly would have had an appeal to German princes, who were in the middle of a plague epidemic in the summer of 1580.[19]

As we saw in chapter 3, a series of successful poison trials on dogs bolstered these claims, especially Landgrave Wilhelm IV's widely reported experiment using multiple dogs and poisons. Berthold's terra sigillata was sold out by November 1580, and the Countess Anna appears to have been lucky to obtain a sample.[20] Because several months passed between Countess Anna's purchase of the antidote and Count Wolfgang's interest in testing it on a convict, it seems likely that they were waiting for a suitable opportunity. Executions were not particularly common in Langenburg in the later sixteenth century. While there may have been one additional person sentenced to death in 1580, that man, Georg Dinkel, was a resident of Langenburg; his crime was somewhat complicated; and it is unclear whether the death sentence was carried out.[21] Wendel Tümler, in contrast, was not a Langenburger and had no family in town to protest his treatment. He represented an opportunity too good to pass up.

The Horse Thief

Tümler's path toward becoming a test subject began with an accusation. "Wendel Tümler of Schmalfelden was arrested this Tuesday, December 20, at night, in the inn in Unterregenbach . . . on account of having stolen one thaler from Michel Wirt of Oberregenbach . . . which was found on [Tümler]." In addition to this petty theft, Tümler had also allegedly committed three other crimes, two of them horse thefts.[22] His profile—a thief—put him in line with many other criminals used for poison trials, such as Ambroise Paré's silver-stealing cook and Claudius Richardus's "strong young man" condemned for thievery. Unlike most of the criminals we have encountered in this book, however, Tümler was not anonymous. Not only do we know his full name, we also know something of his heritage. The official who penned the accusation noted that he was the son of Simon Tümler (or Dümler or Thumbler) of Schmalfelden and Barbara Altenkörener of Lindlein, and that he had been married for seven years to Margaretha Zöllner, daughter of Michel Zöllner, a physician in Rimbach. In some ways, then, Tümler represents a somewhat surprising choice for a poison trial. In contrast to the Corsican mer-

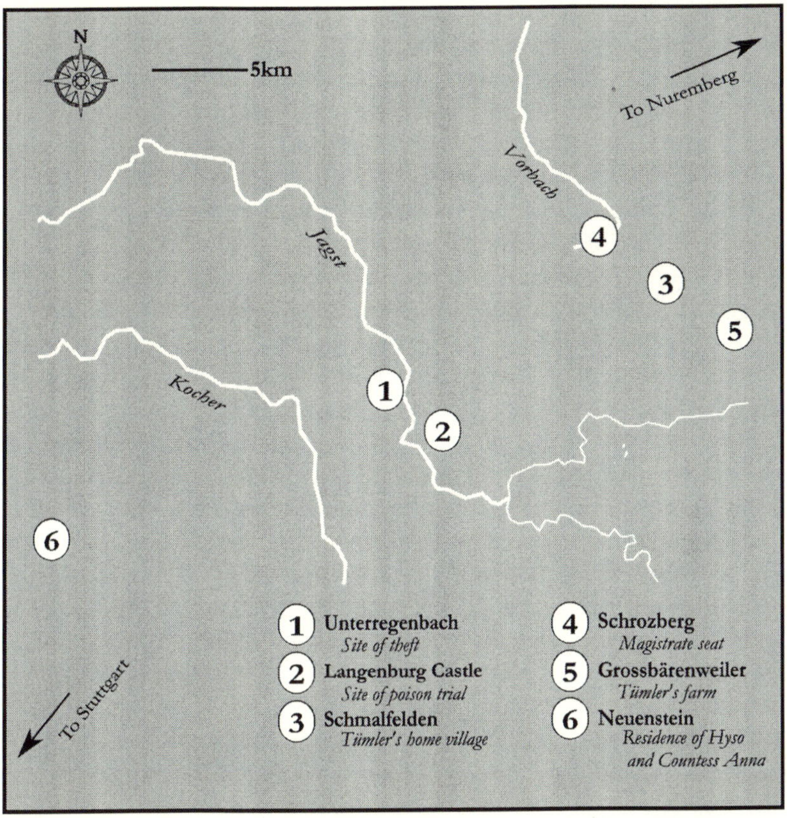

4.3 Map of the Langenburg area in Hohenlohe, southwest Germany. Courtesy of Sarah Rankin.

cenaries in Pope Clement VII's trial, he was not an outsider. He lived locally: although he was not a citizen of Langenburg, his hometown of Schmalfelden lies only nine miles away. His parents were middle-class burghers, and his wife's father was a doctor. There is no question that Tümler had a solid social foundation.

What went wrong? Why, all of a sudden, did Tümler desert his family and allegedly start stealing horses? The letter to Count Wolfgang from Jörg Heber, the chief magistrate of Schrozberg, fills in some details. Six years prior, Heber explained, Tümler had inherited a small farm in a hamlet just outside of Schmalfelden called Grossbärenweiler (see fig. 4.3). He was not a good farmer. Through his own sloth, the farm had failed, and Tümler owed enormous sums of money to the friends and neighbors who had unwisely lent him help. Without saying goodbye to his wife and children or selling his farm, he ran away from his debts and entered a life of crime. In Heber's account, he stole the widow Kälberbach's best horse from her field, rode away, and sold it in Weinsberg, a forty-mile journey to the west. The horse was returned after Heber appealed to the magistrates of Weinsberg, but Tümler remained at large. There were rumors of other horse thefts as well.[23]

There are other documented cases of young men feeling trapped in small town life in southwest Germany: both the father and the brother of astronomer Johannes Kepler left established families to become ne'er-do-wells.[24] In Tümler's situation, the debts he owed likely would have landed him in prison anyway; stealing may have seemed like a preferable option. Once he had been caught, however, his choice to abandon his family and farm made the Schrozberg magistrate little inclined to show him mercy. "Without a doubt," Heber concluded, "if he is tortured, more evil deeds will be brought out of him." Heber's letter, dated December 22, 1580, and addressed to Count Wolfgang, was written a mere day after Tümler's arrest. It attests to Count Wolfgang's avid interest in finding a case likely to end in capital punishment. After receiving Heber's response, Wolfgang then asked Zacharius Hyso to oversee the matter of torture. He asked Hyso to have the executioner arrive in Langenburg on the following Tuesday, which would have been December 27.[25]

His recommendation met with resistance. In a letter dated December 26, Hyso, who was responsible for overseeing legal procedures in Hohenlohe, urged his lord to tread carefully. He cautioned that the only certain crime, stealing a thaler, was very minor, and that the victim, Michel Wirt, was uninterested in pressing charges. Standards of torture in the Holy Roman Empire, he noted, required hard evidence of severe crimes, not merely a suspicion. Although the letter from Heber expressed a general concern, it gave no solid proof that Tümler had carried out the Kälberbach horse thefts. Nevertheless, Hyso continued, he had followed Wolfgang's orders to bring in the executioner, with the idea that the man could show Tümler his instruments and explain what he would do to him, in the hopes it might elicit a confession. This "show-and-tell" approach was commonly used as a precursor to torture.[26] If Tümler confessed, and his confessions were substantiated, one could then proceed with torture. If not, however, Hyso deemed it best to leave the matter be.[27]

Even in the presence of the executioner, a man named Gümlich, Tümler initially refused to confess to stealing the Kälberbach widow's horse, and he claimed that he did not know anything about it. He admitted only to a few minor thefts from family and neighbors in Grossbärenweiler and Schmalfelden.[28] In a second questioning on December 28, likely with a good bit more pressure, he confessed to twelve counts of theft, including some horse thefts and larger monetary thefts. Finally, on December 29, the executioner subjected him to a full session of torture, during which Tümler confessed to seventy-four counts of thievery. The jurists Martin Heinrichs and Bartel Ermans witnessed and recorded his confession. There now seemed to be a cut-and-dried case for a death sentence.[29]

But the legal counsel still hesitated. In an extraordinary display of caution, Heinrichs and Ermans followed up on many of the confessed crimes. They scribbled marginal notations on the long list of Tümler's confessions to state whether they were factual or false (fig. 4.4). "This is true" appears next to a number of the charges. Quite a few, however, were deemed fabrications. In one instance, Tümler claimed to have stolen, in the dead of night, ten liters (*Maß*) of wine and four pounds of meat from the cellar of a farmer named Michel Vogt. The marginal notation reads, "[Vogt] says that he never put wine in [his cellar] and also lost no meat." Tümler's confession that he stole four sheep from Jörg Klencken of Beurenweiler was also refuted, for Klencken "never lost any sheep." A number of notations simply stated that the purported victim knew nothing of the crime.[30] These careful attempts to substantiate Tümler's confessions hint at a continued concern on behalf of the council about the use of torture in this case. In 1532, Emperor Charles V had introduced a new criminal law code, the *Constitutio criminalis Carolina*, which included a careful procedure regarding torture. While the code stated that confessions under torture had to be repeated without torture in order to count as credible, it did not require the court to follow up on these confessions.[31] Under the efforts of Heinrichs and Ermans, nearly half of the seventy-four confessions were found to be either completely false or highly exaggerated. Nevertheless, a fair number of the admitted crimes were substantiated to their satisfaction. Tümler was now not merely a petty thief but a severe criminal. The only fitting punishment was death.

"If it works . . . on a person the same as on animals"

In early January 1581, a new phase of the Tümler case began. Two weeks after the thief's arrest, with a death sentence all but assured, Wolfgang introduced the idea of testing the Silesian terra sigillata to his councilors. Wolfgang's original letter is lost, but the response from Hyso, written on behalf of all of Wolfgang's councilors, gives valuable insight into the

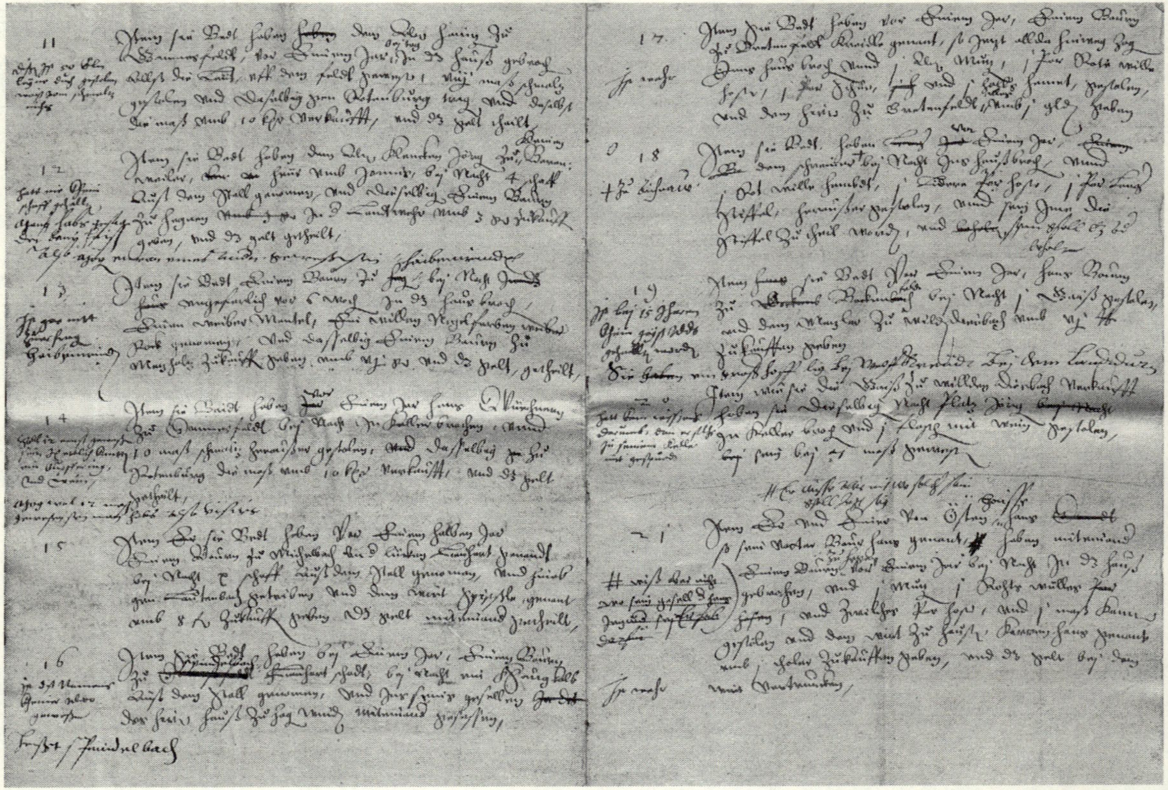

4.4 Tümler's confession, with marginal notations, December 1580. With kind permission of the Hohenlohe Zentralarchiv Neuenstein.

perceived pros and cons of conducting a trial of the antidote on Tümler. In particular, the councilors weighed the potential benefits of testing the antidote against the problems inherent in using a condemned criminal for a medical trial.

Wolfgang and his mother, Hyso's letter indicated, wished to test the Silesian terra sigillata on Tümler in order to discover "if it works against poison on a person the same as on animals . . . so that one might be able to help other people in the future." Hyso found these twin objectives very worthy—in fact, he had already conducted a limited test of the antidote himself. After reading the account of the Hessian dog trials, he explained, he had obtained some of the drug and a description of how to use it from Countess Anna. He had then tried it twice in an attempt to still the buzzing in his ears, with mixed results. The first time he had sweated profusely—a sign that it was working—but he had broken no sweat at all the second time. Given this inconclusive outcome, he too was eager to find out "if it has the same virtues as is claimed."[32] A trial using poison would bring more clarity than self-testing on a minor complaint.

The question at the heart of the Tümler trial, then, was whether a drug proven to be effective on animals could also work on humans. The results of Landgrave Wilhelm's test using dogs were accepted as definitive, but they left open the question of whether the terra sigillata also worked on humans. The extent to which animals could stand in for humans was (and remains) one of the knottiest problems in medical testing. Animals were used extensively for anatomical investigation in antiquity and the Middle Ages, but the problem with this approach—the differences between animal and human structure—spurred the first human dissections in the fourteenth century and more vociferous criticism from sixteenth-century anatomists like Andreas Vesalius. Nevertheless, animal dissections continued to be used in the sixteenth century, and they saw a resurgence in the seventeenth century.[33]

There are fewer examples of systematic attempts to use animals for drug tests before the seventeenth century, but a few prominent examples suggest that it was not uncommon. The radical Swiss doctor Paracelsus (1493–1541) noted the soporific effects of vitriol on chickens, while his more traditional countryman Conrad Gessner (1516–65) claimed to have tried remedies on numerous animals.[34] Nevertheless, most knowledge of drug therapy came from clinical encounters with human patients. As we saw in chapter 1, Avicenna warned against using animals to evaluate drugs, and medieval scholars who assimilated and amended Avicenna's drug testing rules generally assumed that such tests would be done on people.[35] In most cases, testing involved giving a remedy to an ill patient and observing the effects, and physicians shared their experiences with using specific drugs on sick patients. In addition to the printed case study genres discussed in chapter 3, physicians touted their experiential knowledge in private consultations. Dr. Johann Göbel, a personal physician to Elector August of Saxony, assured the elector that he had cured "many people" of stomach and intestinal complaints "through the orderly means" of a succession of drugs, for which he sent recipes. To emphasize the success of this method, Göbel included a list of names of thirty-two people he had cured between 1557 and 1584, including many burghers, a butcher's wife, a pastor, a young nobleman, and "a boy from Bohemia."[36] For drugs not deemed harmful, this approach allowed physicians to take patients' differing humoral complexions into account.

Yet the taboo against testing dangerous drugs on humans remained. Gessner explicitly stated that "dangerous remedies" should be tried on "beasts" before they were tried on humans.[37] The thirteenth-century physician Bernard de Gordon, as we saw in chapter 1, recommended that physicians test a new remedy on a sequence of subjects to avoid unexpectedly encountering a poison.[38] This hierarchy of test subjects tried to protect humans against dangerous substances—although some humans were deemed more worthy of protection than others. Significantly, Bernard did not mention condemned criminals in his list, and he suggested using animals as a buffer against accidentally poisoning humans. His careful approach, echoed in Gessner's warning, followed the general principles handed down from Galen's time: one should not use humans for fatal tests.

Some Renaissance physicians viewed condemned criminals as an exception to this rule. As we saw in chapter 2, the eminent Italian physician Antonio Musa Brasavola asserted in 1552 that new or uncertain medicines should be tried on "scoundrels" who had been condemned to death before they were tried on the broader population.[39] In practice, however, the poison trials appear to have been the main focus of such testing, and even they were limited. All of the poison trial accounts suggest enough unease about the practice that special assurances were required. Increasingly, those measures included permission from the criminal himself.

"Without any tricks or force"

How did condemned criminals feel about serving as test subjects? The earliest poison trial accounts demonstrate very little interest in this question. Let us briefly return to Caravita's oil and the trials conducted on the command of Pope Clement VII in 1524. The Corsican prisoner Ambrogio was given a large quantity of poison and left to die in agony. His countryman, Gianfrancesco, fared a little better: after he survived the poison, he was condemned to the galleys rather than executed. It appears that neither man knew that the marzipan they were eating was poisoned; in fact, the testers made deliberate attempts to disguise their intentions. There is some suggestion that the prisoners nevertheless suspected something. Ambrogio (the more scurrilous prisoner) ate his marzipan cake only haltingly, while Gianfrancesco gobbled his cake and advised Ambrogio to trust in God. However, both men rejoiced when the executioner was sent away, a sign that they did not realize what was facing them. Similarly, the Mantuan prisoner named Antonio, who drank the eggs poisoned with arsenic, was not told that he was being used as a test subject, although he figured it out once he started noticing the effects of the poison.[40]

Other Italian poison trials demonstrated a similar lack of concern about the prisoner's point of view. The test conducted in 1540 in Bologna on Bastiano di Gianfrancesco was done "without the knowledge" of the prisoner.[41] As we saw in chapter 2, Gabriele Falloppio received a condemned criminal from Cosimo I de' Medici, with the understanding that he could put him to death in any way he liked before dissecting him; there is no indication that the prisoner was consulted.[42] The Medici poison trial on Jacopo Gemignano similarly made no mention of whether the prisoner's consent was sought or not. In addition, no prisoner appears to have come out of the early Italian cases (1524–40) with a full pardon; the two who did survive were sent to a miserable life in the slave galleys. Pope Clement explicitly stated that the prisoners still had to pay for their crimes.[43] As I argued in chapter 2, the main concerns in the early poison trials had to do with procedure.

A more careful attention to the prisoner emerged out of the cluster of poison trials that took place in the 1560s. The series of prominent tests conducted by the Holy Roman Emperor Ferdinand I and his son, Archduke Ferdinand II of Tyrol, in 1561 introduced two new elements. The prisoner's *consent* was sought, and the trial took place with the understanding that the convict would *go free* if he survived. The issue of consent appears in nearly every account of poison trials at Emperor Ferdinand's court. Mattioli's narrative of the failed test in Prague claimed that the prisoner agreed to the trial willingly, as he preferred to die in prison rather than in public. Georg Handsch's German translation of Mattioli explained even more vociferously that the prisoner "took the poison willingly and gladly, as he said he would rather die (should it come to that) in a quiet place, among few and honest people, than be hanged in public, in front of everyone."[44] As we saw in chapter 3, French surgeon Ambroise Paré echoed Mattioli's sentiment nearly exactly in his account of a poison trial at the court of King Charles IX. The prisoner, Paré claimed, preferred "to die of the aforementioned poison in prison than be hanged in the view of everyone."[45] Significantly, these statements accompanied cases in which the antidote did not work and the prisoner died, suggesting that the issue of consent went beyond a concern that a surviving prisoner might protest his treatment. Even in cases where the criminal died, some additional justification was needed.

Can we believe these portrayals of the prisoners as willing—even eager—participants? The records of these cases were all written by people in positions of power over the criminals, and we should certainly read them with a large grain of salt. Yet there are some reasons why a criminal condemned to hang might have been willing to suffer the gruesome effects of poison in place of a public execution. Hanging was seen as a low and dishonorable form of capital punishment in early modern Europe, in contrast to the more honorable death by beheading.[46] It was applied almost exclusively to lower-class men who had been convicted of theft or larceny, and, particularly in northern Europe, it brought shame upon the criminal's entire family. The shame derived partially from the fact that the executioner, seen as a dishonorable and defiled individual, laid his hands on the criminal in order to help him up the ladder, which was compounded by the added disgrace of having the body left to decompose on the gallows.[47] The frequent assertion that the prisoner preferred to die "in prison rather than in public" likely reflected a very real consideration about the horrors of a public execution.

Although this tendency was stronger in the German cases, some Italian prisoners also expressed a preference for poison over a public execution. In his 1502 treatise on anatomy, physician Alessandro Benedetti noted that condemned criminals slated for dissection sometimes requested to be put to death by a physician rather than the executioner.[48] The Mantuan murderer Antonio allegedly rejoiced once he figured out he was being poisoned in the second trial conducted by Pope Clement, as he had heard of Gianfrancesco's pardon and hoped to receive the same reward (as indeed he did).[49] Falloppio's prisoner survived the first dose of opium, and the man then asked to be given a second dose in the hopes that he would survive again and be set free.[50]

Beyond the fear of bringing dishonor to oneself and one's kin, then, the poison trial offered a chance at freedom. The condemned man had the rare possibility to escape being executed at all. Mattioli mentioned that the prisoner at Ferdinand's court agreed to the test not only for its privacy but also because he hoped that he might survive and be set free. This offer of freedom was no empty promise. Prisoners who survived actually were set free and sometimes even given an additional reward. As we saw in chapter 3, Claudius Richardus noted that the prisoner in Prague, who survived a test of aconite when he took bezoar, was set free and given "a monetary reward." Handsch's notebook clarified that the reward included six thaler and a new suit of clothing.[51] This practice of pardoning the prisoner appears to have been most common in northern Europe, but it sometimes occurred in Italy as well. In June 1567, Cosimo I de' Medici tested "a certain antidote" on two prisoners condemned to hang. At least one prisoner survived the poison and was rewarded with his life in thanks.[52] Given the certainty of death and dishonor otherwise, there were good reasons why a condemned criminal might have chosen the option of poisoning.

At the same time, there were also good reasons why the princes and their physicians might have wanted to highlight the prisoner's consent, even in unsuccessful tests. Hyso's reservations about deviating from regular procedure spoke to a wider problem with poison trials. However useful they might have seemed to physicians, they interrupted the usual process of executing a condemned man for his crimes. The emphasis on the prisoner's agreement can be seen as a signal to the reader that he was not being ripped wantonly away from the expected course of events. Poisoning was especially reviled because it was a secret crime: the victim did not know who had attacked him.[53] By obtaining permission from the condemned criminal, the testers could demonstrate that there was no secrecy or deception and that the trial had proceeded "on his own free will, without any tricks or force." In theory, the condemned criminal was supposed to welcome the public execution as a purifying path to salvation.[54] The poison trial accounts suggest that criminals were more than willing to circumvent this process and hope for a reprieve. At very least, the prisoners' consent provided a convenient justification for the testers.

A Good Death

As gruesome as the spectacle of public execution seems to us today, capital punishment fulfilled an important cultural and religious purpose in the early modern criminal trial process.[55] While the Caroline criminal code was intended to make criminal justice less arbitrary, in practice trials increasingly took place behind closed doors in an inquisitorial fashion. Public involvement was limited to the very end, when the verdict was announced and the punishment decreed and carried out. The punishment phase was the only direct involvement the public had with the criminal trial, and it therefore acted as an important confirmation that justice had been done.[56] A strict blueprint guided this process. The court's verdict was to appear fair; the prisoner, who had to confess before he or she could be convicted, was to show repentance and go willingly to death; the executioner was to complete his job efficiently. Should the punishment go awry, it was usually seen as a sign that a false judgment had been pronounced. Accordingly, a prisoner who recanted his confession on the scaffold or an executioner who bungled his task caused a public uproar and often a riot.[57]

Far more than merely a guarantor of justice, the public played a crucial role of helping the "poor condemned" to a good death. The *ars moriendi*, or art of dying, had held important religious significance since the later Middle Ages. In order to die a good death leading to salvation, a person had to go into his final moments joyfully and without fear. It was the task of the living to help prepare a dying person for death, part of a wider connection between the living and the dead.[58] The ritual of execution was intended not only as punishment and a warning, but also as a way to help the condemned criminal to a good death and, consequently, to salvation. Hence the traditional procession of the criminal to the gallows or block usually involved hymns and prayers. As we saw in chapter 2, the Florentine confraternity San Giovanni Decollato was dedicated specifically to helping criminals die a good death on the gallows, both by comforting the condemned and by holding up images of the crucified Christ as the criminal ascended the ladder to the gibbet or awaited the sword.[59]

In the Protestant context in which Tümler was embedded, the concept of a good death remained just as important—if not more so, given the guarantee of salvation in Lutheran theology, as long as a person died in the full belief of Christ. Luther himself wrote a *Sermon on Preparing to Die* (1519), and the description of Luther's own good death became a best-seller.[60] Public processions of the "poor condemned" to the scaffold, accompanied by prayers and religious songs, were particularly common in Protestant territories. Songs and prayers usually continued until the moment of death. More than just spectacle, the execution was a crucial way to guide the "poor condemned" to heaven.[61] At stake was not merely a fitting punishment but the very soul of the criminal.

Poison trials interrupted this process. In all of the antidote tests examined thus far, the prisoner died not in front of a crowd, but in prison in front of a few physicians. Not only did these tests remove the prisoner from the public view, and thus from the normal course of justice (and a good spectacle), they also prevented the populace from helping the criminal to a good death or even knowing whether a good death had been achieved. Compounding these problems was the perception that poison was a particularly horrific way to die. Viewed as an "abominable crime," poisoning was frequently connected to witchcraft. The acts of killing with poison, herbs, or evil spells had been linked since the early Middle Ages.[62] Poison was not a substance that fostered a good death.

Perhaps for this reason, most poison trial accounts contained subtle assurances to the reader that they were helping the prisoner to a good death. As we saw in chapter 2, the authors of the *Testimonium* describing Pope Clement VII's poison trial went to great lengths to emphasize that the usual procedure was being followed in every way, from the presence of the Roman senator at the trial to the confraternity members comforting the condemned men with prayers.[63] It is likely no accident that Handsch portrayed the Prague prisoner as having died quietly, "as if he were falling asleep," a sign that a good death had taken place.[64]

The central importance of a good death can be seen in the one case in which it was markedly lacking, French surgeon Ambroise Paré's test of bezoar on a cook condemned for stealing silver. For the most part, Paré was very mindful of his own reputation in his description of this trial. He emphasized that the impetus for the test came from King Charles IX; that the king himself had asked the cook's permission, with a promise that he would go free if he survived; and that the prisoner expressed his preference for taking poison in the privacy of prison. An apothecary's apprentice provided the poison, and Paré portrayed himself as merely viewing the test rather than presiding over it. He arrived late, to find the cook "ambling on all fours, like an animal," with his eyes and face fiery and blood streaming from his ears, nose, mouth, backside, and penis. Paré attempted to give him a drink of oil "to save his life," but it was no use. The poor man died "miserably," crying that the gallows would have been preferable.[65] His was not a good death, and Paré did everything he could to absolve himself of blame.

Nevertheless, this episode came back to haunt him. In 1575, the physicians of Paris brought up a series of charges against Paré, who had long been a thorn in their side owing to his privileged position as the king's personal medic. For the most part, the accusations involved cases of malpractice and unprofessional behavior, including many instances in which he had overstepped his bounds as a surgeon and treated patients for internal ailments. However, the cook's death in the poison trial appeared as one of the counts against him. In an unpublished response to all of the physicians' accusations, Paré included a defense against the charge that the poison test was "detestable" and defamed the good name of King Charles IX.[66] He argued first that the prisoner himself had preferred to "die in prison" than to "finish his days publicly on the gibbet." That willingness, he asserted, made the trial acceptable. In addition, he noted slyly, the trial had not damaged the reputation of King Charles. After all, Pope Clement VII had not been defamed by the tests Mattioli described. Finally, he maintained, the test showed that the bezoar stone could not be counted on and thus served a public good.[67] The twin assurances of consent and public benefit, backed by Mattioli's authority, helped balance out the tragedy of the prisoner dying a death that could hardly be deemed good.

Princes and the Public Good

The argument about the public good was another consistent theme through nearly all of the poison trials. When Pope Clement VII declared that Caravita's oil should be tested on "condemned bodies," he did so "for the benefit of the public."[68] Cosimo de' Medici's 1567 poison trial in Florence similarly claimed to be for the "benefit of all."[69] This notion of public welfare appeared as a rote justification for poison trials at all levels, not merely trials on humans. As we saw in chapter 3, the charlatan Bertolani used "public benefit" as one of his arguments for testing his venom antidote in front of the Venetian Magistrate.[70] In most cases, authors argued that the "public benefit" came from finding a drug that would help large numbers of people. Paré's defense flipped the notion of public welfare on its head: by showing that the bezoar did not work, the trial benefited those who might otherwise have relied on it. Even though it failed, it warranted the terrible death of the poor cook.

The invocation of "public benefit" was not always a façade in princely poison trials. Pope Clement's interest in testing Caravita's oil was spurred by Caravita's success in treating plague victims during the 1523 epidemic in Rome. Similarly, Landgrave Wilhelm IV of Hesse-Kassel's test on dogs took place in the context of a plague epidemic in the summer of 1580. Throughout that summer, Landgrave Wilhelm kept tabs on the number of plague cases and the use of various cures, and following the

4.5 The devil as a tyrannical ruler, holding a cup of poison. Ambrogio Lorenzetti, Allegory of Bad Government, c. 1338. Palazzo Pubblico, Siena. Wikimedia Commons.

success of his poison trials on dogs, he made the Silesian terra sigillata available to local villages in order to treat plague victims.[71] It was not enough, it seems, for a prince to use a prisoner for a poison trial merely to protect his own person. A more altruistic justification was needed, even for powerful princes and popes.

Nevertheless, altruism was only part of the reason why the "public benefit" was so important. As Paré's case shows, it provided a powerful defense against objections to poison trials. These concerns were particularly germane for princes. Since the Middle Ages, there had been a strong negative association between tyrannical rulers and poisoning. Poison was depicted as the epitome of tyranny in a famous fourteenth-century fresco by Ambrogio Lorenzetti on the walls of the public palace in Siena: in a portrayal of bad government, a tyrant wearing a devil's face carries a cup of poison (see fig. 4.5).[72] Rumors of attempted poisonings sullied the reputations of several Renaissance aristocrats, most famously Lucrezia Borgia and her brother Cesare (unfairly, as historians think today).[73] The Englishman Fynes Moryson, who traveled through Italy in the 1590s, framed the propensity of the Italian princes to test antidotes on criminals as evidence of their expertise in the art of poisoning:

In our tyme, it seems the Art of Poysoning is reputed in Italy worthy of Princes practice. For I could name a Prince among them, who having composed an exquisite poison and counterpoyson, made proofe of them both upon condemned men giving the poison to all, and the Counterpoyson only to some condemned for lesse Crymes, till he had found out the working of both to a minute of tyme, upon divers complections and ages of men.[74]

This description greatly exaggerated the use of criminals for poison trials, but the satire worked because it had a grain of truth. Any prince who wished to conduct a trial of poison on a human thus had good cause to tread carefully.

Count Wolfgang had a particular reason to proceed with caution. In comparison with the other powerful men who oversaw poison trials—the pope, the emperor, the king of France—Count Wolfgang II of Hohenlohe-Weikersheim was a fairly minor prince. His power was regional, and although he proved a deft and widely respected ruler, he did not have the same aura of dominance as Cosimo I de' Medici or Emperor Ferdinand I. The documents in the trial of Wendel Tümler reflect this precarious situation. Nearly all authors of poison trial reports imbued their anecdotes with subtle attempts to justify the use of human test subjects, but in most cases the true motivations for these justifications were merely implicit. In the behind-the-scenes direction from Wolfgang's councilors, we can see an explicit, open effort to invoke the twin ideas of consent and public benefit—and a good death—in order to prevent the trial on Tümler from reflecting negatively on Wolfgang himself.

Process und Form

The driving force behind this effort was plodding German bureaucracy. Most German principalities involved some amount of power-sharing between the prince and his council (*Rat*), a relationship that was sometimes fraught but could also be beneficial. Wolfgang appears to have leaned heavily on his advisers and respected them. He particularly valued the opinion of Hyso, who had served his father diligently and who continued to advise Wolfgang from his mother's residence in Neuenstein.[75] In 1578, for example, Hyso provided a very long recommendation on the proper form of address for the future Duke Friedrich I of Württemberg (who was not yet the ruling duke, Hyso opined, and thus should be addressed as a count).[76] All of his surviving documents show a deep attentiveness toward proper procedure and the regular order of things.

The Tümler case underscored that concern in spades. The test of the terra sigillata could take place, Hyso advised the count in his initial letter, but it required careful preparation involving a number of stages. First, he explained, Wolfgang should allow the criminal trial to be carried out to its conclusion, including the public pronunciation of the death sentence. That way, if Tümler lived, he could not complain that due process had not been done in his case or give any cause for unrest among the peasantry or foot soldiers. Only *after* the final verdict was pronounced should Tümler be told, by some third party, "that Your Grace has received a new medicine that is supposed to be good for poisons and has been tried on dogs and found to work. And that Your Grace wishes to try it on a human, for the good of other people." Tümler should then be given a choice: he could take the poison and then a potion containing the antidote, with the promise that if he lived he would be banished rather than executed, or he could reject "such mercy" and be hanged as planned.[77]

This clear desire to avoid insurrection was a prudent nod to the historically fraught relationship between the Hohenlohe peasantry and the nobility. The counts of Hohenlohe and their subjects had frequently been at odds ever since the widespread peasant revolts of 1525, and the rulers constantly feared further turmoil. The years prior to Wolfgang's reign had been rife with struggles between princes and peasants over tax levies, a problem that Wolfgang managed shrewdly during his time in power.[78] Hyso's recommendation that the case be left in the legal system until the point at which the death sentence was pronounced was a safeguard intended to prevent Tümler from accusing the aristocratic court of impeding the course of justice "in the case that he would come away alive." Tensions between the ruling classes and peasants were starkly apparent in his caution. Hyso noted somewhat sourly that complaints about justice were "the custom and habit of this frivolous peasantry" and reminded Wolfgang that "years ago the foolhardy peasants" had made many untenable demands of the rulers, possibly a reference to the revolts in 1525. In addition, he cautioned, "agitated people will bring forth all sorts of evil talk."[79] While Hyso expressed no sympathy for Tümler himself, he did worry that going outside regular procedure would foment unrest.

The solution, then, was to create a new protocol. So pressing was the desire to make sure the test would not cause disruption that, a few days after Hyso's initial letter, Wolfgang's councilors together drew up a revised and far more detailed "Process and Form [*Process und Form*] on How One Should Proceed with the Giving of Poison and the Trial of the Terra Sigillata on the Prisoner in Langenburg" (fig. 4.6). Four trusted

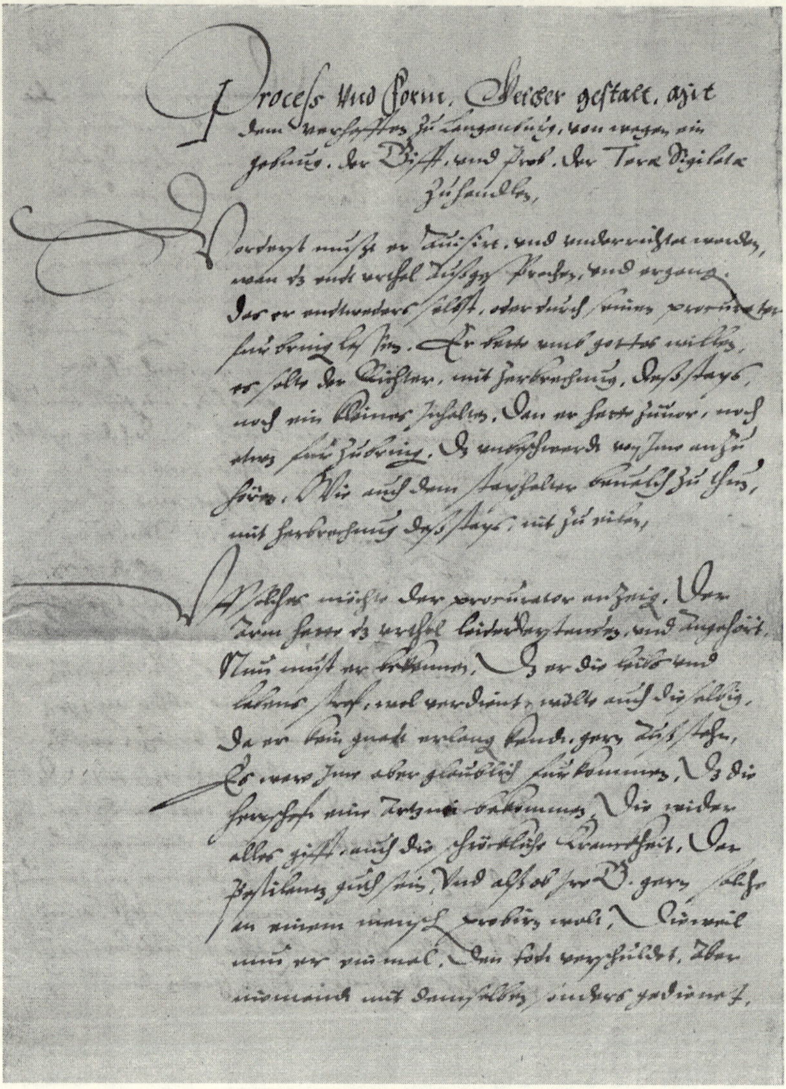

4.6 The *Process und Form.* With kind permission of the Hohenlohe Zentralarchiv Neuenstein.

advisers signed it.[80] This document laid out, step by step, the proper actions of each party involved in the case—Tümler, the judge, and the rulers of Hohenlohe. It reads like the stage directions for a play.

First, the councilors wrote, someone should advise Tümler to plea for a postponement of the execution. Tümler should state that he understood the verdict and knew that he deserved the sentence of death, but that he had heard a credible report that the lords of Hohenlohe had a medicine that was supposed to work against all poisons, and that Their Graces wished to try it on a human. Although he had certainly earned his sentence, Tümler should say, his death would in principle help no one in particular—whereas if the medicine were tried on him and found to be effective, it could help many people. He should state that he would be willing to undertake such a trial "out of his own free will, without any tricks or force, also without being convinced or told to do it." If he were to die, he would suffer no worse fate than already at hand, and he would at least be able to die "with the blessing of the Almighty." The judge should then agree to run the matter by Count Wolfgang before carrying out the execution.[81]

Following this plea, Count Wolfgang should let it be known that His Grace had considered the petition, but that he was greatly hesitant to grant the thief's request given the severity of his crimes. On the other hand, a number of goodhearted people had pleaded that the medicine would "come to the comfort and rescue of many good, pious, God-fearing people." Because of this potential to help "many a good Christian," Count Wolfgang should say, he was open to trying the poison on Tümler, especially since "it was far from certain that he would escape with his life." However, the count should make clear he was pardoning him only because of "the good and help of the many pious Christians" who might be cured by the medicine, and because the antidote "necessitated a proper trial." He should also agree to pardon Tümler only on the condition that the thief would leave the county of Hohenlohe immediately if he survived, and that he would swear not to come within ten miles of it as long as he lived. If Count Wolfgang proceeded in this way, the councilors informed him, he would protect himself against all potential objections that might be raised to the plan.[82]

Wolfgang's councilors made explicit the implicit fears lurking in the earlier trial narratives. They stated the root of those anxieties frankly—political unrest or open rebellion. Such fears were no mere hyperbole, as riots and insurrections around botched executions were known to occur.[83] It would be easy to imagine a crowd viewing a poison trial as just such a bungled execution. Conversely, the councilors also feared unrest should Tümler survive and complain. Their careful balancing act echoed Pope Clement VII's determination that criminals still needed to be punished for their crimes.[84] By appearing reluctant to proceed with a trial that he desperately wished to conduct, by obtaining Tümler's consent, and by agreeing to banish Tümler should he live, Wolfgang could mitigate all of these fears. The prescriptive nature of this charade, with each actor given a specific role, was crucial, as it supplanted the equally regimented script of the execution.[85]

The councilors' script went even beyond Tümler's consent—they wanted Tümler to be seen as suggesting the poison trial. They also urged Count Wolfgang to portray the test as a benefit both to the broader population and to Tümler himself. The "God-fearing" public needed a "proper" trial to ascertain the effectiveness of the antidote on humans, in order to help the wider populace. In the councilors' opinion, Count Wolfgang's motivation must be seen as "the good and help of many pious Christians," a mantra they repeated numerous times. This focus on the "comfort and rescue of many good, pious, God-fearing people" was very different from the question of whether "the medicine worked as well on a human as on an animal," which initially drove the terra sigillata trial.[86]

In the councilors' portrayal, Tümler himself would also benefit, no matter the outcome. If the test was unsuccessful, he still would be able to die "with the blessing of the Almighty," that is, a good death. The horrors of the poison not only served a greater good but also caused no spiritual detriment to the prisoner, despite the physical pain. In addition, Tümler was to be told the hangman "did not need to be involved"—a move that would preserve his family from shame even if he died, a clear reference to the widespread fear of such dishonor and defilement.[87] By agreeing to the trial, Tümler could save his own soul, preserve his family's honor, provide a potential benefit to others, and possibly even walk away a free man. The poison trial, they insinuated, could be portrayed as having some significant advantages over the regular ritual of execution.

In short, the letter by Wolfgang's councilors explicitly and cleverly addressed the many potential concerns about using humans for poison trials: it interrupted the regular process of execution; it involved a horrific way to die; it might be seen as an overreach of the prince's power; it might prevent the criminal from dying a good death. One by one, the councilors recommended ways to allay these concerns, consistent with practices in previous poison trials. How successful were they in this endeavor?

"A perfit triall"

After these extensive recommendations, the archival record of Tümler's trial goes silent. Our only report of the outcome comes in the official testimonial letter from Wolfgang appended to Andreas Berthold's book on the wondrous virtues of the Silesian terra sigillata, published in Latin in 1583 and translated into English in 1589. Written from Wolfgang's perspective, the document described the process of obtaining and testing the Silesian terra sigillata. A careful examination of this printed, public letter allows us to judge the effectiveness of the council's carefully laid plans.

Wolfgang, the document claimed, had learned of the Silesian terra sigillata from "my well beloved friend Andreas Berthold of Oschatz." Gone was the role of his mother, the Countess Anna, who had read Landgrave Wilhelm's account of the Hessian trial, obtained the terra sigillata, and urged a human trial. Instead, Wolfgang portrayed a direct exchange between himself and Berthold, in which the peddler's valorization of the drug and the success of Landgrave Wilhelm's dog tests "made me also desirous to see the triall of it." As it so happened, the thief Tümler, called Thumblandt in the printed version (perhaps to prevent besmirching his family's name), had just confessed to numerous felonies and been condemned to be hanged.[88] Unsurprisingly, Wolfgang gave no inkling of the behind-the-scenes discussions about Tümler's torture and conviction.

After his official condemnation, the unfortunate Thumblandt, "being yet detained in prison," learned that there was a medicine rumored to work "against sundrie sicknesses and the most deadly poisons." He quickly made a "humble request" to his parents and friends, "of which there were present no small number," that he would like to "have given unto him the most deadly poison that might be devised," in order that a "perfit triall" could be made of the terra sigillata. Thumblandt's motive for undergoing this trial, according to the testimonial letter, was a desire not merely for "the mercie of God" and "the respect of his poore life," but also "for the help and benefit of all Christendom."[89]

These words echoed the recommendations of Count Wolfgang's councilors, nearly word for word in some places: after the pronouncement of the sentence, the criminal learned independently of the antidote, and he requested—on his own initiative—that he be given a deadly poison to test it. He made this appeal via his parents and friends, a sign that he had a broad support network and would not be turned loose onto an unwelcoming population should he survive. According to the testimonial, Thumblandt viewed this trial as a means of receiving the mercy of God, as well as a way to help "all Christendom." It would bring him a good death while serving the public good.

The testimonial letter then described the trial itself. Unlike all of the earlier antidote tests, it was not conducted in private, but "in the presence of all our Nobilitie and Commons," including Count Wolfgang and his twelve-year-old son, Count Georg Friedrich. It thus preserved the public spectacle of execution without the executioner; instead, a physician and an apothecary provided the instruments of death. Thumblandt received a dram and a half of mercury sublimate, mixed with conserve of roses. Immediately thereafter, he drank a dram of Berthold's terra sigillata dissolved in wine. Everyone waited breathlessly for the outcome:

> And albeit the poison did, in the judgement of our learned physician Georg Pistor, doctor in physic, and John Lutzen our Apothecarie, who were with him all the while, extremely torment and vexe him: yet in the end the medicine prevailing overcame it, whereby the poore wretch was delivered, and being restored to his health was committed to his parents.[90]

As in the case of Wilhelm of Hesse's dogs, the terra sigillata once again proved itself worthy, and it provided a spectacle to boot. Tümler/Thumblandt escaped with his life, and Berthold had an extremely valuable letter to add to his impressive testimonials of success. Meanwhile, the steps Count Wolfgang's councilors suggested became inscribed in the official printed record.

Of course, Count Wolfgang's power over the narrative should make us question whether the trial proceeded quite as smoothly as depicted. Perhaps Tümler's recovery was less complete or less certain than portrayed; we have no indication of how long he lived once he left Langenburg. Perhaps the dose of poison was not truly fatal. Perhaps the simple statement that the poison "vexed" him masked the bouts of extreme pain, vomiting, and diarrhea that affected other subjects. The point of Wolfgang's testimonial letter was to show the easy success of the terra sigillata rather than describe a medical phenomenon, unlike the cases written by physicians that we examined in chapter 3. As far as the reader knew, the test had progressed smoothly and logically from the moment of Tümler's arrest to his survival and banishment—and the antidote had worked.

Ethics and Poison Trials

What are we to make of this contrast between the public and the private record? On a basic level, it shows that a public historical narrative of an event—any event—rarely reflects the true complexity. Other poison trial accounts, composed after the fact, display subtle hints of underlying anxieties about human experimentation. The letters from Count Wolfgang's councilors, written *in medias res*, provide a smoking gun. Obtaining and using a human being for a fatal test was not as easy as it sounded, even when using the body of a criminal already condemned to die. As test subjects, living convicts posed far greater hurdles than dead ones. While there was significant cultural opposition to dissecting the cadavers of deceased criminals, this practice did not endanger the soul of the prisoner, or position the prince as a poisoner, or return a dangerous criminal back to society. The prince's reputation had little to lose from dissection; not so the poison trials. That is the essence of why Tümler had to be seen to participate "without any tricks or force" and why his poison trial needed to have some perceived public benefit.

I do not mean to suggest that sixteenth-century experimentation on condemned criminals ran aground on moral and ethical concerns. As we have seen, most Italian cases did not appear to require the criminal's consent, and many poison trials took place in northern Europe despite the hesitation. Even in the cases that hinted at underlying discomfort with poison trials, the fears revolved around public perception rather than the poisoning itself. Despite their heavy-handed stage directions, the Hohenlohe councilors stated outright that they saw "no real problem" with the trial. As we saw in chapter 3, Georg Handsch's German translation of Mattioli minimized the Italian physician's role in a failed poison trial. Yet Mattioli's reputation appears to have been the main force behind Handsch's caution in print, not ethical unease about the prisoner. In his private notebook, which includes brutally honest self-criticism in other places, Handsch recorded several poison trials without any indication that they were problematic.[91] The Parisian Board of Physicians condemned Paré's poison trial as "detestable," but again, the outrage focused more on the reputation of King Charles IX than on the treatment of the silver thief.

Still, the fact that public perception was a concern at all strongly suggests that poison trials transgressed expected notions of how physicians and princes were supposed to behave. The special treatment of all surviving criminals reinforces this notion. For men convicted of less dangerous crimes like theft, the reward of freedom was remarkably consistent, and even the worst criminals had their sentences lessened. Pope Clement VII relegated his surviving prisoners, convicted of violent crimes, to the slave galleys rather than executing them. A few years after Tümler's trial, Landgrave Georg I of Hesse-Darmstadt (1547–96) conducted a trial using Berthold's terra sigillata on a mercenary who had been apprehended after a long manhunt and condemned to be broken on the wheel for "many terrible crimes." The prisoner survived and quickly recovered his full health, and Landgrave Georg "benevolently" decreed that his sentence be commuted to beheading. The wheel was a horrific and shameful way to die, while beheading was the most honorable of the various methods of execution. As in Pope Clement VII's case, Georg gave the mercenary a reward (of sorts) for his efforts without letting a murderous criminal go free.[92] In cases in which the crimes were too grave to warrant a full reprieve, the surviving criminal still received recognition for taking part in an extraordinary event.

The promised pardon for nonviolent criminals was particularly noteworthy. In sixteenth-century penal systems, the mitigation of death sentences was not common (although less rare in Germany than elsewhere).[93] This tantalizing offer of freedom went hand in hand with the attempts to receive the prisoner's permission, thereby neatly assuring the convict's compliance. Presented with the possibility of life, how many people would choose certain death? In post-Nuremberg ideas of ethics and informed consent, this basic situation would be viewed as a form of coercion in and of itself. The Nuremberg Code stipulated that the test subject "should be so situated as to be able to exercise free power of choice, without the intervention of any element of force, fraud, deceit, duress, over-reaching, or other ulterior form of constraint or coercion."[94] In 1978, the United States National Institutes of Health severely restricted the use of prisoners as medical test subjects owing to the power differentials that made true consent difficult to determine.[95] Modern bioethics are not very useful in understanding the early modern stakes, however, especially given the generally harsh dictates of the absolutist state.[96] Given the certainty of death and dishonor facing condemned criminals, and given that several surviving criminals were indeed released, the promise of freedom represented a rare opportunity. It was also fleeting. While the public good continued to be a frequent argument in favor of medical tests on humans, the issue of consent rarely surfaced from the decline of human poison trials until the nineteenth century.

We do not know what happened to Tümler after he departed the scene; he disappears from the historical record, presumably with a thankful prayer at his survival. As he faded into obscurity, so too did human poison trials. Tümler's case represents the last extensive record of a poison trial on a condemned criminal that I have found, although the practice continued at least until the end of the sixteenth century.[97] For the most part, as the public face of poison trials faded, the focus returned to animals. Nevertheless, the idea of using condemned criminals for potentially fatal tests continued to have its proponents. A century after the Paris medical faculty condemned Paré's poison trial as "detestable," Denis Dodart, a doctor of the same medical faculty, advocated the use of condemned criminals for testing poison. Writing in 1676, he used the same argument for the public good as Brasavola had in 1552.[98] Although the bulk of testing shifted to animals, the question of which humans should participate in potentially deadly tests remained a problem for centuries to come. Similarly, the idea that part of the answer lies in procedure and protocol has stayed with us to this day.[99]

Tümler's case sits at a crossroads of a sort. On one hand, it represented the culmination of the avid interest in poison trials from 1524 to 1581, as the last of three major trials of the Silesian terra sigillata in the 1580s. On the other hand, purveyors of marvelous antidotes increasingly presented the drugs as useful for a panoply of illnesses, not just poisoning, as in Berthold's portrayal of the German terra sigillata. Poison trials provided dramatic proof of usefulness against poison. They were less helpful in demonstrating efficacy on a broader range of diseases. As we will see in the next two chapters, the rising frenzy for wonder drugs in the latter half of the sixteenth century helped other methods of testing antidotes eclipse poison trials.

PART THREE

Wonder Drugs

FIVE

Powerful and Artful Substances: Bezoar Stone

In 1589, a provincial German physician named Johann Wittich published a small book with an enticing (and very long) title. *Report on the Wondrous Bezoar Stone That Is Good against All Poison*, the heading announced in bold red ink, and it continued, in smaller blackletter: "Also on Other Stones That, from Their Hidden, Imbedded Nature and Virtues, Can Produce Untold and Unbelievable Effects." But that was not all. The book told not only of healing stones, but also of "the new Armenian Balsam"; "foreign wonder herbs, barks, and roots"; and "the new Silesian Terra Sigillata." All of these marvelous substances, he noted, had been "unknown to the old and newer scribes" and had come into European knowledge only in the past thirty years "from the Oriental and Occidental Indies . . . never before translated into German." Wittich had undertaken the arduous task of translation to provide all "high potentates, industrious natural historians, and lovers of health" with an overview of new and marvelous cures.[1] He explicitly tied these wondrous medicaments to the practice of testing, claiming an urgent need for "powerful and artful substances that have been proven adequately in a trial [*proba*]" (see fig. 5.1).[2]

Wittich's *Report* brought together a flurry of writing on wonder drugs from the 1560s–1580s. In 1563, a Portuguese physician living in Goa, Garcia da Orta (d. 1568), wrote of the useful properties of many Indian drugs, including an influential account of bezoar stone. Orta's work influenced another Portuguese physician, Cristobál Acosta (1525–94),

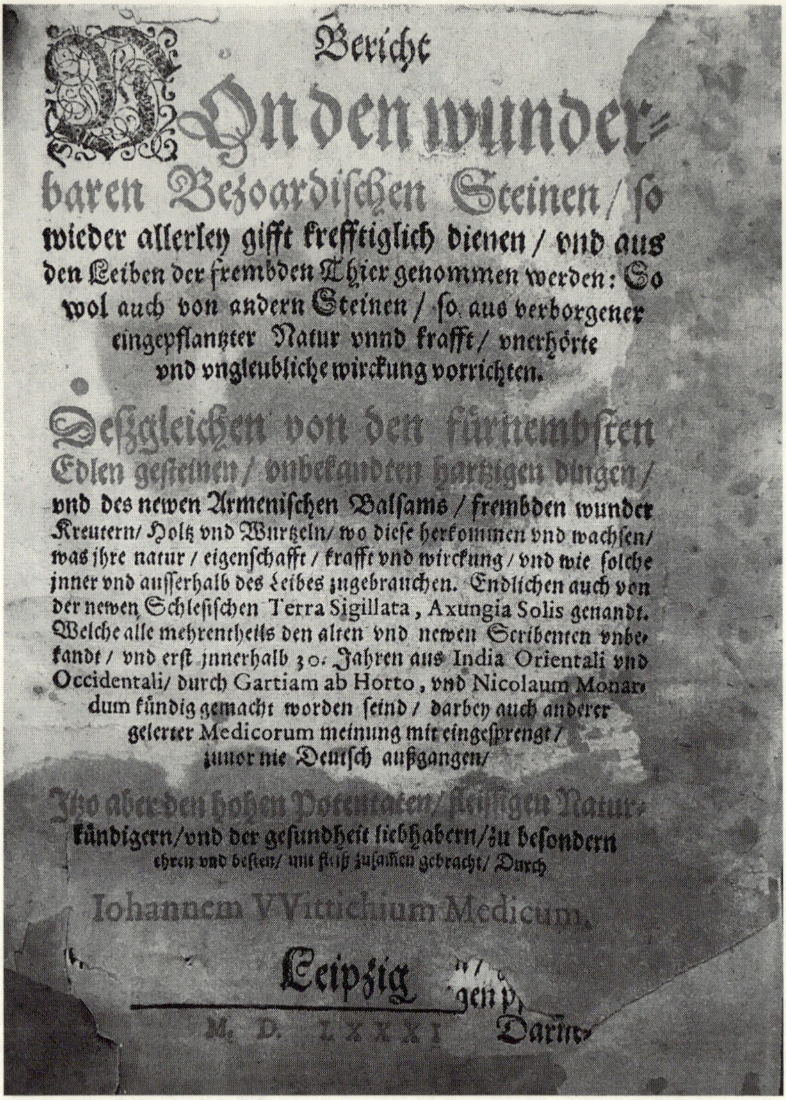

Bericht
Von den wunder-
baren Bezoardischen Steinen/ so
wieder allerley gifft krefftiglich dienen/ vnd aus
den Leiben der frembden Thier genommen werden: So
wol auch von andern Steinen/ so aus verborgener
eingepflantzter Natur vnnd krafft/ vnerhörte
vnd vngleubliche wirckung vorrichten.

Deßgleichen von den fürnembsten
Edlen gesteinen/ vnbekandten hartzigen dingen/
vnd des newen Armenischen Balsams/ frembden wunder
Kreutern/ Holtz vnd Wurtzeln/ wo diese herkommen vnd wachsen/
was jhre natur/ eigenschafft/ krafft vnd wirckung/ vnd wie solche
jnner vnd ausserhalb des Leibes zugebrauchen. Endlichen auch von
der newen Schlesischen Terra Sigillata, Axungia Solis genandt.
Welche alle mehrentheils den alten vnd newen Scribenten vnbe-
kandt/ vnd erst jnnerhalb 30. Jahren aus India Orientali vnd
Occidentali/ durch Gartiam ab Horto, vnd Nicolaum Monar-
dum kündig gemacht worden seind/ darbey auch anderer
gelerter Medicorum meinung mit eingesprengt/
zuuor nie Deutsch außgangen/

Itzo aber den hohen Potentaten/ fleissigen Natur-
kündigern/ vnd der gesundheit liebhabern/ zu besondern
ehren vnd besten/ mit fleiß zusammen gebracht/ Durch

Iohannem VVittichium Medicum.

Leipzig

M. D. LXXXI

5.1 Johann Wittich, *Report on the Wondrous Bezoar Stone* (1589), title page. Courtesy of Houghton Library, Harvard University.

who published his own account of Indian drugs in 1578. In 1565, Spanish physician Nicolás Monardes published a book called *Dos Libros* (Two Books), which contained a treatise on marvelous drugs from the New World and a treatise on bezoar and other poison antidotes. Wittich learned of all of these authors through the best-selling Latin translations of Dutch physician Carolus Clusius (1526–1609), who completely rearranged the original texts and added his own details.[3] In addition to these works from Iberian authors, Wittich included an account of tobacco from French diplomat Jean Nicot (1526–1604); the letter on bezoar that Claudius Richardus had composed for Archbishop Olahus; and an excerpt from Andreas Berthold's book on the Silesian terra sigillata. He also gave details on marvelous gems from medieval Arabic and European authors, and he included examples from his own experience as a city doctor in Arnstein and personal physician to the counts of Schwarzburg.[4] Wittich's *Report on the Wondrous Bezoar Stone* acted as a virtual cabinet of curiosities for the German-reading public, a *Kunstkammer* full of wonder drugs.

Wittich's book was emblematic of a growing frenzy for drugs that "can produce untold and unbelievable effects," and it highlights a crucial point about poison antidotes: they did not just cure poison. The most valuable antidotes had long been seen as substances that could cure nearly any ailment. Theriac remained the most prominent example in Europe, but exotic substances like bezoar, unicorn horn, and emeralds also were thought to possess expansive healing powers. The word "wonder" was used frequently to describe antidotes in sixteenth-century texts, as it was in Wittich's book.[5] As we saw in chapter 1, these items appeared prominently on inventories of the treasures of princely households in the Middle Ages. The drastic expansion of the Portuguese and Spanish drug trade in the sixteenth century made some of the most exotic drugs more widely available in the marketplace and added new ones, as Europeans imported unfamiliar materia medica from both east and west.[6] Concurrently, European healers began to market their own antidote-cure-alls, like Caravita's oil or the Silesian terra sigillata. In the second half of the sixteenth century, drug treatises promising wondrous effects, many of them written by physicians like Monardes or Wittich, began to make waves in print culture. A spate of recent scholarship has shown just how valuable antidotes like bezoar were as objects of curiosity and wonder.[7]

Antidotes' increasing connection with wonder drugs puts poison trials in a somewhat different light. Whatever their contribution to new experimental techniques, poison trials were also closely tied to the drug trade and the demand for wonder cures. Caravita's oil, a secret new antidote from a local itinerant healer, had commercial as well as medical meanings; we should not overlook the rich reward Caravita received for his efforts.[8] Mattioli blatantly used poison trials to advertise his "miraculous" scorpion oil, as we saw in chapter 3. Moreover, the value associated with poison trials went beyond bare monetary accounting. Any prince who had an effective antidote had a potent weapon against both attempted poisoning and the plague epidemics that regularly swept through Europe. Wittich explicitly mentioned "high potentates" in his call for "powerful and artful substances that have been proven adequately in a trial."[9] In this light, poison trials had a symbolic as well as an experimental significance.

Yet "proven adequately in a trial" had multiple meanings that only rarely indicated the use of poison on condemned criminals. As this chapter demonstrates, the appeal of "powerful and artful substances" came not only from their purported ability to prevent poisoning, but also from their powers to cure disease. Accordingly, all of the new wonder drug treatises included dramatic anecdotes of healing drawn from experience as proof of the medicaments' powers. In addition to Richardus's trials on condemned criminals, he described the use of bezoar on two "upstanding men" who had become deathly ill. Monardes claimed to have witnessed nearly a dozen specific instances of bezoar's amazing success, but none of them were poison trials; most were cases of illness, not poisoning.[10] While poison trials provided potent evidence of efficacy, only powerful princes had condemned criminals at their disposal, meaning that very few medical practitioners could actually use them as evidence. Healers situated outside of the princely courts thus tended to use anecdotes drawn from clinical experience rather than contrived poison trials on criminals to emphasize the efficacy of antidote cure-alls.

Another kind of poison trial did appear frequently in wonder drug literature, however: quick-and-dirty marketplace tests using animals. These tests sometimes were used to demonstrate a drug's efficacy, but often they functioned as proof of authenticity. Hand in hand with "powerful and artful substances" came fears about fraud, along with a constant suspicion of the foreign merchants who delivered exotic drugs to eager European consumers. To reassure readers, authors of wonder drug literature turned to marketplace tests. While courtly poison trials focused on the symptoms of poisoning, these marketplace tests were outcome-driven. A successful trial on pigeons or dogs added evidence that a marvelous drug could be trusted. Ironically, most of the authors describing these marketplace poison trials were learned physicians, who, as we saw in chapter 3, generally had little patience for charlatans' poison shows. Eyewitness reporting (by physicians) and the testimony of reliable individuals helped serve as markers of authenticity and a protector against fraud.

The last two chapters of this book reexamine poison trials in the context of antidotes' capacity as "powerful and artful substances" that could cure nearly any disease. As the sixteenth century wore on, the focus began to shift from curing poison to wonder drugs that would work on disease in general—both in traditional learned Galenic medicine, as we will see in this chapter, and in alchemical medicine, as the next chapter will demonstrate. Although poison trials represented a potent proof of success through the eighteenth century, human poison trials did not remain a central part of the narrative. The ethical issues raised in chapter 4 may have been a factor in this shift, but I argue that the transition also had to do with the nature of the cure. Antidotes expanded into much broader wonder drugs, and then, in the hands of alchemists, true panaceas. This chapter demonstrates that learned physicians played a central role in that transition.

Antidotes as Wonder Drugs

Wonder drugs were substances perceived to have remarkable medical effects, either because they healed a particular ailment extremely well or because they could cure nearly any ailment. Although the specific term "wonder drug" is a modern idiom, the concept of a wonder drug is age old, as Francis Brévart has noted.[11] German treatises specifically used some version of the word "wonder" (*Wunder*) to describe marvelous medicaments, while Italian and Latin texts often described such cures as "miraculous" (*mirabile* or *miracoloso*). Numerous treatises on wonder drugs had circulated in medieval Europe, many of them infused with magical and religious significance. Some of these substances, such as vulture's flesh, derived from ancient remedies; others, like vervain, first appeared in early medieval texts. Often the wondrous powers of the drug related only tangentially to health matters in the first instance; vervain, for example, was lauded primarily for its effectiveness in warding off demons, protecting against magic, and revealing the outcome of an illness.[12] Over time, these mystical properties were often eclipsed by medical uses. In 1539, the influential German herbalist Hieronymus Bock (1498–1554) complained that vervain was "collected for magic more than medicine," but he also noted that it was good for "almost all internal complaints," especially problems with the liver, spleen, and kidneys.[13] By 1582, Adam Lonicer omitted any mention of magic and described vervain as useful against myriad illnesses and conditions, including poison.[14]

Poison antidotes figured prominently as wonder drugs, owing to their powers to cure poison alongside a wider variety of ailments. The ubiquitous theriac and its forerunner mithridatium were prime examples. Although primarily antidotes to poison, they had been viewed since antiquity as useful for countless conditions as well as a safeguard against illness. Ancient authors also praised terra sigillata as a cure for stomach ulcers, wounds, and bleeding in addition to its antipoison properties. Between the fourteenth and the sixteenth century, both the number of these cure-alls and the supposed powers of many existing antidotes expanded. In the wake of the Black Death, as we saw in chapter 1, physicians began to use terra sigillata as a plague cure, and subsequent authors recommended it for many other ailments. By the time Andreas Berthold marketed his German terra sigillata, it had become a virtual cure-all. A similar shift occurred with other antidotes, including emerald, Armenian bole, bezoar stone, and unicorn horn. Most of these substances, especially bezoar and unicorn, were precious items that had long figured among princes' most valued treasures. In the sixteenth century, their transformation into cure-alls went hand in hand with a modification in their material worth: still exotic and valuable, they nevertheless became attainable medical commodities.

Bezoar stone provides a particularly potent example. As *Harry Potter* readers might remember, bezoar is not a stone at all but rather an animal product.[15] It is a sort of calcified hairball—a hardened mass of indigestible material found most often in the stomachs of ruminants like goats (although it can occur in most animals, including humans). Bezoar appears to have been unknown or uninteresting to the Greeks, but it had a long history as a prized antidote in India and the Middle East. It was mentioned in pre-Islamic Hebrew texts and appeared in Arabic sources from at least the eighth century. The Persian polymath al-Biruni (c. 973–1050), who traveled widely in India, praised bezoar particularly effusively. He maintained that true bezoars should be "the costliest among stones," since they had the power to heal the body.[16] In the early twelfth century, the respected Spanish-Arabic physician Avenzoar (ibn Zuhr) claimed that bezoar could expel poison "miraculously."[17] This endorsement was particularly influential and appeared regularly in early medical writings on poison, from medieval European scholastic authors to Monardes.

Although cited in texts, bezoars as physical objects were exceedingly rare in the European Middle Ages, found mainly in princely and ecclesiastical collections.[18] Guglielmo de Marra briefly praised "the stone that is called *Bezehar*" as a "medicine of medicines for all poison" in his treatise written at the Avignon papal court in the 1360s.[19] More expansive sections on bezoar appeared in fifteenth-century poison treatises by Sante Arduino and Antonio Guaineri, suggesting that awareness of the antidote was growing.[20] New commercial opportunities helped magnify this interest. When the Portuguese arrived in southern India in 1498, they reportedly received a bezoar as a gift from the king of Cochin, which alerted merchants to a new source for the coveted item. In the sixteenth century, the Portuguese trade in bezoars became a highly lucrative endeavor, and Spanish merchants soon stepped into the fray, making the product somewhat more attainable. Nevertheless, bezoars remained coveted, expensive items: they were packed together with diamonds in one seventeenth-century shipment from Goa to Portugal.[21]

The rising demand for bezoar appears to have spurred a renewed medical interest. Mattioli briefly mentioned (and praised) bezoar as "an infallible antidote owing to its special virtues against all poison" in his 1544 *Commentaries on Dioscorides*, and Italian physician Giovanni Battista da Monte and Portuguese physician Amato Lusitano both included a description of bezoar in their commentaries on ancient authors published in the mid-1550s.[22] Lusitano's description was particularly influential, as he mentioned actually encountering bezoars via Portuguese trade networks. Between 1563 and 1565, Portuguese physician Garcia da Orta, Spanish physician Nicolás Monardes, and the imperial surgeon Claudius Richardus all wrote laudatory descriptions of bezoar. Significantly, these last three authors treated bezoar as a cure-all rather than merely as an antidote to poison. In 1578, Portuguese physician Cristobál Acosta provided a particularly long list of bezoar's uses, following Orta and Monardes. He described the stone as "a universal antidote against all poisons" that also helped against poisonous diseases and long-standing ailments; in quartan and other fevers that were difficult to eradicate; and in leprosy, scabies, smallpox, measles, choler, and old itches. It had great powers against plague and other contagious diseases, and it was similarly useful against melancholy and in melancholic ailments. It healed epilepsy, and it could even preserve against wasting, as it helped reinforce any meat taken into the body.[23] Importantly, the increased medical attention followed the upswing in trade, rather than vice versa.

Trade networks similarly drove the trajectory of unicorn horn, another exotic animal product seen as a powerful weapon against poison. The healing power of the mythical unicorn dated back to the writings of Greek physician Ctesias (5th century BCE), whose description of

India included a one-horned beast that had the power to make water potable with its horn. The animal appeared occasionally in classical and medieval texts, but no authors mentioned properties against poison.[24] It seems likely that the horn's supposed powers against poison derived from Arabic legends of strange, one-horned animals that could ward off poison—possibly based on the rhinoceros, the horn of which was also prized as an antidote to poison. These associations appear to have traveled along trade routes rather than in medical treatises.[25]

Like bezoar, unicorn horn was a precious commodity, but unlike bezoar, it came from a mythical beast rather than a flesh-and-blood animal. Debates about whether the unicorn actually existed abounded but did not slow the popularity of the drug. Along trade routes from Asia and Scandinavia, various exotic substances, including narwhal tusk and rhinoceros horn, began to filter into Europe under the name of unicorn horn.[26] By the fourteenth century, unicorn horns appeared among other poison antidotes as valuable items in princely and (especially) ecclesiastical collections. The most famous specimens belonged to cathedrals, such as the nearly seven-foot-long horn belonging to the French royal abbey of Saint Denis. The Basilica San Marco in Venice had two unicorn horns, one of which had its usefulness against poison inscribed on a silver gilt handle. In both cases, the local populace could view the horn once a year and, upon petition, had access to its medical virtues.[27]

As in the case of bezoar, unicorn horn had long been seen as a cure for plague, but it did not become a topic of extensive medical discussion until the middle of the sixteenth century. When it did, authors gave it powers that extended beyond poison. In his influential *History of Animals* (1551), the renowned Swiss physician and naturalist Conrad Gessner included a striking entry on the unicorn. In addition to his musings on the beast's existence, appearance, and diet, Gessner included an extensive section on the medical uses of its horn. He cited it as good for poison, plague, pestilential fevers, worms in the body, and epilepsy.[28] Stand-alone treatises on the unicorn by Andrea Bacci and Ambroise Paré (an early skeptic of the unicorn) followed in the 1560s and 1570s, mirroring the pattern of bezoar.[29] Unlike bezoar, however, the unicorn was treated as a whole, not just the medicinal properties of its horn. With the exception of Paré, most physicians agreed that the animal existed and that its horn was indeed a powerful antidote, although they tended to focus more narrowly on its efficacy against poison and plague than was the case with bezoar.

It was no coincidence that the antidotes imbued with the greatest exoticism and mystery also were seen as having the greatest healing powers. As discussed in chapter 1, poison usually was thought to harm humans by means of an occult (hidden) property known as specific form or total substance. Antidotes functioned through similar occult properties, and, as Katharine Park and Lorraine Daston have established convincingly, their power as medical wonders came from their exotic nature. Italian physician Marsilio Ficino (1433–99) specifically described their powers as impressed by the heavens.[30] These occult properties remained even as the objects themselves became more attainable.

Treasures of Health

Antidote wonder drugs were hybrid objects, imbued with value both material and medicinal.[31] As such, they lent themselves particularly well to the increasing trend of collecting and displaying natural wonders that spread through sixteenth-century Europe. Along with bezoar and unicorn horn, new antidotes such as rhinoceros horn and Maldives coconuts (also called Seychelles nuts) began to filter into collections.[32] Many antidotes figured among the items in cabinets of curiosities assembled by physicians and apothecaries, especially in Italy, where they functioned as objects of research (at least potentially) as well as exotic natural specimens. The collections of Ulisse Aldrovandi in Bologna, Francesco Calzolari in Verona, and Ferrante Imperato in Naples all boasted marvelous antidotes such as unicorn horn, bezoar, and coconut, as well as terra sigillata from various locations (see fig. 5.2).[33]

Whatever these items' intangible value as cures, they also held significant worth as commodities. An inventory of Calzolari's collection appeared in a book by Gian Battista Olivi in 1584, intended to praise (and advertise) the marvels of Calzolari's "Museaum." Numerous antidotes appeared among the rarities, including bezoar, emerald, amethyst, Indian balsam, Armenian bole, St. Paul's earth, and a piece of unicorn horn received personally from Mattioli. The list also included many nonmedical wonders, such as marine specimens, fossils, fungi, and preserved crocodiles, as well as other exotic medical cures such as tobacco and a number of distilled oils and waters.[34] Many of the medical rarities also happened to be items on sale in Calzolari's shop. An introductory letter from the late Mattioli explicitly praised the many poison antidotes that Calzolari carried in his shop, particularly his famously authentic theriac. Mattioli noted that the apothecary also sold "my poison antidote and my oil of scorpions" as well as other marvelous antidotes such as unicorn horn and hart's heart bone.[35] The letter, dated 1568, was not

5.2 Calzolari's Musaeum, as pictured in Benedetto Ceruti's 1622 *Musaeum Franc. Calceolarii*. Courtesy of Houghton Library, Harvard University

written specifically for Olivi's book (Mattioli had died in 1578), but its inclusion emphasizes antidotes' delicate balance of material and medical worth. A web of traders and merchants from near and far stood in the background of these material objects.[36]

At the princely courts, prized antidotes held an even stronger status as symbols of both health and wealth. Often set in delicately crafted filigree or studded with gems, they merged art and nature, wealth and craftsmanship. Marnie Stark has demonstrated the widespread practice of adding gold, silver, and jewels to bezoar, rhinoceros horn, and coconut. Bezoars, in particular, masked their dull appearance with beautiful cloaks of gold filigree and gems (see figs. 5.3 and 5.4).[37] In 1580, the Spanish nobleman Sebastian de Santoyo, chamberlain to King Philip II of Spain, sent a massive shipment of valuable items to Francesco I de' Medici, presumably on Philip's behalf. Along with jewels, delicately painted furniture, textiles, and finely wrought weapons, Santoyo sent "three bezoars from the Portuguese Indies, two adorned in gold filigree and one without ornament," as well as "five large bezoars from the Castilian Indies as well as four or five small ones." He also sent "a coconut from the Maldives adorned in silver."[38] Not all of the bezoars had precious settings, underscoring their inherent value even without adornment, but those that did were worthy of mention.

Many of these priceless and beautifully crafted antidotes found their way into the princely Kunstkammer (chambers of arts), which began to take shape in the second half of the sixteenth century. The Hapsburg imperial Kunstkammer became famous for its impressive display of bezoars and rhinoceros horn, especially under Emperor Rudolf II, although the first imperial bezoars were likely acquired before Rudolf's reign.[39] Saxony's royal Kunstkammer in Dresden, founded by Elector August I in 1560, displayed a unicorn horn and bezoars among its most important treasures, while Duke Albrecht V of Bavaria had a table full of bezoars in his Munich Kunstkammer, founded in 1565.[40] Most princely collections in the Renaissance also contained richly decorated bowls and drinking vessels intended to either neutralize poison or warn the user if poison was present.[41] One particularly impressive example from the Dresden collection, created by the goldsmith Kaspar Widmann in the 1550s, was a delicately crafted gold-gilded bowl, out of which rose a large agate and in which four pieces of unicorn horn were embedded (see fig. 5.5). Already a marvelous item through its expert craftsmanship, the bowl held the power both to indicate poison (agate) and neutralize it (unicorn horn). The imperial Hapsburg Kunstkammer contained numerous drinking vessels made of bezoars, as well as a golden bowl attached to a bezoar by a chain (see fig. 5.3).

This connection between antidotes and display went beyond exotic items obtained from afar. In the sixteenth century, Saxony became the European center for mining serpentine stone or snakestone, another mineral valued as an antidote to poison. Very rare until the discovery of serpentine mines in Zöblitz in the mid-fifteenth century, it remained a prized commodity. If worked soon after mining, serpentine could be turned on a lathe like ivory and then polished to a sheen once the material dried. Saxony's talented turners soon began to create elaborate serpentine drinking vessels, some fashioned by Elector August, an expert turner himself. Like Widmann's bowl, the serpentine vessels were masterpieces of craftsmanship in addition to wielding power against poison.

5.3 Two bezoars from Goa, one in gold filigree (late sixteenth century) and one attached to a golden bowl (seventeenth century). Kunsthistorisches Museum, Vienna. Photo by the author.

5.4 Spanish bezoar set in gold and studded with emeralds and rubies, late sixteenth century. Kunsthistorisches Museum, Vienna. Photo by the author.

5.5 Bowl set with agate and unicorn horn, termed a *Giftprobe* (poison trial). Kaspar Widmann, c. 1556–58. Grünes Gewölbe, Staatliche Kunstsammlung, Dresden. Photo credit: bpk Bildagentur, Grünes Gewölbe, Paul Kuchel, Art Resource NY.

The Dresden Kunstkammer displayed a number of examples of turned serpentine, and Elector August sent them as gifts to other European monarchs.[42] Wittich's description of marvelous antidotes as "powerful and artful substances" aptly reflects the status of these valuable objects.

That value should be understood materially as well as metaphorically. Many precious antidotes were kept not in the Kunstkammer, but in the treasure chamber, or *Schatzkammer*. Mark A. Meadows has argued that the earliest Kunstkammer were *primarily* treasuries, rather than collections of art or curiosities.[43] The mid-sixteenth-century literature on antidotes supports this point of view. Gessner began his discourse on the unicorn horn by noting that the two specimens held at the Basilica San Marco in Venice were kept among the rubies, emeralds, and pearls of its treasure chamber. He claimed that the king of Poland also had two unicorn horns in his Schatzkammer, and that there was once a full unicorn horn in the treasure chamber at the Strasbourg Cathedral until a canon sawed several inches off its end, markedly decreasing its value.[44] One of the three inalienable treasures of the Hapsburg dynasty was its unicorn horn. Bezoars were officially part of the treasury even if they were displayed in the Kunstkammer, which was also true of the bezoars and unicorn horns displayed by the dukes of Bavaria in Munich. Even most of Elector August's turned serpentine was kept in the Schatzkammer, not the Kunstkammer.[45]

Testing Princely Power

In the early 1560s, when Emperor Ferdinand I conducted his poison trials on bezoar stone, it was still an exceedingly rare object. These tests predated the major flurry of publications on bezoar that would come in the mid-1560s. Bezoar remained rare enough, indeed, that even seeing one represented a newsworthy point of discussion among physicians. In 1564, Gessner noted that the imperial physician, Johann Crato von Krafftheim, had seen a Spanish (i.e., Peruvian) bezoar at Emperor Ferdinand's court. No doubt, Gessner noted, the stone had come via the Fuggers—a reference to the wealthy Fugger trade network that controlled many imports from Spanish America.[46] Claudius Richardus's description of Ferdinand's tests on bezoar thus represented an attempt to provide new knowledge about a genuinely unknown substance with powerful potential. We should not, however, assume that scientific curiosity was the only motivation behind the tests. Successful poison trials provided the princes who oversaw them with a doubly potent weapon. They drew attention to the monarch's priceless antidotes and suggested that these objects' properties against poison had actually been proven—thereby giving the prince an aura of wealth and invincibility.

This projection of princely power made a particularly strong statement in an era of rampant religious and political strife. Until the 1580s, all of the poison trials on condemned criminals took place in Catholic regions. The trials of Caravita's oil in 1524 were initiated by Pope Clement VII himself, and several cardinals of the Church conducted their own tests (chap. 2). Emperor Ferdinand I oversaw poison trials on bezoar stone, unicorn horn, and the archducal poison powder in the context of his avid efforts on behalf of the Counter-Reformation. The only poison trials conducted by Protestant princes were the tests of the Silesian terra sigillata (chap. 4). At the same time, the German physician Wittich, a devout Protestant, identified the need for "powerful and artful substances that have been proven in a trial" specifically as a means for German princes to defend themselves against the "devil's servants, poison-skulkers, thief-assassins, and murderers in Italy, Spain, and France," the three main Catholic powers. Reformation religious struggles represented an important backdrop to the drive to test these "powerful and artful substances."[47]

The glowing letter on bezoar from Richardus to the Hungarian archbishop Nicolaus Olahus contained a strong subtext of both imperial power and Catholic sovereignty in general. Richardus presented bezoar as a "kingly medicine" proven by tests conducted "at the command of our true sacred Catholic Imperial Majesty."[48] The choice of Olahus as recipient of this glorious news was no accident. Known for his strident Counter-Reformation efforts, Olahus was one of Emperor Ferdinand's most faithful servants.[49] Similarly, the two "upstanding men" who had fallen ill while visiting the imperial court and received bezoar as a cure both happened to be active in the Counter-Reformation as well. The first of the sick patients, a seventy-five-year-old Transylvanian nobleman named Lord Scander, had fallen into a "dangerous illness . . . with the reduction of all strength and the pulse, and a weak and unruly stomach, and . . . pain around the ribs." On the emperor's command, Richardus gave him "five grains of this stone bezoar in wine" and carefully recorded the patient's long recovery. The second patient, a prominent Counter-Reformation scholar named Friedrich Staphylus, had been sick "for seventeen whole days of an acute and deadly fever." Once again on Emperor Ferdinand's command, he received seven grains of bezoar. He made a full recovery at the house of a friend, Peter Canisius—another Counter-Reformation theologian.[50] This confluence of figures active in Catholic reform was almost certainly not coincidental.

Richardus's account of poison trials on condemned criminals directly followed these marvelous instances of healing. The terrible symptoms he described in all four cases (the two ill patients and the two poisoned criminals) made the recovery of the four men seem miraculous. Yet Richardus attributed the eventual cure not to God's favor but rather to powers inherent within the bezoar itself. He claimed the stone should be lauded as "good and pleasant," on account of its "wonderful power, virtues, and efficacy as ascertained by daily experience." It could be used "on men and women, young and old, whatever complexion and nature they might have." It even worked "in the middle of the dog days [of summer], when all other medicines do nothing," and it could be taken "without any annoyance, harm, or danger. For it does not alter or weaken the body, but rather helps its nature and strengthens the whole body." He closed his letter with the hope that reading about the "powers and efficacy of this bezoar stone" had given Olahus "peace of mind" and that its myriad "heroic virtues" would help him to a "lucky and happy path."[51] Richardus's letter not so subtly positioned bezoar as a powerful aid for those faithful to Catholic reform, with Emperor Ferdinand the key to discovering its potency.

Several other Catholic princes also tested their exotic treasures. According to Amato Lusitano, Duke Cosimo I de' Medici had overseen a successful test of unicorn horn on two condemned criminals in Florence.[52] As we have seen, King Charles IX of France led another attempt to test bezoar (unsuccessfully, in that case). When tests on bezoar and unicorn horn took place, they had a special import owing to the objects' rarity and pricelessness. While few authors explicitly tied poison trials to Catholic supremacy, the fact that Richardus connected the two hints at the perceived power of an antidote proven in a trial on a human subject.

Yet this kind of test represented only a small subset of poison trials. Many other princely poison trials using condemned criminals took place on substances with less tangible material value, especially new drugs created by local practitioners. There was already a hint of this trend in Guglielmo de Marra's fourteenth-century manuscript, in which he recommended a "secret" remedy against plague and poison created by Master Benedicto of Mantua and tested at the court (chap. 1).[53] Gregorio Caravita's distilled oil (an alchemical remedy) and Andreas Berthold's Silesian terra sigillata (an alchemically prepared clay) both promised local solutions to poison and plague. There was a low risk/high reward calculation for princes in testing these substances: if they worked, the prince had a valuable new antidote, but if they did not, the prince could send the itinerant healer away with little fuss.

Quite a few of the antidotes tested in poison trials, however, came from the princes themselves. It was common for Renaissance princes to have a signature remedy against poison, often a so-called "poison powder" to be mixed in wine or other liquid. Elector August I of Saxony's poison powder (*Giftpulver*) became particularly famous, and he sent it as gifts to other monarchs, although there is no record of his ever testing it. Both Archduke Ferdinand II and Cosimo I de' Medici tested their poison powders, as we have seen, with mixed success. In Italy, at least two different monarchs tested their own oils, both likely derivative of Caravita's oil. Pope Clement VII had his Oleum Clementis tried at least once (chap. 2), and the Medici grand dukes tested their scorpion oil multiple times (chap. 3). These princely antidotes were attached to aristocratic men nearly exclusively. As I have argued elsewhere, noblewomen also compounded many highly regarded cure-all remedies and demonstrated an interest in collecting and recording cures for poison. Only princes, however, had their names connected with prominent poison antidotes.[54]

There was a risk in testing these princely poison cures. Sometimes the test failed and the prisoner died, as in the case of Mattioli's trial of Archduke Ferdinand's poison powder. In most cases, however, failed tests did not find their way into printed documents. The Oleum Clementis appears to have been used in two unsuccessful tests on condemned criminals in Ferrara and Bologna, but neither appeared in print, and neither archival record specifically identified the oil as Pope Clement's. When failures were described in print, they tended to be sanitized. Mattioli explained the failed test of Archduke Ferdinand's poison powder by claiming that the batch was old and had lost its potency.[55] As we saw in chapter 2, Gabriele Falloppio described a test of the Oleum Clementis in which the prisoner eventually died, but he portrayed the man's survival for a week as a great success. In contrast, Ambroise Paré described the failed test of bezoar under King Charles IX as unequivocal proof of the antidote's worthlessness. In that case, the bezoar came from a Spanish trader and was not yet connected to the king himself.[56] Princes and their physicians thus found ways to publicize successful antidote tests while preventing failures from reflecting poorly on the monarch.

Poison trials on condemned criminals were intended to be exceptional. Only princes could command trials using humans as test subjects, and they tested drugs that were hard to come by in some way or other, whether exotic rarities or local specialties. Princely physicians did not test the vast number of common European plants also thought to have powerful properties against poison, such as angelica, gentian, and garlic.[57] Even so, not all poison trials at the princely courts were intended as demonstrations of princely power. In 1588, eight years after Landgrave Wilhelm IV of Hesse-Kassel oversaw his extensive trial of the Silesian terra sigillata on eight dogs (chap. 3), there were questions about its effectiveness. At the landgrave's request, his physicians conducted a second test on two dogs. This trial was not written up in a formal *Observatio*, nor did it involve a drama-filled spectacle. The physician Albert Hyperius merely wrote a letter to Landgrave Wilhelm describing a small, private test. Although the letter now has a large, barbell-shaped hole ripped in the middle, making it difficult to read, Hyperius appears to have deemed this follow-up test a success.[58] Clearly, both the prince and his physicians continued to value even a relatively brief and discreet poison trial as a form of evidence.

They were not alone. Although poison trials on human subjects were not an option for most physicians, tests using animals appeared frequently in medical writings. These trials also tended to focus on exotic substances like bezoar or unicorn horn. Rather than dramatic demonstrations of princely power, however, they generally acted as a sideshow—a quick demonstration, often to prove that a particular specimen was authentic.

The Local, the Foreign, and the Fraudulent

Expensive, exotic, and novel poison antidotes came with a big caveat: fears about fraud. As Wittich put it, "daily experience shows what powerful effects precious stones and other [antidotes] have, as long as they are not falsified."[59] This anxiety was hardly new in the sixteenth century. Galen himself was intrigued by exotic and hard-to-obtain drugs, but he was so concerned about authenticity that he embarked on several long, arduous voyages to examine prized remedies himself. His travels included trips to Cyprus and Palestine for rare ingredients and two separate voyages to the island of Lemnos in order to find out the truth of the rumors he had heard about terra sigillata. The first of these voyages failed, as the ship put in at a town on the wrong side of Lemnos, but he found the correct town, Hephaestias, on his second trip. After interviewing several locals about the terra sigillata and its uses, he came away satisfied—and claimed that he bought twenty thousand of the small medallions to take back with him.[60] Galen also stressed the importance of authenticity in compounding theriac, a drug that was always fraught with potential for either error or fraud.[61] It should come as no surprise that theriac became the focus of the ancient and medieval poison trials described in chapter 1.

Galen's initial trial using poison on wild roosters aimed to show the marvelous efficacy of theriac, but nearly all other poison trials in ancient and medieval medicine focused specifically on the authenticity of a given batch of the drug. Most physicians assumed that theriac would work if one had a good batch but that fraud and incorrect compounding were rampant. Galen himself proclaimed that "there is much trickery practiced about the drug by tricksters," which made it necessary to "test the drug to see if it is adulterated."[62] When theriac trials appeared in medieval Arabic or Latin literature, they tended to mimic Galen's poison test on roosters: one should feed poison to two birds (usually roosters or pheasants) and give the antidote to only one.[63] If that bird survived, the theriac was good. If it died, the particular batch was tainted. That theriac worked was unquestioned.

In these trials, the ability of theriac to counteract a poison or purgative confirmed its status as an antidote and cure-all writ large. If it was effective against one particular poison, it would presumably work against other poisons and myriad ailments. Authenticity and efficacy thus reinforced one another in a tidy circle. Authentic theriac would work against poison and many diseases. A trial would demonstrate whether the theriac was good. If the trial worked to counteract poison, the theriac was authentic. That theriac could then be used with confidence for a variety of prophylactic and curative purposes. In the Middle Ages, theriac trials were inevitably theoretical; they described a test that *could* be carried out, not necessarily one that *had* been carried out. In many cases, they simply acted as a recognition of the concerns about fraud.

Anxiety about adulterated or fraudulent theriac only grew in the Renaissance, and with it came a spate of actual theriac testing, although only rarely via poison trials. The physician and naturalist Ulisse Aldrovandi tested theriac by poisoning pigeons in the 1570s, but for the most part testing was done through what we might now call quality control.[64] In many cities, the sale of theriac became highly regulated, and several of them, including Venice, Lyon, and Nuremberg, required open-air ceremonies in which apothecaries laid out their ingredients for inspection. Compounding was allowed only once a year, in the spring, and was done publicly, with much pomp (see fig. 5.6).[65] This process was particularly important in Venice, seen as the supplier of Europe's finest theriac. Because a number of components of theriac were expensive, exotic, and hard to come by, substitutions or *succedaneum* were often allowed. Apothecaries known for having the best theriac usually boasted of having the fewest substitutions.[66]

Despite these official measures intended to guard against trickery, there were constant concerns about inferior theriac. The tightening control over the compounding of theriac was a symptom of a larger struggle between physicians, who diagnosed patients and prescribed remedies, and apothecaries, who compounded and sold drugs. Physicians frequently accused apothecaries of selling lower-quality or fraudulent drugs and of diagnosing and treating patients on the sly.[67] They also raged against empirics who sold theriac and similar drugs in the marketplace, as we saw in chapter 3.[68] In 1519, German physician Lorenz Fries ridiculed empirics who claimed to know how to make theriac, when the process was so complicated that it challenged even physicians. Without a physician's knowledge of specific conditions, he advised, promises of wonder cures were fallacious.[69] As we have seen throughout this book, theriac was an important linchpin in physicians' attempts to distinguish themselves from empirics.

Theriac was far from the only poison antidote to raise suspicions of fraud. Because of its rarity, terra sigillata raised constant concerns about authenticity, a skepticism that had inspired Galen's arduous voyages to Lemnos. After the Ottomans took over the island in 1479 and embarked on a deliberate project to increase the terra sigillata trade, Western Europeans faced a conundrum: they now had access to a lauded ancient antidote,

5.6 Public compounding of theriac in Nuremberg. Hieronymus Brunschwig, *Liber de arte distillandi de Compositis*, 1512. Wellcome Collection.

but they had to trust the Turks, and supplies remained limited.[70] As the popularity of terra sigillata grew in the Renaissance, a number of newly discovered European earths began to compete with the Lemnian variety. One of the earliest and most successful of these new earths was the Maltese variety widely known as the "grace of St. Paul" or St. Paul's earth, which the *pauliani* snake handlers sold in Italian marketplaces (chap. 2). The Maltese earth was specifically compared to—and sometimes confused with—the Lemnian terra sigillata from at least the 1520s.[71]

That confusion came in part from the Maltese earth's material form. Like the Lemnian terra sigillata, it began to be sold in small medallions stamped with a seal (see fig. 5.7). Purveyors of other, newly discovered earths soon followed suit, leading to an array of small, stamped earthen medallions sold as poison antidotes. The variety heightened concerns about fraud. The Italian physician Antonio Musa Brasavola praised the properties of true Lemnian earth in 1537, but he noted that most specimens were fraudulent.[72] That sentiment was echoed in medical texts published throughout much of Europe, although fraud was usually blamed on the duplicity of the Turks rather than on the growing selection of European earths.[73] When French traveler Pierre Belon visited the Ottoman Empire in the 1550s, he found dozens of different varieties and, following in Galen's footsteps, decided to travel to Lemnos himself in order to sort out "the true from the false" by carefully examining both the material appearance of the medallions and their inscriptions (see fig. 5.8). He claimed that nearly all terra sigillata sold in the city of Constantinople was fraudulent.[74]

Bezoar stone and unicorn horn were similarly the subject of concerns about fraud from their first appearance in medieval texts. Arabic treatises on bezoar frequently contained thorough descriptions of the stone's appearance, feel, and smell, in hopes of guiding the reader toward an authentic specimen. The description by al-Biruni also suggested a number of tests to help determine whether a bezoar was authentic or a fraud.[75] After the Portuguese expansion of trade made bezoars more available in Europe, concerns about fraud skyrocketed. By the seventeenth century, the Dutch East India Company (VOC) refused to trade in most bezoars because of the high risk of fraud.[76] Unicorn horn provoked even greater anxiety, especially since the very existence of the animal was uncertain. Gessner gave a thorough account of the many fraudulent practices in

5.7 Specimens of Maltese or St. Paul's earth. Pharmazie-Historisches Museum, Basel. Photo by the author.

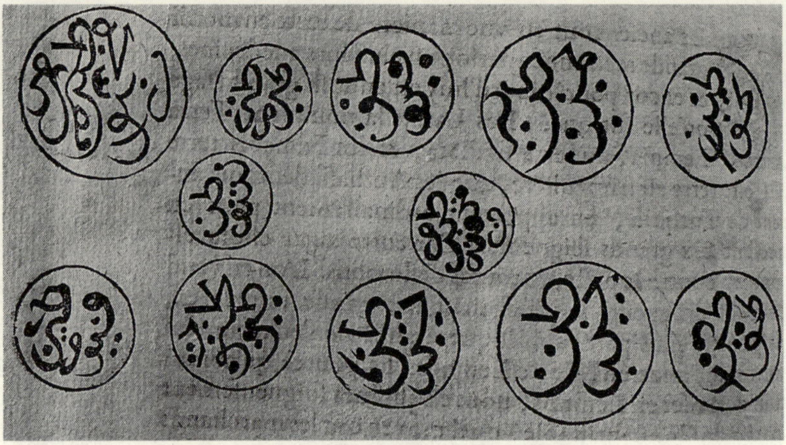

5.8 Pierre Belon, images of supposed Lemnian terra sigillata. Courtesy of Houghton Library, Harvard University.

creating false specimens of unicorn horn before he even mentioned the healing properties.[77]

Antidotes were not the only drugs to face concerns about fraud. As Valentina Pugliano has pointed out, the issue of authenticity loomed large in sixteenth-century discussions of materia medica of all sorts, as distinctions such as "true apium," "legitimate cardamom," or "false balsam" became increasingly common. Her research has suggested that apothecaries, as a group, were particularly concerned about establishing authenticity.[78] The apothecary Calzolari boasted of having found "true balsam, true Jerusalem rose," and "true galingale" as well as "very rare things that are not in Italy, like true cinnamon."[79] Any drug could provoke suspicion of falsification, but the mistrust tended to be far stronger with foreign imports, especially valuable antidotes.

These concerns about fraud helped spur a backlash against both compound drugs (like theriac) and foreign imports (like bezoar or terra sigillata) in the 1530s and 1540s. The hostility was especially pronounced in Germany, where humanist botanists promoted local herbs, but it had prominent French and English constituents as well.[80] There were three main arguments against compounds and foreign drugs: (1) many foreign drugs sold at the apothecaries were fraudulent; (2) simple herbs were superior to compounds; and (3) local drugs were superior to foreign drugs. The physician and botanist Hieronymus Bock, for example, complained that the many allowances for substitutions in apothecaries meant that sick people were now sold "mouse shit as pepper." He praised the powers of local herbs as cure-alls and poison antidotes and claimed that he would far rather have angelica, masterwort, lovage, and gentian than "the Arab's bezoar."[81] Gentian, in particular, was "the most useful herb in Germany." It was so efficacious, according to Bock, that hawkers simply mixed it with a few other herbs and honey and credibly passed it off as theriac.[82] The message was clear: foreign drugs were suspect, expensive ingredients were probably fraudulent, and Germany boasted of many marvelous herbs. Why not use them?

This contrast between "indigenous" and "exotic" drugs has been studied extensively by historian Alix Cooper, who identifies it as a narrative that lasted from the Renaissance well into the eighteenth century. The indigenous-exotic divide derived in part from an animosity toward Arabic sources, especially Avicenna, who, in the eyes of many humanists, had corrupted Greek medicine with foreign ingredients and complex compounds. Bock's scorning of "the Arab's bezoar" did double duty as a humanist critique of Arabic scholarship and a slur against foreign drugs. In the era of both Reformation and global expansion, promoting local resources was an important goal.[83] One German physician who closely followed Bock's example, Tarquinius Schnellenberg, argued that "we have so many good and priceless *Simplicia* in our German nation, which are much closer and more suited to our nature than those from across the ocean or far away."[84] Rather than undermining the interest in wonder drugs, authors like Bock and Schnellenberg displaced the wondrous properties onto local herbs.

The indigenous/exotic debate shifted significantly in the middle of the sixteenth century. Although attacks against foreign drugs remained a trope well into the eighteenth century, they were joined by an increasing number of writings extolling exotic substances like bezoar and unicorn horn. To get around the criticism of the exotic, authors turned to experience—and particularly their own experience of using wonder drugs in practice. Despite the ever-present concerns about fraud, which only increased as time went on, the constant stream of promising new wonder drugs arriving from both the New World and the East was too tempting to ignore. One way to instill confidence in wonder drugs was to test them.

Marketplace Trials

In the middle of the sixteenth century, physicians began to praise both bezoar and unicorn horn as poison antidotes and general remedies based on their own firsthand experience. The earliest widely published accounts came from the Portuguese physician Amato Lusitano (João Rodrigues de Castelo Branco, 1511–68), who came from a family of converted Jews, or *conversos*. Lusitano spent time in Antwerp and Ferrara before settling in the coastal papal state of Ancona, where his family ran a commercial venture. In 1553, Lusitano published a commentary on Dioscorides that drew on his personal knowledge of materia medica, both from his medical practice in Antwerp and Ferrara and from his access to Portuguese trade networks. Based on firsthand knowledge, he corrected some perceived errors in Mattioli's *Commentaries*, a stance that earned him Mattioli's lifelong enmity. In 1558, Mattioli published a public and personal denunciation of Lusitano and used his powerful position to exploit Lusitano's vulnerability as a Jewish convert. Lusitano spent his last years in de facto exile in the Ottoman Empire.[85]

Mattioli's hostility almost certainly curtailed Lusitano's status as a botanical expert after 1558. Nevertheless, his work remained widely read and respected. With his connections to Portuguese trade networks, his details about exotic drugs outstripped Mattioli's.[86] His entry on bezoar stone is a prime example. Although he paid tribute to Avenzoar and other authors in his account of bezoar, he also related direct eyewitness accounts from "our Portuguese, who tell us much from India," and explained that the "fabulous stone" was taken from the stomachs of animals. The best dose, he stated, was three grains taken in oil, wine, or vinegar (depending on the specific illness). Finally, he described a large bezoar he had seen in Venice, owned by Doña Beatrix de Luna (1510–69), a fellow *converso* who had followed a similar path through Antwerp and Ferrara. He noted that Doña Beatrix had purchased her bezoar from the viceroy of Portugal for the staggering sum of one hundred and thirty gold ducats. The value was not overblown: Lusitano claimed that its effectiveness against poison resembled that of unicorn horn.[87] This comparison put bezoar, still very rare in Europe, on a plane with a familiar priceless remedy against poison. In a later chapter, he mentioned that the two unicorn horns kept in the treasury of the Basilica San Marco in Venice, "each the thickness of an arm," were displayed to the public once a year on Ascension Day, along with other precious items.[88]

Yet Lusitano could not recommend unicorn horn unequivocally. He cautioned that many specimens sold as unicorn horn were actually whale bones or artificial creations manufactured out of lime. To avoid being hoodwinked by a fraud, one had to know the proper method of testing a specimen. "Do not listen to those who test unicorn," he warned, if the tests involved infusing the shavings with water to make them sweat.

This kind of test could be faked. Instead, he asserted, one could prove whether a sample was genuine unicorn horn by the following trial: Take two puppies or chickens and give them poison to drink. Give one of them an equal or greater weight of the unicorn horn and give nothing to the other. If the unicorn was genuine, the first animal would live, and the second would die. This poison trial put a version of Galen's rooster test into practice in the marketplace—and unlike accounts in medieval texts, Lusitano's made it clear that he had, in fact, carried out this test as described. "Last year in Venice," he wrote, "I made a trial on a horn, two pounds in weight, for which the merchant sought two thousand ducats. [I tested it] on two pigeons, to whom I gave arsenic. The one that did not drink the antidote died within the hour. The other that drank it lived five hours." Although the second pigeon eventually died, Lusitano saw the test as proof that "the horn was optimal and genuine," since arsenic was such a deadly poison.[89] As already mentioned, Lusitano also referenced Duke Cosimo's successful test of unicorn horn on two condemned criminals.

Marketplace tests were nothing new. Many prized foreign imports had long been seen as susceptible to fraud, especially the exotic spices that formed the basis of European pharmacy—cinnamon, nutmeg, cloves, black pepper. The first-century Greek physician Dioscorides recommended a method "to test cinnamon" by breaking it in half. If it immediately produced a strong scent, it was the proper cinnamon; if not, it was inferior.[90] Most of the time, however, marketplace trials involved carefully examining the substance, which required a certain knowledge of how the substance was supposed to look, smell, or taste. Substances sold primarily as poison antidotes, in contrast, opened the possibility of poison trials. While medieval tests had suggested using poison trials to test the authenticity of theriac, sixteenth-century physicians described *actual* tests on various exotic antidotes.

Marketplace poison trials soon appeared in other works on exotic substances—all from authors who had read Lusitano. In his *Colloquies* on Indian drugs (1563), Portuguese physician Garcia da Orta related a tale he had heard from "persons worthy of credit" regarding the use of unicorn or rhinoceros horn against poison: "Arsenic was given to two dogs, one receiving a double dose. To the one given a double dose, the scraping of the horn was given to him to drink. This one lived; the other that had taken a smaller dose of arsenic died."[91] Spanish physician Nicolás Monardes wrote in 1565 of a Venetian merchant who, in order to validate a piece of unicorn horn he was selling, conducted a trial using two chickens poisoned with black hellebore. One chicken was then given some unicorn horn water to drink. As in Lusitano's test, the trial was only semisuccessful: the chicken given the antidote died after two days, while the chicken given only poison died within fifteen minutes.[92] Both stories echoed Lusitano's and may be derivative. Unlike Lusitano, however, Orta and Monardes emphasized not the issue of authenticity but rather the question of whether the antidote worked.

Later treatises on bezoar stone made fraud a more central point, and they specifically blamed Portuguese, Indian, and Jewish traders for fraudulent goods. Testing represented a way to protect against deceit by foreigners and religious outsiders. German physician Christoph Hybele or Hieblin drew heavily on his own personal experience in the Portuguese colonies of Goa and Cochin in his 1598 treatise on bezoar, and he portrayed testing as a necessary protection against counterfeit. He told of a "rich Jew" in Cochin named Moyses Villela, who had perfected the art of creating false bezoars, a process so secret that he refused to teach it to Hybele, "even though this Jew was a great friend of mine." However, "all those who understood and tested it did not buy it." Hybele recommended three different tests of the bezoar, none of which were poison trials, but he reinforced the importance of testing to guard against fraud.[93] In other cases, poison trials were presented as an easy method to assess the authenticity of a dubious substance. Wittich warned that the Lisbon merchants "do not want to let people test [bezoars] to see if they are authentic or not," but he suggested that one nevertheless conduct a trial by passing a needle and thread through some hellebore and then through the leg of a dog. If the dog survived, "[the bezoar] is authentic. But if the dog dies, it is false."[94] Poison trials represented one potent weapon in the arsenal of tests against fraud. As we have seen throughout this book, they appeared to give a definitive answer.

"The best medicine of all"

Poison trials may have been a handy tool to uncover fraud, but they were used sparingly as evidence of efficacy in medical treatises. For the most part, arguments about the virtues of marvelous medicaments came from clinical experience on sick patients. This pattern was especially true in the case of bezoar stone, which was particularly lauded for its dual role as poison antidote and cure-all. (Unicorn horn, in contrast, tended to be recommended more narrowly for poison and poisonous illnesses.)

An extensive account of bezoar's effects against myriad illnesses can be found in the writings of Garcia da Orta, who, like Lusitano, came from a Jewish *converso* family. His family had fled the Spanish Inquisition for Portugal in 1492, although Orta returned to Spain to study at the universities in Salamanca and Alcalá de Henares. His departure for Goa in 1534 was almost certainly related to the increased reach of the Inquisition in Portugal, but he embarked on the journey in a prime position as the personal physician to the Admiral of the Indian Sea, Martim Afonso da Sousa.[95] Toward the end of his life, at the urging of friends, he penned a thorough treatise on South Asian drugs, *Colloquies on the Simple Drugs and Medicinal Herbs of India* (1563), the first European work on the topic and only the third book ever published in Goa.[96] The book was presented as a series of dialogues between Orta and a highly learned (and likely fictional) visiting Spanish doctor named Ruano; in the last colloquies, a friend of Orta's, Dimas Bosque, joined the discussion. Part botanical, part anthropological, and part medical, the treatise professed to portray simple drugs as they were used in Goa and nearby territories. The discussions in the dialogue continually emphasized the superiority of knowledge drawn from experience (represented by Orta) over text-based knowledge taken from Pliny and Dioscorides (represented by Ruano).[97]

As a professed student of experience, Orta generally was far more interested in correcting European misconceptions about Indian drugs than in introducing new wonder drugs. For the most part, he focused more on the appearance, habitat, and properties of the simples than on their medicinal uses. However, concerns about poison and discussions of effective antidotes permeated the book. In his colloquy on cardamoms, Orta claimed that the "kings, or rather tyrants, in these parts are much in the habit of poisoning their brothers," and he explained they highly valued a theriac made with cardamom as an antidote.[98] He gave a more thorough list of a variety of available antidotes in a section about a disease called morxi (likely cholera), and he particularly praised bezoar. When asked by Ruano whether there were any effective drugs one could use to treat morxi, Orta replied that theriac, *pau de cobra*, unicorn horn, and a powdered root called *contra erva* all could be used effectively, "but the best medicine of all is three grains of bezar stone, which the Persians call pazar. It is of such use that it almost miraculously dilates the powers of the heart. I have had many patients who said to me after taking it, not knowing what it was, that the medicine . . . had given them renewed force and made the soul return to the body."[99] Given the restraint with which Orta discussed most materia medica, this validation was high praise indeed.

Orta later included a chapter specifically on bezoars, which included extensive information from his own experience. He claimed that they were taken from a he-goat called a Pazam, the best of which came from Persia and Khorosan, and that he had seen one of them, a large, reddish beast, in Goa with his own eyes. The bezoar was formed by twisting around a fine straw in the goat's stomach, "forming a rind like that of an onion." He noted that people used bezoars to counteract poison, all poisonous diseases, and all diseases caused by melancholy, and they were extremely expensive in Goa for that reason. Wealthy locals, he said, also used them in the spring and fall as a preservative, "and they say that . . . it preserves their youth." Orta had heard of its successful use on old melancholic diseases such as "prickly heat, leprosy, itch, ringworm," and he himself had tried it on patients for "other diseases and found it to be very good." He also thought it would be useful as a preservative in times of epidemic disease like plague or measles.[100] Once again, this effusive description of bezoar as a near cure-all presents a stark contrast to most of Orta's careful text.

Unusually for Orta, he also relied on hearsay in his description of some antidote cure-alls, and in one case he introduced evidence of drugs he had not tried himself: rhinoceros horn and unicorn horn. In colloquy 42, he noted that he himself had never seen a rhinoceros, but that the people of Nizamoxa would pay over two hundredweight in gold for "a little tested unicorn [*unicorno experimentado*]" and even more for rhinoceros horn. While he and Ruano both agreed that no one had ever seen a unicorn, Orta could attest that the rhinoceros actually existed. In 1512, he stated, the "king of Cambay" had sent a rhinoceros as a gift to the "king of Portugal," who had then sent it to the pope (see fig 5.9).[101] In addition to this historical example, Orta noted that he personally had seen the horns for sale in the market still attached to the skin, a sign that they came from a flesh-and-blood animal and were not manufactured. This evidence was followed by the poison trial on dogs conducted by "persons worthy of credit," likely a veiled reference to Lusitano.[102] But the poison trial represented only a parenthesis in the discourse on rhinoceros and unicorn, and Orta's short section on the horns was dwarfed by his extensive account of bezoar, which contained no hint of a poison trial.

Orta did, however, place weight on dramatic instances of healing poison. When his friend Dimas Bosque joined the conversation in his *Colloquies*, Bosque brought up a different antidote to poison, similar to

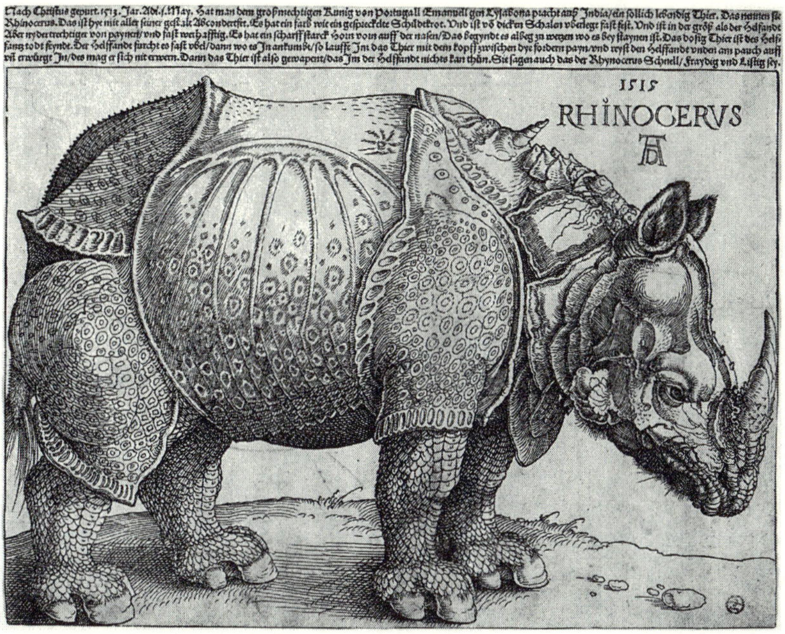

5.9 Dürer's Rhinoceros, 1515. National Gallery of Art, Washington, DC. Wikimedia Commons.

bezoar but found in porcupines in Malacca. Bosque had personally seen the stone's marvelous effects on two men who had been poisoned and were very ill. He gave them some of the stone, and they recovered. Thus, Bosque proclaimed, "I give you my experience as the result of an experiment." Orta responded that he would have many opportunities to repeat that experiment, as poisoning was "very common in this land."[103] While Bosque's anecdote did not describe a contrived trial, it still gave evidence of a sudden recovery from poison. Neither Bosque nor Orta, however, dwelled on the symptoms of poisoning in the same way as the authors of poison trial reports discussed in chapter 3. Orta's interest was in conveying the successful recovery.

Monardes's Poison Anecdotes

That pattern can also be seen in the writings of Nicolás Monardes, who became famous especially for his laudatory descriptions of New World drugs. As Marcy Norton points out, Monardes's embrace of New World substances put him at odds with most humanist physicians and likely derived directly from his mercantile connections to the West Indies.[104] Although he never traveled to the colonies himself, Monardes lived in Seville, an important port city for ships coming from the Spanish holdings overseas. He became involved in colonial trade as early as 1533, when he participated in a business that sent slaves to the New World in exchange for shiploads of medical commodities and exotica, a venture that left him bankrupt in 1567.[105] Before the bankruptcy, however, Monardes published his book on foreign medicines, *Dos Libros* (Two Books), in 1565. It became so popular that he was permitted to write additional works on New World drugs, despite his precarious financial situation, which almost certainly influenced his glowing descriptions of the drugs' wondrous properties.[106] He eventually composed three separate books on materia medica from Spanish America, published together in 1574 and eventually translated into numerous languages.[107] The success of these books—and Monardes's growing fame around Europe—helped keep the Sevillian authorities at bay until he could pay off his debt.

Monardes did not just write about New World drugs, however. The second book of his initial publication, *Dos Libros*, was a short treatise "on the marvelous medicines that work against all poisons." He focused on two poison antidotes: bezoar and a Catalonian herb he called *escuerçonera*.[108] Monardes claimed he had chosen to highlight bezoar because it could "remedy and cure so many diverse infirmities" in addition to being useful against poison.[109] He drew his arguments in favor of the bezoar's virtues from a variety of Arabic and Western European authors, especially emphasizing ties to Spain in both cases. While it is unclear how well he knew Orta's work, he repeated Lusitano's account of bezoar nearly word for word.[110] Monardes mentioned the problem of fraud, but like Orta, he did not fixate on it. He claimed it was easy to determine a bezoar's authenticity by conducting a simple marketplace test. One merely had to break it open to see if there was a powder inside: if not, it was false. Indeed, he personally had seen one bezoar stone hawked by an Indian merchant that was filled with grain or seed when broken open, "whereby the Indian deceived many."[111] Rather than dwell on instances of fraud, however, he moved quickly on to describe bezoar's virtues.

Monardes cited numerous scholarly authorities to emphasize the long-standing belief in bezoar as a marvelous antidote and cure-all, but as in his works on New World drugs, he used firsthand experience to underpin his arguments. He dedicated the poison treatise to the duchess of Bejár, who Monardes credited with introducing him to the bezoar when she requested a specimen in a desperate attempt to cure her son of epilepsy. According to Monardes, the boy's dramatic recovery convinced him of bezoar's virtues, and he included an extensive account of this success among the many examples of bezoar's healing powers "that I have seen by experience," both in illness and in cases of poisoning. In one instance, Monardes described using bezoar to cure a gentlewoman named Doña Maria Catano of a long-standing illness. Shortly thereafter, he was called to an emergency: a scholar named Luis de Cuerva had accidentally eaten something poisonous. Monardes ran back to Doña Maria's house, found the bezoar still in her hands, and was able to run back in time to save the learned man. The same de Cuerva, who was apparently accident-prone, had a similar experience after drinking from a pool of water full of venomous worms while on a hunting trip with his patron. Luckily, his patron had a bezoar on hand and was able to rescue him. All in all, Monardes reported ten cases of curing patients from illness or accidental poisoning using bezoar.[112] Like Orta, he did not include a poison trial.

The dramatic descriptions of bezoar cures in *Dos Libros* had an immediate effect that would have an impact on both global trade and local ecologies. Monardes later claimed that a gentleman of Peru, having read his account in *Dos Libros*, had sought and found bezoar stones in Peruvian animals. In the man's estimation, these newfound bezoars were similar—if not superior—to those that came from India via Portugal. Monardes called the discovery of bezoars in Spanish-controlled lands a "most marvelous thing, and most precious"—for now they would be more readily available and "more certain and authentic." His former conviction that bezoar fraud was easily detectable quickly altered once the Spanish had a way to enter the bezoar trade. East Indian bezoar, he now claimed, was so fraudulent that "for every ten true ones you have one hundred false ones."[113] Monardes elaborated on Peruvian bezoars later, in his second book on New World drugs, in which he added evidence that they were comparable to Persian bezoars owing to the similar conditions in which the Andean animals lived. He claimed the Peruvian stones were good not only against all poison but also against fevers, pestilence and plague, melancholy, leprosy, and worms, among other complaints.[114] While he included several anecdotes in which Peruvian bezoar had been tried in various illnesses, he once again cited no poison trial.

Monardes's reports on bezoar are prime examples of the ways in which medical literature, experience, and trade reinforced each other. Lusitano's experiential description of bezoar prompted Monardes's glowing report on it—which, in turn, spurred Spanish readers to find a South American substitute. Marcia Stephenson has studied the impact of the bezoar craze in Peru and found that the desire for precious items (and profit) had profound costs: the utter decimation of flocks of wild vicuñas and guanocos, which in turn had an impact on the indigenous peoples who depended on them.[115] Monardes's boasts of readily accessible Peruvian bezoars are chilling in this light. This carnage also underscores the high financial stakes of the bezoar trade, in which success led to fortune. Physicians' anecdotal stories of treating sick patients with bezoar, which attributed near-miraculous qualities to the drug, could have real-world repercussions.[116] In the grander scale of things, poison trials were rather tangential to this intersection of medicine and commerce.

Wonder Drug Tobacco

Yet Monardes was by no means averse to using poison trials as evidence. As we have already seen, he included a marketplace test on unicorn horn in his poison antidote treatise. Even more significantly, a poison trial appeared in the famous entry on tobacco in his second treatise on New World drugs, first published in 1571. Monardes boldly trumpeted tobacco as a wonder drug and cure-all, good for nearly two dozen conditions, including "poisons and poisonous wounds."[117] As in his description of bezoar, he used anecdotes drawn from experience to illustrate his point. A number of historians have pointed out that his dramatic claims about tobacco rested largely on knowledge appropriated from Amerindians, and his description of its uses against poisonous wounds provides a case in point.[118] Monardes described how certain "Caribbean Indians who eat human flesh" frequently used poisonous arrows to kill "Indians or Spaniards, as they find them." In one incident, the Caribs came in canoes to attack San Juan, Puerto Rico, and they killed and wounded a number of "Indians and Spaniards." Lacking the herb usually used to combat poison, Spanish and Indian healers decided to try the juice and leaves of tobacco. It revived their ailing men and healed their wounds wonderfully. Thereafter they did not need to fear the Caribs: they simply brought tobacco leaves wherever they went.[119]

Monardes's tales of tobacco's success against poison did not only take place in Mesoamerica, however. In Spain, he noted, tobacco had also been found to be very useful in combatting the herb called black hellebore (*ballestero*), which he described as "a most potent poison that kills without remedy." The king wished to probe tobacco's powers against the herb, and he commanded his underlings to "make an experience" of it. Led by a physician named Dr. Bernardo, they wounded the throat of a little dog, put the herb into the wound, and then covered the wound with tobacco juice and leaves.[120] The dog lived, to the amazement of all who watched. According to Monardes, Dr. Bernardo had reported, "I made this experience on the command of His Majesty. I wounded the dog with a knife, and then I put the herb hellebore into the wound, and the [tobacco] was selected, and the dog took the herb, and the dog returned to health after [taking it]."[121] The trial appeared significant enough that Monardes essentially repeated it twice, once in Bernardo's own words.

This test very closely resembled the marketplace poison trials mentioned by a number of authors, going all the way back to Bernard de Gordon (chap. 1). An animal was wounded and poison was put in the wound, followed by an antidote. Yet King Philip II's poison trial was different. Most poison trials made a broader statement about the antidote they tested. In marketplace trials, success usually indicated that the drug was valid, which meant it would also work against myriad poisonous diseases. But Monardes made no claims about tobacco's general powers against poison. Instead, he focused on the specific external application on poisonous wounds. Unlike bezoar or unicorn horn or terra sigillata, tobacco was not, in principle, a poison antidote that also had other virtues. Instead, it was a wonder drug that also happened to have strong powers against poisons delivered externally. King Philip II's poison trial thus tested only one thing: whether tobacco would work against wounds poisoned by black hellebore.

Beyond this important difference, tobacco also *worked* differently than antidote cure-alls like bezoar. It did not cure through its "hidden, embedded nature and virtues," the long-standing theory of occult forces or specific form. Instead, it worked through its humoral properties. Monardes described tobacco as a hot and dry drug, a classification almost certainly drawn from Amerindian categories, as Norton has argued.[122] Applied externally, it healed bodily complaints that came from a cold cause, like certain headaches, aches and pains, stomach conditions caused by cold wind, and problems with the uterus. A syrup of tobacco put on the navel could expel worms. The leaves also helped against cold swellings, cold sores, toothache, and wounds, and the leaves and juice together drew out poison. Nearly all of the applications were external, save for his description of Amerindians and African slaves using tobacco smoke to help with weariness and other bodily and emotional conditions. Monardes did not claim, however, that tobacco cured plague or poisonous fevers, standard for most antidotes. It was a wonder drug for numerous quotidian, mainly nonpoisonous complaints. Tobacco represented a novel cure-all that did not evolve from the poison antidote tradition, and it was emblematic of a greater interest in panaceas in the later part of the sixteenth century, as we will see in the next chapter. Even so, it counted the power to cure poison among its many virtues.

Physicians and Wonder Drugs

In theory, learned physicians rejected the notion of drugs that could cure multiple ailments, especially when they were sold by empirical practitioners. In practice, they endorsed them. Just as physicians were at the center of testing poison antidotes at Renaissance courts, so too were they instrumental in spreading word of marvelous cures that, to modern ears, sound far-fetched at best. The Iberian authors who published books on drugs from the East and West Indies were all learned physicians. Dutch physician Carolus Clusius translated and reworked these Portuguese and Spanish texts, which brought news of marvelous foreign substances to a Latin-reading public. Published by the Antwerp-based printer Christoph Plantain, Clusius's books on exotic foreign medicaments became best-sellers. His original 1567 edition of Orta sold out in a year, a staggering success for a Latin book.[123] Four other editions followed, as well as new editions of Monardes and Acosta and several editions combining the authors.[124] As a number of historians have pointed out, these works were not really translations. Clusius reorganized them, left parts out, and added his own comments. He depicted materia medica from both the East and West Indies in increasingly exoticized terms, blurring the distinction between East and West into a general concept of the Foreign. Yet he also muddied the distinction between these foreign drugs and new European plants.[125] The consistent thread through all of his works was a focus on wonder.

Clusius produced these works in the context of the aristocratic interest in marvelous objects and natural wonders described at the beginning of this chapter. Florike Egmond has noted that Clusius's patron Charles de Saint Omer oversaw an extensive Kunstkammer, one of the earliest and largest collections north of the Alps.[126] A later patron was Wilhelm IV of Hesse- Kassel, the German prince who tested Berthold's terra sigillata on dogs, who oversaw his own Kunstkammer.[127] With his strong interest in natural history, Clusius focused more on the plants and other substances he described than on their medicinal uses.[128] At the same time, he also emphasized eyewitness accounts and verification. He included the poison trials mentioned by Orta in his translation and added a note about a marketplace test to authenticate bezoar, but they took a backseat to observation and experience in instances of illness.[129]

Yet poison trials remained central to the validation of unfamiliar and exotic poison antidotes. This point can be seen particularly clearly in the herbal (*Kräuterbuch*) of German physician Adam Lonicer, first published in 1557. Like many sixteenth-century herbals, Lonicer's book contained information about more than just plants. He included animals and minerals, as well as a section on distillation, and he continually updated the book throughout his lifetime. In his final edition, published in 1587 (the year after his death), Lonicer added an entry on bezoar, "a stone against all poison," which he highlighted as "more powerful than any other poison antidote or theriac." He listed the various learned men who had written about bezoar's potency, both Arabic authorities such as Rhazes and Avenzoar and the sixteenth-century authors Julius Scalinger, Johannes Lange, and Mattioli. In addition to being the most effective remedy against poison and plague, bezoar could also kill worms in the bodies of children; heal stomach cramps that were otherwise impossible to heal; drive out "evil and long-standing fevers"; help melancholy and all melancholic conditions, "whatever the cause"; protect people against many complaints if taken prophylactically; and even preserve youth and ward off old age. Owing to these many powers, Lonicer claimed, the Turkish emperor took it many times a year. But, he cautioned, "great fraud and falsification is perpetrated with this stone," as other stones and artificial materials were frequently sold as bezoar.[130]

Lonicer reassured the reader that one could separate true from false through three tests. First, one could stick the bezoar with a glowing-hot needle or iron spike. If the substance let off smoke, it was authentic; if not, it was false. For the second test, "one can give poison to an animal like a hen, pigeon, goose, or human, and then a spoonful of this stone, powdered and dissolved into a pleasant water." The last test was far simpler: one merely had to spit into a cloth, rub the stone with the cloth, and if color came away, the stone was authentic. Lonicer listed these three tests matter-of-factly and gave no qualitative differentiation between spitting into a cloth and giving an animal (or a person) poison, but he made it clear that testing was integral to the use of bezoar. In comparison, his entry on unicorn horn, included since his 1557 edition, noted a similar problem with fraud but gave no information about testing. The entry on unicorn also focused more narrowly on poison and plague, rather than the expansive powers credited to bezoar.[131] In the thirty years between his book's editions, the purview of exotic antidotes had expanded, and the concept of testing to ensure authenticity had become ingrained in the medical literature.

"Proven adequately in a trial"

Although poison trials had become common in medical writings by the 1580s, they were not portrayed as either the only evidence or the definitive word on an antidote. I would like to return full-circle to Wittich's *Report on the Wondrous Bezoar Stone*, with which I began this chapter. As we have seen, Wittich presented his book chiefly as a collection of poison antidotes, and he explicitly cited a need for "powerful and artful substances that have been proven adequately in a trial [*proba*]." A *proba* frequently denoted an experiment, a deliberate contrived event. In another work, Wittich used the term to describe a test to see whether a woman was pregnant, or whether her milk was good.[132] Yet as we saw in chapter 3, sixteenth-century terminology around experimentation was notoriously imprecise, and Wittich most certainly did not mean that antidotes needed to be subjected to a poison trial, or indeed any sort of contrived experiment. Instead, he weaved together a range of forms of validation from the sources on which he drew.

For the most part, Wittich used the testimony of sixteenth-century physicians as evidence that a drug worked, whether or not he described a specific test. On bezoar, he claimed that "many learned men" praised the stone's efficacy against poison and plague, and he cited Mattioli, Falloppio, and Guinter of Andernach. He included copious evidence on the bezoar's appearance from both Orta and Monardes (via Clusius), often beginning with the phrase "Gartia says" or "Monardus says." He also gave information from local sources, especially the noblewomen whom he served. In his entry on gritstone, used for kidney pain, Wittich remarked that the duchess of Bavaria, after suffering from kidney pain on three occasions, had set one in an armband and was fine from that point forward. His patron, Duchess Katharina of Nassau, had used it on her husband, who was plagued with gout, and it had helped. Duchess Katharina had told him that one could acquire the stone "from the Portuguese in Andorf, if one pays enough."[133] This statement gives a valuable sense of the pathways along which information and goods flowed. Gossip networks helped prop up trade networks, touting various kinds of healing successes. Wittich made it clear, however, that the testimony of physicians represented the gold standard.

Wittich's evidence also included some poison trials. As already mentioned, he suggested a marketplace test to authenticate bezoar, drawn (and embellished) from Clusius's commentary on Orta, and he gave a prominent position to Claudius Richardus's letter on bezoar. Similarly, he included Garcia da Orta's anecdote about Dimas Bosque's validation of Malacca porcupine stone.[134] In one glaring instance, however, Wittich left out poison trials he easily could have included. At the end of the book, he added a description of the "German terra sigillata" from Andreas Berthold's 1583 book, translated more or less faithfully. He included Berthold's description of the thirteen different ailments that the terra sigillata cured, numbered just as Berthold numbered them; he repeated Berthold's arguments about the superiority of German terra sigillata over the fraudulent Turkish variety; and he included Berthold's methods of administering the drug. He even called the drug "tested and proven," in a number of places. Wittich did not, however, relate Berthold's poison trials or, indeed, make any mention of Berthold at all. He merely stated that the account came from a "Latin book published in Frankfurt am Main in 1583," and he prominently credited the Silesian physician Johannes Montanus with discovering the terra sigillata. In so doing, he left the strong impression that the account came from Montanus, a respected learned physician trained in Italy (albeit an alchemically inclined one).[135] This elision of Berthold appears deliberate. As impressive as Berthold's poison trials were, Wittich gave them a pass, likely to avoid highlighting the success of an empiric.

Physicians' claims of expertise regarding marvelous remedies reached well into the seventeenth century, as did the debates about proof, testimony, and fraud that went hand in hand with their efforts. In 1625, the Swiss physician Caspar Bauhin (1560–1624) published an erudite study, in Latin, on bezoar stone. His work represented a masterful assimilation of previous scholarship on bezoar, drawing from over 150 authors ranging from Avenzoar to Lusitanus to Monardes to Orta to Wittich to seventeenth-century physicians Daniel Sennert and Johannes Hartmann. His book took the reader through the varying ideas about how bezoar was formed, what it looked like, its relationship to other animal wonder drugs like unicorn horn and hart's heart bone, and its powers and efficacy. Near the end of the book, Bauhin included a chapter with numerous examples (*historia*) of bezoar's success in both cases of illness and cases of poisoning, most of which were drawn from Mattioli, Monardes, and Richardus. He separated out cases of illness and poisoning, but he did not distinguish contrived poison trials from cases of accidental poisoning.[136] As with Wittich and Clusius, examples drawn from experience appeared as valid evidence, however they had occurred. Bauhin followed these successful cases with instances in which bezoar appeared unsuccessful, including Paré's failed trial, but these examples were less numerous and generally less dramatic than the tales of success.[137]

In an appendix to his book, Bauhin noted that he had not had time to discuss a different category of drug: artificial bezoar. He included a brief description of deliberately crafted "bezoardics," or drugs intended to have the same properties as bezoar but made artificially using animal or mineral material. The expense of bezoar and fears about fraud had driven physicians to create legitimate fakes, bezoar-like substances that were open about their counterfeit. Bauhin did not mention (and possibly did not know about) the elaborate Goa stones created by Jesuits in Portuguese India, which came to rival true bezoar in their beautiful mounts.[138] He did, however, include several recipes for artificial bezoar, including from German physician Daniel Sennert and Italian physician Francisco Pona.[139] Bauhin gave bezoar enough credence as a wonder cure that even artificial bezoar was worthy of note.

Bauhin's text demonstrates the legacy of sixteenth-century physicians' efforts. By claiming "powerful and artful substances" as their territory, physicians like Clusius and Wittich designed a category of wonder drugs legitimized by European learned physicians. This process followed the same pattern as Mattioli's deliberate attempts to recast poison trials as a learned endeavor in the 1540s–1560s. While poison still played a central role in the discussion of wonder drugs in the later sixteenth century, doctors increasingly emphasized the broad healing powers of substances like bezoar or tobacco. Poison trials remained a powerful validation when they occurred, but they were just one tool among many. Because antidote tests on humans were so rare, dramatic tales of successful healing offered a more accessible form of proof.

Yet by attaching themselves to wondrous remedies, physicians in some sense made it harder to distinguish themselves from empirics, who had long claimed near-miraculous effects from their cures. This project became even more difficult with the increasing interest of alchemists like Montanus and Berthold in poison antidotes and wonder drugs. Bauhin never once mentioned alchemy in his description of artificial bezoar, but all of the recipes he included required alchemical techniques. As we will see in the next chapter, physicians soon found themselves faced with a major conundrum from alchemical empirics: a true panacea.

SIX

A Universal Cure: The Panacea Amwaldina

In the wake of Andreas Berthold's triumphant poison trials with the Silesian terra sigillata, a Bavarian alchemist named Georg am Wald invented his own version of the drug. In 1581, am Wald published a pamphlet advertising a plague remedy and poison antidote called the Terra Sigillata Amwaldina. Like other varieties of terra sigillata, am Wald's was formed into sealed medallions, stamped with the letters of his name (fig. 6.1). Like Berthold, am Wald described it as a German terra sigillata, a plague remedy, a poison antidote, and a cure for many different illnesses. Unlike all other examples of terra sigillata, however, the Terra Sigillata Amwaldina was not dug from the earth. Am Wald had made it himself, through a secret alchemical process that he alone had discovered. He deliberately placed himself outside of both the Galenic and the Paracelsian medical traditions, portraying himself as following only "truth." As the title page prominently touted, his terra sigillata was a true "universal medicine."[1]

How could the reader trust these dramatic claims? Am Wald only briefly mentioned a poison trial. If anyone doubted its usefulness, they could just try it on a "dumb animal" using poison, but this possibility was merely an aside.[2] Instead, he preferred a different kind of approbation. There were some "so-called doctors," he complained, who did not recognize the virtues and efficacy of his terra sigillata. He had silenced these "naysayers" by stuffing their mouths full with "many *Exempla* . . . from the many thousands

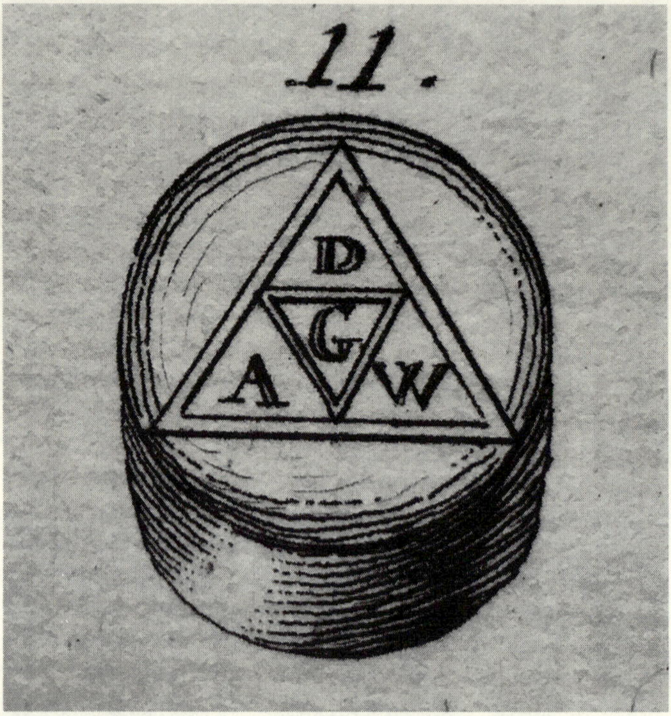

6.1 Terra Sigillata Amwaldina. In Christian Gottlieb Ludwig, *Terrae Musei Regii Dresdensis* (Leipzig, 1749). Courtesy of Houghton Library, Harvard University.

who gave me their testimony." Bohemians, Bavarians, Swabians, Austrians, Palatines, Allgäuers, Bodenseers, and residents of many other regions had all sent word of the drug's success.[3] Testimonials from happy patients, not poison trials on "dumb animals," were the best means of ascertaining success.

While Berthold's terra sigillata became a prized commodity after the success of his poison trials, am Wald's had a more limited impact. His pamphlet garnered some initial local notoriety, quickly going through a second 1581 edition and two in 1582, but he attracted no widespread attention.[4] In the 1590s, however, the fortunes of the two men began to flip. Berthold was summarily ousted from the Silesian terra sigillata business after a falling-out with the drug's discoverer, physician Johannes Montanus, and he struggled to fend for himself without access to his signature remedy. Because his terra sigillata was tied to Silesia rather than to Berthold himself, his poison trials could not protect him from commercial disputes. He slid into increasing hardship and obscurity until he died in poverty in 1610.

Meanwhile, am Wald's star began to rise. In 1591, he published a pamphlet advertising a new medicament called the Panacea Amwaldina. The panacea resembled the Terra Sigillata Amwaldina, he claimed, but it was *chiefly* a universal medicine and could cure 116 different diseases and ailments. Poison, plague and poisonous diseases, and the bites of poisonous animals were the first specific conditions am Wald mentioned and among the ones he discussed most extensively, but to these he added a dizzying array of illnesses (including dizziness) that the panacea would cure. The list encompassed diarrhea, dysentery, constipation, lunacy and frenzy, toothaches, diseases of the womb, any sort of internal bodily pain, the French disease (syphilis), sciatica, dropsy, sore throat, cough, and many more. It would promote breast milk, drive out dead fetuses, restore memory, sate insatiable hunger, stop bleeding, and prevent people from becoming werewolves.[5] It was specifically an alchemical drug, in the Paracelsian tradition. This time, it did make a splash. A second edition quickly appeared in 1592, and there was enough demand for the drug (and the books) that am Wald was able to purchase a knight's manor house outside the city of Feuchtwangen in 1593. There he remained until his death in 1616.[6]

The Panacea Amwaldina merged poison antidotes with another storied tradition of marvelous remedies, alchemical elixirs. In an understated way, alchemical medicaments had composed one important subcategory of poison antidotes from the early sixteenth century. We have encountered some of them throughout this book, especially the Silesian terra sigillata and the various varieties of scorpion oil attached to Gregorio Caravita, Pope Clement VII, Pietro Andrea Mattioli, and the Medici family. Am Wald, however, made wild claims about his "universal" cure that went far beyond the purview of most antidotes, and he tied the drug to alchemy in a central way. His panacea adroitly highlighted the shift, in the later sixteenth century, from poison antidotes with cure-all properties to drugs advertised *primarily* as cure-alls. This trend was nascent in the examples of bezoar and tobacco discussed in the previous chapter, and, not coincidentally, tobacco was described explicitly as a "panacea" in the 1580s.[7] In am Wald's hands, however, the panacea took on an entirely new character. Just as am Wald was the first to use the term "terra sigillata" to describe an alchemically created sealed medallion, so too was he the first, as far as I can tell, to use "panacea" to signify an alchemical cure. This choice would prove influential and lasting, and it was novel enough in the 1590s that it drew the vociferous opposition of learned physicians, particularly the "father of chemistry," Andreas Libavius (c. 1550–1616).[8] His objections served only to heighten the impact of am Wald's panacea.

The panacea's success rested on a method of proof that both resembled and differed from Berthold's poison trials. In his expanded book editions of 1592 and 1594, am Wald included an appendix, longer than the treatise itself, with dozens of testimonial letters from grateful patients. He argued that this evidence was so overwhelming, it was unnecessary to provide "further trials." Nevertheless, he stated, if someone did not believe him, they could go ahead and try it on "sick dogs, horses, hens, or other dumb animals."[9] Poison trials remained an option, but they were but a lowly method of proof. Instead, patients provided the most important testimony of the drug's success. Berthold also turned to alchemy to bolster his claims for the Silesian terra sigillata, but he continued to use his poison trials as irrefutable evidence of his own method of preparing and sealing the drug. Both methods of evidence involved testimonial letters, but the forms of proof within the letters (contrived trials vs. patient testimonies) were very different.

This chapter uses the parallel stories of Berthold and am Wald to highlight the complicated role alchemy played in the fascination with wonder cures described in chapter 5. As in the case of exotic poison antidotes, alchemical wonder drugs drew the attention of physicians. A number of learned doctors approached the Panacea Amwaldina with the same interest that they brought to wonder drugs like bezoar or tobacco. For both Berthold and am Wald, the testimonial letter represented an important form of proof, and, significantly, learned physicians did not always reject this kind of evidence out of hand.[10] At the same time, they remained particularly skeptical of alchemical empirics with grand claims and dubious credentials. No one type of evidence, I argue, was viewed as inherently better than another. Instead, the question of proof was closely bound up with the particular circumstances of the practitioner and his drug. Physicians' approval, moreover, did not always correlate with commercial success. As we shall see, Berthold's terra sigillata benefited from the greater support of learned doctors, but am Wald's panacea proved far more lucrative.

Pseudo-physicians

Although both am Wald and Berthold practiced medicine without a license, neither fit the typical portrait that physicians tried to paint of empirical practitioners. Both men came from rather isolated parts of the Holy Roman Empire, Berthold from a small town in Saxony called Oschatz and am Wald from the southeastern Bavarian city of Passau. It appears that both came from humble backgrounds. Am Wald was the son of a poor Protestant bookseller named Jacob Baldinus or Jacob Amwald, and Berthold's itinerant lifestyle does not suggest wealth or family connections.[11] Berthold first appears in the historical record nearly two hundred miles from his birthplace, in the remote northern city Stettin (now Polish Szczecin), and he later somehow ended up in the mountains of Silesia.[12] Both men, however, managed to receive an education. Am Wald studied law at the University of Basel and received his degree as a legal licentiate in 1573. Although he was later mocked for his insufficient Latin, his legal education required at least a competent level of Latinate knowledge.[13] Berthold's educational background is unclear, but he initially described himself as a calligrapher and a schoolteacher, and he published most of his books in Latin. Later, he became a mine foreman, a respected position. Intriguingly, both men were also devout Lutherans and wrote religious books early in their careers.[14] They may have lacked a traditional Galenic medical education, but neither man was an ignorant marketplace mountebank.

Both Berthold and am Wald, then, fell into a gray area of medical empirics. They were both reasonably learned and upwardly mobile, and indeed, they helped themselves along on that front. Berthold signed his letters "Andreas Berthold von Oschatz." The "von" in German could simply mean "from," but it could also be a marker of nobility. Am Wald ennobled himself even more blatantly, expanding the surname Amwald to the more aristocratic-sounding "am Wald," and, eventually, the rather pompous "am und vom Wald." This simple subterfuge appears to have been rather effective. The Hessian testimonial letter attesting to the success of the Silesian terra noted that the drug had been brought to the court by "the honest man Andreas Bertholdus von Oschatz (*Andream Bertholdem ab Oschatz*)."[15] Although several physicians taunted am Wald for his attempt to appear noble, many patients, princes, and physicians accepted the name at face value. He married into a wealthy family in Memmingen, and when his first wife died, married the daughter of a Donauwörth patrician.[16] To the casual observer, both men appeared to be utterly legitimate.

Despite the parallels in their biographies, there were some significant differences. Most crucially, Berthold never claimed to be a physician. In all of his letters and books on the Silesian terra sigillata, he identified himself as a mine foreman in the town of Kupferberg in Silesia. Nevertheless, he also did not contradict the impression that he was a doctor when the opportunity presented itself.[17] Am Wald was far bolder in claiming medical expertise. Despite his law degree, he never practiced as a lawyer, but instead set up practice as a physician in the Bavarian cities of Memmingen and Donauwörth. In 1578, he received a dubious doctorate in medicine from the University of Padua. As a Protestant, he had to receive the degree from a nobleman rather than from the university, and it was well-known that many of these "Count Palatine" degrees were bought rather than earned. Given that am Wald enrolled in 1577, returned to Memmingen to get married, and briefly returned to Padua to receive his degree in 1578, it seems likely that he fell into that category. Nevertheless, he was once again practicing as a physician in Donauwörth by 1580. In 1583, he tried his luck as a doctor in Augsburg, with disastrous consequences: the authorities refused to accept his Padua doctorate as valid, and he failed the required examination. He was asked to leave the city and returned to practice in Donauwörth, where he lived until driven out for inciting religious unrest in 1591. After a brief period in the small town of Schwabach near Nuremberg, he was able to purchase the manor of Thürnhofen (or Durnhoff), an estate to match his noble name.[18] There he ran his pharmacological practice until his death in 1615 or 1616.[19]

Did Berthold and am Wald know about each other, or perhaps even know each other? I have found no direct evidence of any interaction, but they certainly knew some of the same people. Johannes Posthius (1537–97), a poet and the personal physician to the Prince-Bishop of Würzburg, wrote laudatory poems for both of them, as did the theologian Johannes Lauterbach. I suspect that am Wald, at least, followed Berthold's career with interest. His marketing strategies very closely followed Berthold's, in a manner that seems more than coincidental given their geographical proximity and shared social contacts. Whatever the case, the two men, at around the same time, both hit on alchemical wonder drugs that were primarily poison antidotes but also allegedly cured many ailments beyond poison and plague.

Alchemical Wonder Drugs

Berthold and am Wald were far from the first people to offer alchemical wonder drugs. One of the main aims of medieval European alchemy was medical, linked to the notion of the philosopher's stone. While the stone, sometimes called the elixir, was most often cited as an agent for transmuting metals, there had long been an assumption that it also had cure-all properties, and interest in its healing powers strengthened in the thirteenth and fourteenth centuries.[20] Roger Bacon, a thirteenth-century English Franciscan friar, emphasized the elixir's abilities to restore youth and extend human life, while a text associated with Arnald of Villanova similarly argued that the philosopher's stone could preserve health, restore youth, and cure illness.[21] An influential fourteenth-century treatise falsely attributed to Ramón Lull (c. 1232–c. 1315), a Catalan Franciscan friar, described a "stone" or "universal medicine" that could transmute metals, cure all disease, and stimulate growth in plants.[22] Another Franciscan friar, John of Rupescissa (c. 1310–c. 1370), wrote extensively on a distilled medicinal liquor called *quinta essentia*, or fifth essence, which had wondrous medicinal properties.[23] In various ways, then, medieval authors strengthened the association of alchemy with a marvelous healing elixir.

By the late fifteenth century, texts attributed to Bacon, Arnald, Lull, and John circulated throughout Europe. As these works spread in Latin and various vernaculars, elixirs and quintessences became newly attractive to healers who gave little thought to transmuting metals.[24] A surgeon and pharmacist from Strasbourg named Hieronymus Brunschwig, for example, printed two influential and richly illustrated books on distillation and distilled medicines in 1500 and 1512. The latter publication drew heavily on Rupescissa and included extensive chapters on quinta essentia and other, related alchemical drugs. Brunschwig explicitly noted that his interest in alchemy was solely for medical purposes.[25]

Intellectually, these distilled wonder cures came from a very different tradition than the poison antidote cure-alls that have been the focus of this book thus far. Nevertheless, they had a number of things in common. Like poison antidotes, alchemical drugs operated through occult (hidden) powers and worked on people of all humoral complexions. They also flourished in a similar context: the princely courts. Pope Clement IV was a patron of English alchemist Roger Bacon, and Bacon's works on prolonging life were taken seriously by later popes and their physicians. Distilled medicines appeared at the papal court by the fourteenth century, one of the earliest places they can be documented in Europe. Arnald of Villanova reported that the cardinals at the court of Boniface VIII regularly drank potable gold, an alchemical remedy.[26] For the most part, however, the alchemical elixir and poison antidotes like theriac, bezoar, and unicorn horn remained separate traditions.

From the early decades of the sixteenth century, the two traditions slowly converged, as medical alchemy began to be used to create remedies for poison and plague. In the 1520s, a Florentine goldsmith named

Lorenzo da Bisticci rose to fame for his distilled elixir that he called the "Christ of medicines" (*Cristus medicinarum*), which could cure many ailments, including plague.[27] As we saw in chapter 2, the Italian surgeon Gregorio Caravita compounded his own oil against plague, tested in the first Renaissance poison trial in August 1524.[28] In the German city of Leipzig, a physician named Caspar Kegler (c. 1461–1537) similarly began to sell his own electuary and aqua vitae against plague and poison, both created using alchemical techniques. Unlike Caravita, Kegler kept the secret of making his most prized remedies to himself, and he and his heirs cleverly used print to advertise them (an approach that both Berthold and am Wald later exploited as well).[29] Like Brunschwig, none of these practitioners drew explicitly on philosophical alchemy, but they created poison and plague remedies using alchemical techniques. These efforts merged the alchemical tradition with the interest in poison antidotes, plague remedies, and wonder drugs.

The spread of Paracelsus's writings in the mid-sixteenth century provided a new theoretical dimension to ideas about alchemy and poison. An iconoclast who derided Galenic medicine, Paracelsus made poison a central element of his medical theory. Everything, he argued, had a poisonous part and a healing part, and medicine's aim should be to separate out the healing part through alchemical means.[30] This concept was very different from the universalizing theory of specific form, which the Galenists used to explain the action of poison (see chap. 1). In Paracelsus's view, every disease was particular and thus required a particular process to separate out the poisonous and healing elements. In his influential *Archidoxa* (published 1569), he touted the quintessence and the elixir as cures for all diseases. Like Brunschwig, however, he did not imagine one substance that would cure every ailment, but instead a variety of quintessences and elixirs. The quintessence of antimony would cure leprosy, for example, while the quintessence of coral was good for spasms. In this manner, he posited a cure for every disease but not a cure-all.[31]

As Paracelsus's works became widely known across Europe in the 1560s, the Rupescissan and pseudo-Lullian traditions also remained influential.[32] While some aspects of these traditions were highly theoretical and esoteric, one practical result was an increasing focus on alchemical drugs as wonder cures. A popular German version of Rupescissa, first printed in 1555 and often falsely attributed to Paracelsus, explicitly advertised the quintessence as a wonder drug in its title, *Wonder Medicine for All Bodily Ailments and Acquired Diseases*.[33] As I demonstrated in my earlier work, many German noblewomen became known for their distilled aqua vitae (waters of life), which they touted as a cure for dozens of ailments, including plague.[34] King Philip II of Spain, an avid pseudo-Lullian, hired alchemists to produce a universal cure in the 1570s.[35] Even as Europeans sought exotic antidotes and wonder drugs from afar, they also looked for sources closer to home to produce them.

Alchemical Artisans

This growing demand opened new entrepreneurial doors. One of King Philip II's alchemists was the Italian surgeon Leonardo Fioravanti, who rose to fame in the 1550s for his "new way of healing" and his concept of a universal cure, as William Eamon has shown.[36] Although Fioravanti was influenced more by the pseudo-Lullian tradition than by Paracelsus, he adopted Paracelsus's strong stance against learned Galenic medicine. As with Paracelsus, Fioravanti's concept of a universal cure involved not one single medicine but rather a number of cures—in his case, strong emetics and purgatives created by alchemical means. Unlike Paracelsus (but like Kegler), he fiercely marketed these drugs with what we might now call "brand names": Precipitato, Mighty Elixir, Blessed Oil, Quintessence, Liquor, The Philosopher's Stone.[37] He enjoyed enormous success. All of his books went through numerous editions in Italy, and several were translated into English.[38] Although his work did not break into the German publishing scene until the seventeenth century, his marketing strategies may have spread, as we shall see.[39] Whether or not he was a direct influence on German alchemists like am Wald and Berthold, Fioravanti demonstrated the great commercial potential of alchemical drugs.

Another successful medico-alchemical entrepreneur, who certainly would have been known to both Berthold and am Wald, was a Swiss goldsmith from Basel named Leonhard Thurneisser. Alongside his apprenticeship in the 1540s, Thurneisser also worked for a medical professor named Johannes Huber, who taught him botany and introduced him to Paracelsus's writings.[40] As a professional goldsmith he became interested in metallurgy, assaying, and mining, and he traveled widely. He became known in aristocratic circles for his expertise in botany and alchemical medicine and published two well-received books on alchemical cures.[41] In 1570, through various twists of fate, he found his way to Berlin and, the following year, became personal physician to the newly crowned Elector Johann Georg of Brandenburg. Johann Georg granted him an old Franciscan cloister, known as the Gray Abbey (*Grauen Kloster*), as a laboratory for making his alchemical medicines. It was an alchemist's dream job, and Thurneisser took full advantage of it. He not only created and sold a

6.2 Leonhard Thurneisser's *Sigillum Solis* in his *Historia vnd Beschreibung* (1578). Courtesy of the Economic Botany Library of Oakes Ames, Harvard University.

steady stream of costly elixirs, especially potable gold and medicines made from precious gems, but also established a printing press in the Gray Abbey and printed his own books.[42] His manuscript recipe books were full of entries for poison antidotes, elixirs, wondrous oils, and the "Philosopher's Stone."[43] Like Fioravanti, Thurneisser also had an extensive mail-order business, and he gave medical advice to correspondents, both aristocratic and common.[44] Through all of these activities, he amassed a large fortune (which he later lost in a marital dispute).[45]

Along with medicines to be taken internally, Thurneisser also created a variety of metal coins stamped with a seal to be used as amulets, which became popular among German aristocrats. Some of these *Taler*, composed of seven different metals, drew their powers from the sun, moon, and planets and could be used for a variety of illnesses. Thurneisser also created talismans that harnessed both astrological and metallic powers at a specific moment. He claimed, for example, that he made his *Sigillum Solis* (seal of the sun) by hammering gold into a thin sheet at sundown (see fig. 6.2).[46] The idea of stamping that informed Thurneisser's talismans and *Taler* also intrigued alchemists who became interested in terra sigillata around the same time.

Alchemical Earth

Since the time of Galen, terra sigillata had been viewed as an effective poison antidote by the most traditional of physicians. It did not fit the usual idea of an alchemical drug, as it was a clay dug from the earth rather than a distilled elixir. The inspiration for recasting terra sigillata as an alchemical cure came from an offhand reference in one of Paracelsus's earliest printed books, his *Great Surgery* (1536). In discussing remedies for poisonous animal bites, Paracelsus had briefly praised a kind of terra sigillata or "*auxungia solis*" as a remedy for the bites of toads and spiders as well as poisons like arsenic and mercury. This terra sigillata, formed into small medallions and stamped with a seal, drew its powers from the metallic ore it contained. Paracelsus noted that many called it "St. Paul's Earth" (sold by the *pauliani* we met in chapter 2), but that it was rare to find an authentic specimen. In typical Paracelsian fashion, he derided the substances generally sold as terra sigillata in the apothecary shops as "nothing but a swindle."[47] He thus added his voice to those of the many other authors complaining that most terra sigillata sold in Europe was fraudulent (see chap. 5).

Paracelsus's brief endorsement of terra sigillata as a metallic cure proved influential on a Silesian physician named Johannes Schultz, widely known as Johannes Montanus. An avid Paracelsian who owned an extensive collection of Paracelsus's manuscripts, Montanus claimed to have discovered the Paracelsian terra sigillata around 1550 in abandoned gold mines in the hills on the outskirts of Striga (later Striegau and now Strzegom). He called this new earth Silesian terra sigillata (*terra sigillata silesiaca,* sometimes also called *terra sigillata strigonensis*).[48] In many ways, Montanus made the clay more explicitly alchemical than Paracelsus had done. He argued that Silesian terra sigillata drew its powers from gold, viewed by alchemists as the most perfect of metals and thus the most medicinally powerful. While Paracelsus described the auxungia solis as white, the Silesian terra sigillata was yellowish, emphasizing its golden virtues. It had been made "from the chimical arts," with God and Nature acting as the alchemists. Through divine Providence, Nature had "transmuted" the earth into a cure for many diseases, especially poison.[49] Montanus intimated that the terra sigillata might even have virtues approaching that of the philosopher's stone.[50]

Montanus claimed to have kept the new earth secret from all but a few close friends for thirty years, but there is evidence that he was using it by the 1570s. In 1576, he wrote a letter to Thurneisser and signed it "Johannis Trimontanus." The letter merely introduced a fellow Paracelsian and was not particularly noteworthy, but the moniker "Trimontanus" referred to the three hills outside of Striga where the terra sigillata had been found.[51] Apparently Montanus had become known enough for the earth by that point for him to reference it. At some point, the city of Striga became involved in extracting the earth and producing medallions of terra sigillata, which they stamped with an official seal depicting three mountains (see fig. 6.3). This motif would represent the Silesian earth, in various iterations, for centuries to come.

By the late 1570s, as well, Montanus had become acquainted with our old friend Andreas Berthold, who was now working as a mine foreman in the copper mining town of Kupferberg (today's Miedzianka). As we have already seen, Berthold went on a trip across the western Holy Roman Empire in 1580–81 to sell this new German version of terra sigillata, using poison trials as his main method of proof. In 1583, Berthold published a book, in Latin, titled *Terra Sigillatas Recently Found in Germany*, to which he appended a short treatise by Montanus on the Silesian terra sigillata. Berthold's part of the book described the virtues of the new German earth and included the testaments of three poison trials from powerful princes and city officials as proof of its efficacy.[52]

There has long been considerable confusion, both in the sixteenth century and among historians (myself included), over which earth, exactly, Berthold was selling. Sixteenth-century individuals largely saw Berthold's book as an advertisement for Montanus's terra sigillata from Striga, and historians have generally agreed.[53] I have come to realize, however, that there were multiple earths and that Berthold was promoting his own variety alongside Montanus's. He referred to Silesian "earths," in the plural, and he described himself as the "inventor" of the German clay, although he printed Montanus's priority claim to the Strigan variety. Berthold's earth also looked different. In their testimonial letter, Jülich town council members described his terra sigillata as stamped with a seal depicting the sun, moon, five stars, and the letters ABVO (Andreas Berthold von Oschatz). Specimens of both his earth and the Strigan earth still exist today, although the latter is far more common (see figs. 6.4–6.7). In the middle of his text, Berthold invited his readers to come purchase the earth either from him in Kupferberg or from "the brilliant and most learned Doctor Johannes Montanus" in Striga, which suggested that the two varieties were interchangeable. In contrast, a postscript to the text mentioned only Kupferberg. Berthold also planned to leave a supply in Frankfurt am Main with his printer, Christoph Corvinus, every March, along with a variety of stones and gems useful in medicine.[54] Although he was involved with the Strigan terra sigillata in some way, he mainly appears to have been selling his own variety.

Berthold also had his own thoughts about the new terra sigillata. He called it *Axungia solis,* specifically linked it to Paracelsus's terra sigillata, and described it as an oily yellow earth dug out of gold mines and "diligently prepared." There was also a white *Axungia lunae* (grease of the moon) or "mineral unicorn," retrieved from silver mines.[55] Apparently sensitive to the numerous detractors of Paracelsus, Berthold defended this nomenclature, arguing that he had drawn on "the counsell and advise of the best Miners and Doctors of Phisicke" and simply used descriptive titles. The earths had the same virtues as gold and silver, and the silver version replicated the efficacy of unicorn horn.[56] Berthold also praised the terra sigillata as a specifically German cure and thus "of better use for the bodies of our Germains" than Turkish (Lemnian) terra sigillata, a reflection of the widespread argument, embraced both by German humanists and by Paracelsus, that German medicines were best for German bodies.[57]

Despite these attempts to bring the German terra sigillata into the alchemical canon, the drug was still, first and foremost, a cure for poison,

6.3 Silesian terra sigillata from Striga. In Christian Gottlieb Ludwig, *Terrae Musei Regii Dresdensis* (Leipzig, 1749). Courtesy of Houghton Library, Harvard University.

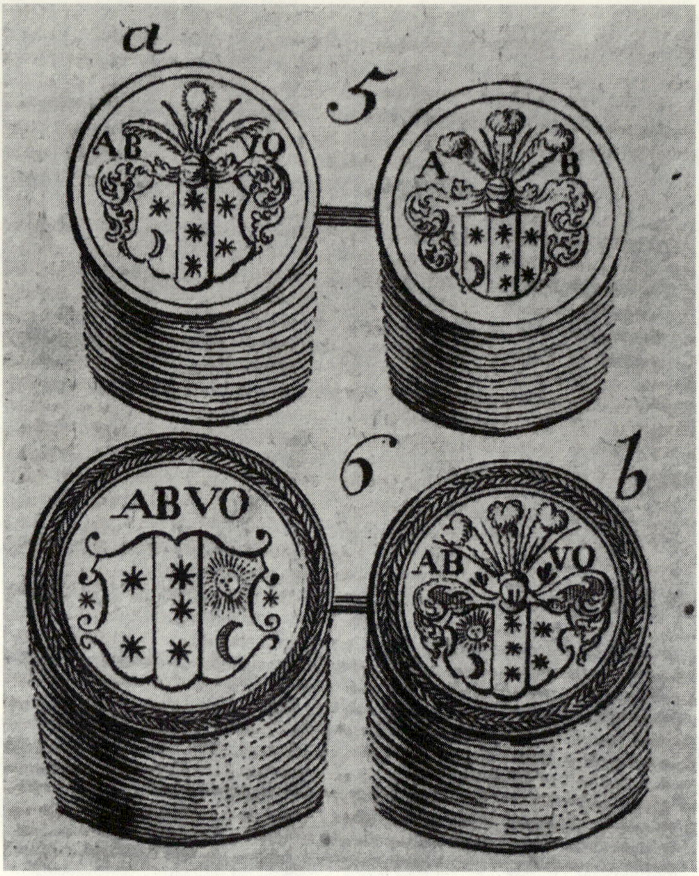

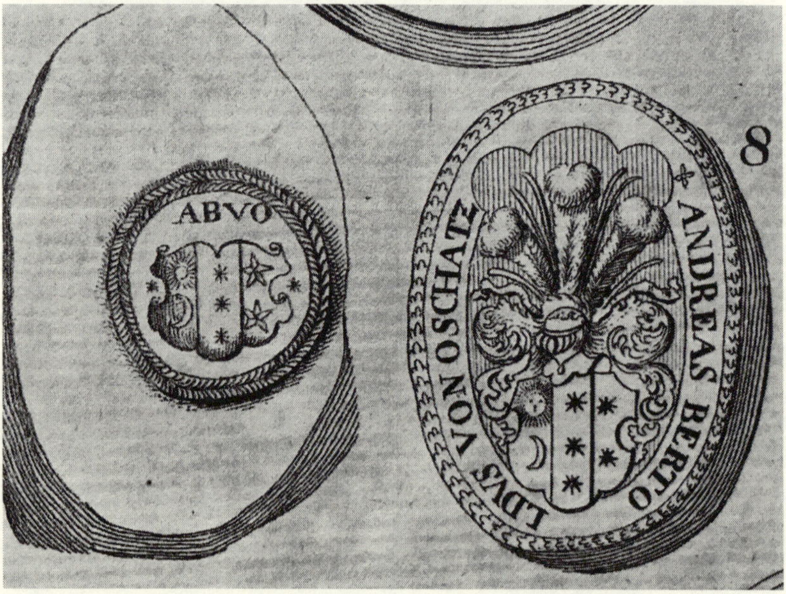

6.4 and 6.5 Two versions of Andreas Berthold's terra sigillata. In Christian Gottlieb Ludwig, *Terrae Musei Regii Dresdensis* (Leipzig, 1749). Courtesy of Houghton Library, Harvard University.

6.6 Specimen of Berthold's terra sigillata. Pharmazie-Historisches Museum, Basel. Photo by the author.

6.7 Specimens of Silesian terra sigillata from Striga. The bottom left medallion depicts Johannes Montanus and includes the seal. Pharmazie-Historisches Museum, Basel. Photo by the author.

plague, and the bites of poisonous animals. It thus fit into the long-standing Galenic tradition of terra sigillata. In detailing the specifics of the drug, moreover, Berthold presented remarkably conventional advice. He claimed that it would expel all poisons and venoms by vomit if administered immediately after the taking or dry them out by sweating if the poison had been in the system for a while. The sweating would also drive out plague if administered early in the illness's progression—but if the plague had already infected the blood, one should use a combination of bloodletting, lancing the buboes, and applying a plaster made of terra sigillata and various herbs to the heart. None of these suggestions differed markedly from the conventional way terra sigillata was understood to work in poison and plague. In other illnesses, Berthold similarly suggested taking the terra sigillata with wine, vinegar, or aqua vitae; applying it externally as a plaster; and using it in conjunction with bloodletting.[58] The *Axungia solis* may have been an alchemical cure in theory, but that aspect nearly disappeared in practice. Berthold's success in marketing the German terra sigillata hinged less on its alchemical nature than on his poison trials, which he used as concrete proof that the antidote worked.

A Cure for Everything

The Silesian terra sigillata quickly had other competition. In 1581, the same year as the poison trial on Wendel Tümler in Langenburg (chap. 4), Georg am Wald published his pamphlet on the Terra Sigillata Amwaldina. There are a number of parallels between his terra sigillata and the Silesian variety. Am Wald presented his remedy primarily as a cure for plague, poison, and the bite of poisonous animals, but he also listed over twenty further conditions against which it was useful, including quartan fevers, all internal pains and complaints, dropsy, jaundice, worms, lameness, the French disease, epilepsy, backache, and gout. It would clean out the womb, help stimulate urination or menstruation, strengthen the whole body, and protect against "acts of God." Like the Silesian version, it was much more than just a poison antidote. Am Wald went even further than Berthold and Montanus and called his terra sigillata a "universal medicine."[59]

Although the concept of a universal medicine was most prevalent among alchemists, am Wald did not specifically describe his remedy as alchemical. Indeed, he declared his independence from all medical ideologies and, in a simple German poem "Against Those Who Begrudge

Me," depicted his medicine as God-given. He followed it with a Latin quotation stating, "Galen is a friend, Paracelsus is a friend, but the greatest friend is truth, which shall prevail."[60] Toward the end of the treatise, he derided "corrupt goldsmiths" who praised alchemy without reason and sold false tinctures as potable gold (quite possibly a dig at Thurneisser). These hoodwinkers, he argued, had given "the good and godly study of medicine" a bad name among learned doctors, "be they Paracelsians or Galenists." Even if one of these alchemists had managed to create a potable gold, that was no great feat. Am Wald himself had already done it, "as subtly as Theophrastus ever could." No potable gold, however, could come close to his terra sigillata in potency and powers.[61] While he did not shy away from alchemy in principle, he made it clear that he did not belong to the ranks of "corrupt" so-called alchemists. Like Berthold, am Wald appeared to be trying to make peace with learned medicine.

Did am Wald know about Berthold's terra sigillata? If he did, he did not let on. He made no mention of any other German terra sigillata in this brief work, although he claimed his remedy was better than other common poison antidotes, such as "unicorn horn, Turkish terra sigillata, bolus armenus, theriac, mithridatium, and the golden egg."[62] Whether the omission of Berthold's terra sigillata was a marketing decision or a sign of genuine lack of awareness is difficult to assess. Despite the excitement about Berthold's earth in some courtly circles, it is unclear whether word of it had reached Bavarian patrician society by 1581. There are a few vague statements that could be references to Berthold in am Wald's text, but nothing overt, unlike his more direct barbs at Thurneisser. At the same time, a completely independent appearance of another new German terra sigillata would have been a very great coincidence.[63] It is possible that am Wald had heard of Berthold but did not feel it necessary to reference him, as the Silesian terra sigillata was just beginning to make its name in 1581. It is also possible that terra sigillata, in general, had been gaining steam in German-speaking regions in the 1570s and that Berthold, am Wald, and Montanus all hit on it independently.

In any case, the Terra Sigillata Amwaldina had a number of important differences from Montanus's and Berthold's varieties. Unlike Berthold, who aimed to reach a learned audience in Latin, am Wald wrote his pamphlet in German. His audience was local, and his drug came cheap, by am Wald's own assertion. It cost only four *Batzen*, or "the same as really bad medicines cost in the apothecaries."[64] Am Wald's terra sigillata had another crucial difference: it was tied not to a place but to a person. The ancient terra sigillata came from a particular mountainside in Lemnos;

the *pauliani*'s earth from Malta; Montanus's and Berthold's from specific mountains in Silesia; and Berthold later promoted new earths from hills in Hesse and Laubach. Although entrepreneurs like Berthold might claim the preparation as crucial, most terra sigillata began as a clay from a known location, as did other medicinal earths such as Armenian bole. Am Wald made no reference to any place or any clay at all. His terra sigillata was man-made.[65]

In describing how the Terra Sigillata Amwaldina actually worked, moreover, am Wald bore out his claims of independence from both Paracelsian and Galenic thought. His explanation did not match any widespread theory of disease and treatment:

> As soon as this terra sigillata comes into the body, it penetrates to all of its appendages, which everyone who pays attention will feel. For some people, the hands and feet grow heavy and tired. For some, [the hands and feet] fall asleep. For some, they tremble, which gives them the creeps, but this all happens without pain, and it does not last long. And where [the terra sigillata] finds something excessive, unclean, or evil that does not belong in the body and causes illness, it attacks it without any harm to the body's limbs. If this *vitiosa* or harmful material has coagulated (as not all illnesses are *humores* or moistures), it loosens it and consumes it with a fire that is no fire, makes no inflammation, neither metallic nor otherwise, no matter what jealous men might say.[66]

In contrast to Berthold, who described his terra sigillata as operating in very typically Galenic ways (provoking sweating and vomiting), am Wald claimed that if vomiting occurred with his drug, it was only because of the poison, not the antidote. Illness was caused by some "harmful material" that was alien to the body, and the Terra Sigillata Amwaldina harmlessly consumed it. This idea was closer to Paracelsian notions of disease as an invasive element than Galenic humoral ideas, but it was far simpler. Rather than Paracelsus's theory that a healing element could be separated from every poisonous, disease-causing agent, am Wald proposed one cure that would work for all diseases.

Like Berthold, am Wald described the actual application of the Terra Sigillata Amwaldina along more traditional lines. He gave diet recommendations and included several recipes for purges to take before taking the terra sigillata. For plague, one could burn the terra sigillata on an ember in order to release the sweet-smelling smoke, a traditional method to ward off the poisonous vapors that caused plague. Am Wald also gave multiple recommendations on how to make the terra sigillata provoke sweating in plague patients and in cases of poison, another standard Galenic response.[67] Despite his claims about the "fire that is no fire," am Wald stuck to methods that would be familiar to his readers. This combination of radical promises and conservative applications would become a hallmark of his alchemical medicine.

Am Wald did not, however, include any hard evidence to support his strong claims. He gestured bombastically to the "many *Exempla* . . . from the many thousands who gave me their testimony," but he did not include any testimonials. Whereas Berthold had officially notarized proof that his terra sigillata had been successful in three different poison trials, am Wald argued that the Terra Sigillata Amwaldina "needs no certificate," as it was "known all over the land."[68] He did, however, describe a trial of a different kind. It was likely, he claimed, that copycats would attempt to mimic the Terra Sigillata Amwaldina. In order to be sure that you had an authentic exemplar, you should put a little bit on a knife's edge and hold it over a flame. If it caught fire and gave off a sweet smell, it was the real deal. He also advised that the Terra Sigillata Amwaldina was rather bitter and that it was easily recognizable by his initials, in triangles, on the stamped medallion (see fig. 6.1).[69] Rather than providing proof of his drug's efficacy, as Berthold had done, am Wald asserted without question that it was effective and instead focused on ways to prove its authenticity.

"Admirable, but hardly believable"

How did traditional physicians respond to the new terra sigillata craze sweeping through the Holy Roman Empire? The reaction varied greatly, but the one constant was considerable interest. Physicians, even vehement opponents of Paracelsian medicine, brought the same fascination with wonder drugs that we saw in chapter 5 to new, local, alchemical remedies. Johann Wittich, for example, remained a staunch Galenist throughout his life, but, as we have seen, he eagerly embraced the Silesian terra sigillata. At the same time, he ignored the existence of Berthold, excised all overt reference to Paracelsus from Montanus's text, and shortened the explanation about the names *Axungia solis* and *Axungia lunæ*.[70] In Wittich's exposition of the terra sigillata, the drug was simply a mineral substance, much like the original Lemnian terra sigillata or other poison antidotes such as theriac. Another traditionalist, the physician Georg Marius, published a book on terra sigillata in 1589 that especially lauded the original Lemnian earth. Marius warned against the recent proliferation of terra sigillata varieties, but he did allow that the Silesian terra sigillata appeared to be a promising substitute for the

Lemnian clay.[71] Marius's student Johannes Posthius, poet laureate and personal physician to the prince-bishop of Würzburg, had an even more positive response. He was so enamored of the Silesian drug that he wrote two panegyric poems praising it and Johannes Montanus.[72]

This generally positive response was rooted in three different factors. First, there was widespread interest in Paracelsian ideas in the Holy Roman Empire, and a great number of German physicians straddled the Paracelsian and Galenic traditions. Although there were staunch Galenists and committed Paracelsians, many physicians could not easily be pigeonholed into one category or the other. Posthius, for example, had trained as a traditional Galenist, but he eagerly sought out promising alchemical cures, and he praised the "wise" Montanus for the "arcana" he had discovered. This praise highlights a second important point in favor of the Silesian cure. Johannes Montanus, despite his avid Paracelsian leanings, was a respected physician trained in Italy, with a legitimate medical degree from Padua. He thus had an easier path toward respect than empirics without the proper credentials. Significantly, Posthius's poem hailed Montanus as the inventor of the Silesian terra sigillata and mentioned Berthold only as the "distributor," much as Wittich simply ignored him.[73] Berthold's attempts to draw attention to his own Silesian terra sigillata largely failed. Yet his poison trials were the crucial third reason why the response to the Silesian terra sigillata was so positive. The wide variety of illnesses Berthold and Montanus purported to cure seemed literally incredible, yet the reader did not need to trust Berthold on his "bare words," as Landgrave Wilhelm IV of Hesse-Kassel noted. Berthold provided hard proof from respected individuals that it worked.

The Terra Sigillata Amwaldina was a different story. Am Wald's *Report* set off an initial flurry of excitement, as with the Silesian terra sigillata. In 1582, Posthius wrote to Joachim Camerarius, court physician in Saxony, asking his opinion of the Terra Sigillata Amwaldina, for which the "Donauwörth city physician" Georg am Wald claimed "admirable, but hardly believable" properties.[74] Despite this initial skepticism, Posthius decided to take a closer look. Barely a month later, he reported to Camerarius that he had paid a visit to am Wald in Donauwörth, even though colleagues in Augsburg had nothing good to say about the man. The visit had not gone well. Am Wald had told him nothing useful. He was arrogant and expressed pride in being an autodidact in medicine; he viewed himself as superior to all learned doctors and was even contemptuous of Paracelsus; and although he continued to tout the virtues of his terra sigillata, he refused to divulge anything about it, even though Posthius had shown himself very eager to learn.[75]

Posthius's damning report provides excellent insight into what it took to be considered legitimate among the learned doctors. He remained surprisingly open to the Terra Sigillata Amwaldina even after he heard the negative opinions of am Wald from Augsburg, enough that he undertook the hundred-mile journey from Würzburg to Donauwörth. Am Wald's behavior, however, had provided no reason for Posthius to trust the drug. Not only did he condemn all kinds of scholarly medicine (Galenic or Paracelsian), he also gave no information in support of the Terra Sigillata Amwaldina. Would Posthius have proven more amenable to the drug if am Wald had handed him certificates from powerful princes? It is hard to imagine that they would not have made a difference. After the disastrous conversation about the terra sigillata, am Wald promised Posthius he would send him a remedy against cough and dizziness. Posthius assured Camerarius that he would take it only if am Wald told him what was in it.[76] That he considered taking it at all underscores how eager physicians were to find effective new remedies, even from dubious sources.

Whether or not am Wald drew inspiration from the Silesian terra sigillata, others connected the two new earths. Posthius, as we have seen, quickly made efforts to examine both new drugs. Physician Johann Weidner expressed great initial interest in the Terra Sigillata Amwaldina and corresponded about it with several physicians in 1583–84.[77] One of his correspondents, Peter Monau, eventually became very skeptical. Monau grumbled to Weidner that much of what am Wald wrote of his terra sigillata smacked of "credulity and superstition." He also reported that he found the Silesian earth useful only for stomach ailments and fluxes, but he speculated that its powers might be increased if one washed it first, as physician Jacob Theodor had touted its efficacy against plague. Another earth with similar virtues as the Strigan terra sigillata, he continued, had been found near the city of Glogau and was, he thought, more effective.[78] All of this chatter among physicians showed a real willingness to engage with and explore the new drugs, alongside a healthy skepticism of the expansive claims made for them.

The Seal of Striga

One physician was not at all happy about the proliferation of German terra sigillata varieties. Johannes Montanus, the discoverer of the Silesian terra sigillata, was rather surprised to see his own essay appear in Berthold's 1583 book. When he received the publication, he struck out against what he saw as attempts to undermine his own terra sigillata from Striga. In 1585, he published a "brief and exquisite" treatise "by the most excellent philosopher and doctor Johannes Montanus of Striga in Silesia on the true, native . . . Terra Sigillata, invented by him, in that place." The emphasis on person and place signaled that Montanus viewed his Strigan earth as the only "true" German terra sigillata. He noted on the title page that he had been forced to publish in order to counter the "mangled and depraved" version of his treatise that Berthold had "appended to [the book] on his earth." Near the end of his text, Montanus similarly complained that Berthold had published his work "without my knowledge" and brought a premature version to the public. Even learned scholars and princes, he complained, thought that Berthold was speaking in his name, with his blessing. This unexpected circumstance had forced him to counter with the present treatise, although he would have preferred to remain anonymous.[79]

What made Montanus so annoyed? He did not deny outright that he had written the treatise that Berthold published in his name, and indeed, his 1585 book was simply a more extended version of that work. (Both versions were far more esoteric and theoretical than Berthold's practical treatise.) One way or another, an early version of the work had ended up in Berthold's hands. While Berthold saw utility in tying his earth to Montanus's, Montanus explicitly did not want the Silesian terra sigillata to be confused with Berthold's earth. This risk of confusion, indeed, appears to have rankled Montanus in particular. On the back of the title page, he placed an admonition to the reader. "We have realized," it read, that "not a few peddlers" were circulating and selling different exemplars of terra sigillata "with various seals and forms," out of "an eagerness for greater wealth." He dismissed these attempts as "diabolic" and warned the readers to purchase only the medallions with the official seal of Striga, authorized by the Strigan Senate.[80]

Berthold, in response, did nothing to clarify the confusion about the Silesian terra sigillata but instead added evidence in favor of his earth, while continuing to mention Montanus. In 1586, he published a new edition of his 1583 book, prefaced by a series of laudatory poems. A new title page listed the same title as the first edition (*Terra Sigillatas Recently Found in Germany*) but added Berthold's coat of arms, surrounded by the words "Andreas Berthold, Explorer and Lover of Nature's Mineral Powers."[81] The panegyrics praised the terra sigillata in general and Berthold in particular, but Montanus made an appearance as well. As already mentioned, the opening set of poems by Posthius honored Montanus and mentioned Berthold ("Barthold") only briefly as a salesman.[82] Most of the rest of the poems, however, specifically sang the praises of Berthold, including a poem addressed to Posthius from the Belgian humanist Franciscus Modius.[83] Berthold thus added laudatory validation from learned men, a common trope in scholarly medical books, to his poison trials. In all cases, he made it easy to conflate Montanus's Silesian terra sigillata with his own variety.

Along with this work, Berthold published a short pamphlet called *Epistola panegyrica*, which may have been intended as a preface to the panegyric poems. The *Epistola* was an open and rather obsequious letter to the three landgrave brothers who ruled Hesse (Wilhelm IV of Hesse-Kassel, Georg I of Hesse-Darmstadt, and Ludwig IV of Hesse-Marburg). It reminded the landgraves of the proven virtues of Berthold's Silesian terra sigillata and announced a new earth discovered "in the mountains of Hesse" that was the equal of the Silesian variety. Berthold claimed he had been motivated to write by the Marburg physician Johannes Wolff, personal physician to Landgrave Ludwig, as well as other "learned men" who had asked him for more information. His new, Hessian terra sigillata had the exact same virtues against poison and disease as the Silesian *Axungia solis*—Berthold called it *Axungia solaris*—but it had the advantage of convenience, as it was hard for people to get to the Silesian mountains.[84]

As before, Berthold used poison trials to advocate for this new earth. He reminded Landgrave Wilhelm that, in 1580, he had tested "my Silesian terra sigillata" in the castle in Kassel and found it "powerful and effective" against poison. He also added a new poison trial to this evidence. Two years ago, Berthold recounted, Landgrave Georg had "made an experiment" of the *Axungia solaris* in the town of Gerau, where a "notorious mercenary" responsible for "many terrible crimes and deeds" had been detained and sentenced to death by the wheel. Recalling the powers of the Silesian terra sigillata, Landgrave Georg called for a delay in the execution, in order to have the Hessian earth tested on this criminal "for the good and health of many mortals." The condemned man was given mercury sublimate, which Berthold described as a terrible and effective poison that would cause certain death. Indeed, the prisoner went through many paroxysms of pain, with vehement spasms, and he seemed sure to die. Once he took the terra sigillata, however, the symptoms miraculously ceased.[85] As mentioned in chapter 4, the prisoner had his sentence commuted to beheading as a reward for his participation.

While much of the *Epistola panegyrica* merely repeated points Berthold had made in his 1583 book, his emphasis on the earth's alchemical nature—and his own role in the process—only increased. He wrote continually of "my earth" and emphasized that the clay (whether Silesian or Hessian) was an arcanum that he had "extracted, tempered, prepared, and sealed" to make it effective. Echoing the words on his coat of arms, he signed his letter with the title "Investigator and Explainer of Nature's Arcana and Minerals." For the devotedly Protestant landgraves, he also added several pages of verses from the Psalms, mostly about the mineral bounties that God had put on the earth. Berthold's message was transparent: his terra sigillata was a God-given marvel sealed within the earth that he, through alchemical means, knew how to unlock. He appealed to the Hessian landgraves in their capacity as devoted fellow "students of arcane things."[86] To emphasize this point, he added additional, handwritten Bible verses to at least one copy of the *Epistola panegyrica*, likely a presentation copy for one of the Hessian landgraves. He embellished one of these verses (from Job 22) with glittering golden ink.[87]

These efforts by both Montanus and Berthold illustrated the problem with terra sigillata. As a recognizable form—a medallion stamped with a seal—it invited imitation and substitution, even though the seal was intended to guarantee authenticity. Montanus's attempts to declare only one legitimate German terra sigillata, bearing the seal of Striga, may have resonated in learned spheres, but he could not stop physicians like Peter Monau from speculating that other German earths might have similar (or even superior) virtues. Berthold, for his part, continued to rely on poison trials as his main method of proof, but he could not escape the widespread impression that the poison trials proved Montanus's Strigan terra sigillata. His attempts to portray the terra sigillata as his own arcanum had only a limited effect. Like Montanus, Georg am Wald initially attempted to use a description of the seal to show the authenticity of his Terra Sigillata Amwaldina. Eventually, however, he hit on a different solution to this problem. He left terra sigillata behind and introduced an entirely new genre of wonder drug.

The Panacea Amwaldina

Am Wald's new cure was called the Panacea Amwaldina, and it was no longer a plague remedy and poison antidote, but instead specifically a universal remedy. In 1591, he published a new book to advertise it, *Short Report on How and in What Manner the Panacea Amwaldina, as a Unified Medicine . . . Should Be Used*.[88] This work lifted entire passages from his pamphlet on the Terra Sigillata Amwaldina, but it contained two crucial differences. Whereas the terra sigillata had been depicted primarily as a plague cure and poison antidote and secondarily as a universal medicine, the panacea had far grander claims. Am Wald proclaimed it a "unified medicine" that would help against 116 different ailments, including poison and poisonous diseases.

Am Wald sharply defended the idea of his panacea as a universal medicine that could heal all disease. Just as a very small quantity of a poisonous herb or other poisonous material could kill you, he argued, so too could a very small quantity of a remedy heal you and keep you healthy. Similarly, just as poisonous air could "go through, infect, and poison all of the body's appendages," a healthy and healing air could penetrate the body and promote wellness.[89] Illness, in his view, originated when the "*spiritus vitæ* or living spirit is aggravated, weighted, burdened, or impeded, so that it cannot fulfill its function and operation." If, then, "all illness has a universal fountain, a common origin and cause," why should there not be a "single, universal, and general medicine" to cure all illnesses? His panacea, he explained, mixed itself with the *spiritus vitae*, strengthened it, and cleansed it. It also purified and cleansed the blood and removed any excess and "anything that harms the body." In this manner, it could heal any sort of illness, "even the most grievous illnesses that the school of Galen believes to be incurable."[90]

This last statement underscores a second major shift from am Wald's previous work. Gone was the moderate tone that had proclaimed "amicus Galenus, amicus Paracelsus." Instead, am Wald revealed himself to be both a sharp critic of Galenic physicians and a fully converted Paracelsian. He complained that "not a few Galenists have, in an un-Christian manner . . . obstructed me and stood in my way," and he insinuated that Galenic physicians were behind two different plots to poison him.[91] Despite his idiosyncratic medical ideas, he insisted that he followed the teachings of "the ancient and true philosopher and doctor Philippi Theophrasti von Hohenheim," and he claimed that Paracelsus himself, had he been alive, would have approved.[92] His panacea was an "astral" cure that had been "cleansed through the highest spagyrical, chymical, and separative arts."[93] Indeed, it was most comparable to the philosopher's stone "because of its great virtues in healing all illnesses."[94] In his conclusion, he engaged in a full-blown defense of himself and Paracelsus against Galenic naysayers.[95] In addition to designating Galenic theory a "mistake," he tried to undermine the historical figures of Galen and Avicenna on religious as well as medical grounds. He cited texts from Galen that seemed to question the omnipotence of God, and he called Avicenna "a great blasphemer and Musselman." Following their examples, he argued, was hypocritical and sacrilegious.[96]

Some of this sudden shift toward Paracelsianism may have reflected a genuine evolution of am Wald's interests. In his later life, he corresponded with a number of alchemically inclined physicians and empirics, and he carried out new chemical experiments in his palace at Thürnhofen.[97] As we have seen, he was also a strict and fanatical Lutheran and had stirred up trouble with his religious writings, so the religious criticism of Galen was not out of character. Despite these very real motivations, the cynical reader might suspect a considerable amount of business calculation in his new allegiance to Paracelsus. As am Wald certainly knew, many wealthy townsfolk and members of the lower and upper gentry were open to Paracelsian ideas and interested in alchemical medicine.[98] Am Wald may have been particularly interested in targeting the gentry, who, as we have seen, proved eager to try new wonder drugs that promised marvelous cures and who had the wealth to provide handsome rewards. Nearly a year before the publication of the *Kurtzer Bericht*, he sent a sample of his Panacea Amwaldina and a letter explaining its uses to Count Gottfried of Oettingen-Oettingen, who was known for his alchemical interests. Although am Wald did not openly criticize Galenic medicine in this letter, he noted that he had heard of the count's love for the "veram Philosophiam, Medicinam, Chymiam, and artem Spagyricam," and he presented his panacea as comparable with the philosopher's stone.[99] In contrast to his less ideological (and only moderately successful) work on the terra sigillata, the treatise on the panacea was crafted to appeal specifically to high-profile, alchemically inclined, Protestant clients.[100]

Am Wald left the connection between his panacea and his terra sigillata ambiguous. The actual text of his *Report on the Panacea Amwaldina* incorporated significant portions of his work on the Terra Sigillata Amwaldina, and he similarly described the panacea as a reddish-brown substance, although he sold it as a powder in small envelopes (*Brieflein*) rather than in the form of a stamped medallion. In an attempt to promote his new medicine without damning the old one, he claimed that the terra sigillata had been a great success "in countless trials in France, German and Italian lands, Jerusalem, and in almost all of Europe," although he provided no proof of this triumph other than to say it was "evident and well-known around the land." At the same time, the terra sigillata was "not to be compared with or considered the same as my panacea."[101] The panacea was, of course, far better. It was also much more expensive, at one gulden per *Brieflein*.[102]

An Unusual Panacea

Am Wald's interpretation of the panacea represented a significant departure from the previous understanding of the term. The Greek word *panakeia* literally means "all-healing cure" and was related to the goddess Panakeia, daughter of the healing god Aesculapius. From classical times through the Renaissance, the panacea was most often connected specifically to herbs, especially a class of herbs known as the "panaces." A number of respected classical authors discussed the panaces, including Pliny, Dioscorides, Theophrastus, Nicander, and Serapion. Although there was little consensus as to which herbs could be counted among the panaces and what properties they had, many were seen as useful against poison (see table 6.1). These differences caused headaches for sixteenth-century humanist botanists trying to make sense of classical texts. Hieronymus Bock, who published an influential German herbal in 1539, grumbled that it was impossible to figure out which herbs were the panaces, since all the classical authors were in conflict on their descriptions.[103]

The panacea also appeared as a metaphor in sixteenth-century religious works. In 1523, Erasmus of Rotterdam proclaimed to King Henry VIII that "no drug has yet been discovered by the physicians that can cure all the diseases of the body, however much they may boast of that famous panacea, which, they claim, is so powerful and yet is quite unknown." In contrast, he argued, God could rid the human of all diseases of the spirit through true faith.[104] The same year that am Wald published his first work on the Panacea Amwaldina, 1591, a northern German pastor named Laurentius Drabitius published a book about a

Table 6.1. The panaces in Pliny and Dioscorides

Plant	Pliny	Dioscorides
Panaces Asklepion	juice of plant has coagulating properties	root is good for snakebite and stomach ulcers
Panaces Heraklion	"good for nothing"	resin opopanax comes from root; good for dozens of things
Panaces Chironion	flower is extremely efficacious	root is good against snake poison
Panaces Centaurion or Pharnacion	root used to flavor wine	root is good for many things, including as an abortifacient
Cunila bubula	called a "panacea" because useful against serpent bites	not mentioned as a panacea
Lovage	not mentioned as a panacea	good for stomach complaints and against poison; called a panacea

"spiritual" panacea "out of the apothecary of the prophet David," which suggested that the only universal cure was heavenly, not earthly.[105] The heavenly panacea continued as a common religious metaphor for years to come.[106]

In the second half of the sixteenth century, there was a rising interest in the goddess Panacea as a symbol of healing, especially in northern Europe. In 1558, physician Philip Michael Novenianus hailed the German noblewoman Dorothea of Mansfeld as a Christian version of Panacea, owing to her talents in making excellent distilled medicines.[107] An image by the Dutch engraver Philips Galle from the 1570s portrayed Panacea as the personification of medicine, holding a urine jar, the symbol of physicians, and a pill box, the symbol of apothecaries (see fig. 6.8). Among the pharmacists' paraphernalia, the engraving included equipment for making distilled medicines. There was thus a conception, at least in parts of northern Europe, of Panacea as a goddess tied to the compounding of medicines, including distillation. Still, I have found no evidence before am Wald of an alchemical medicine being described as a panacea.

The word had, however, been applied to a new drug in one very prominent case. Just a few years before the first evidence of am Wald's panacea, in 1581, the Dutch physician Gilles Everaerts published a book on tobacco called *On the Panacea Herb* [*De herba panacea*]. This provocative title suggested to the reader that with tobacco, the ancient panacea had finally been found. The term "panacea" may have been added by the printer rather than by Everaerts, as the heading throughout the rest of the book reads "On the Herb Tobacco" [*De tobaco herba*], and Everaerts did not refer to the panacea in the text. Despite its alluring title, the book was mainly a compilation of existing writings on tobacco by Monardes and Rembert Dodoens.[108] Nevertheless, the fact that someone decided to use "panacea" in the title suggests that the term was viewed as attractive to potential readers. It was a good marketing choice, as *On the Panacea Herb* enjoyed significant success.[109]

Am Wald invoked both the classical panaces and new drugs from the New World in defending his Panacea Amwaldina. Whoever was not content with his defense of a universal cure, he argued, should go read what Galen and his "forefathers and followers" had written of the "Panace Asclepio, Heracleo, Chrionio, & Pharnaceo," as well as related herbs like Moly, rhubarb, China wood, sarsaparilla, guaiac, sassafras, theriac, mithridatium, quinta essentia, and tobacco.[110] With this statement, he brought the ancient panacea under the same roof as poison antidotes like theriac, the alchemical quinta essentia, and modern wonder drugs

6.8 Panacea Personifying Medicine. Philips Galle after Frans Floris after Marten van Cleve, 1574. Rijksmuseum NL.

from the New World like sassafras or tobacco. He thereby put his own panacea in the canon of cure-alls past and present, many of which were also poison antidotes. Applying the word to his own medicine, am Wald suggested that he had, through alchemical means, created the mythical ancient cure for all diseases.

"That noble and brilliant man"

In presenting his panacea, am Wald took a page from Berthold's book, almost literally. Although he once again began his *Short Report* with a poem against "those who begrudge me," he also added Bible verses and two panegyric poems, in Latin, from learned men. The first poem was a tribute to am Wald that "Mr. Johann Hasler, medical doctor and astronomer in Bern, wrote in my honor thirteen years ago." Johann Wieland, a pastor in Donauwörth, contributed the second, a "*Carmen* in honor of the Panacea or Universal Medicine against All Disease, piously and industriously invented by that noble and brilliant man, Doctor Georg am Wald, legal licentiate and most celebrated doctor and philosopher."[111] Both poems involved a certain amount of subterfuge. Hasler's panegyric had nothing to do with the Panacea Amwaldina, but simply lauded am Wald's medical and juridical talents in general. He had composed it on the occasion of am Wald's marriage to his first wife, Margaretha, in 1578, although am Wald conveniently omitted the lines that addressed the wedding. Wieland's poem directly lauded the panacea, but he was no neutral party; he was one of am Wald's closest friends.[112] Nevertheless, the overall effect suggested that learned men considered am Wald a legitimate and worthy medical doctor and had heaped praise on his panacea.

His tactic worked like a charm. Am Wald's panacea stirred up extensive interest and, at least initially, considerable praise. Johannes Posthius once again showed great fascination with the new drug, despite his prior disregard for am Wald and his terra sigillata, and he even wrote a poem lauding the panacea after reading am Wald's *Short Report* (a decision he would come to regret). Justus Zimmermann published his own "Carmina" in 1592, and the theologian Johannes Lauterbach, who had written one of the panegyrics on Berthold's terra sigillata, published a series of five poems "in favor of the Panacea Amwaldina" that same year.[113] Am Wald was inundated with demand for the drug. Count Albrecht VII of Schwarzburg-Rudolstadt requested and received 134 doses of the panacea in 1591, and the prominent physician Andreas Libavius asked for a sample "for a friend" that same year.[114] The Panacea Amwaldina became popular enough that am Wald printed a form letter to give to his patients, with blanks to fill in the patient's name and the appropriate dose (see fig. 6.9).[115]

The attention and corresponding income were well timed, because am Wald had just been expelled from Donauwörth for inciting religious unrest.[116] In 1592, now living in Schwabach (near Nuremberg), he published a second, expanded edition of his book on the panacea. A third "once again expanded" edition followed in 1594, by which time he had moved again—this time to his Thürnhofen manor house, which he always described as a "castle" (*Schloss*).[117] The proceeds from his pricey panacea had allowed "that noble and brilliant man" to purchase a palace to reinforce his aristocratic persona.

In the expanded editions of his book, am Wald included even more evidence in favor of his panacea. He added new laudatory poems and

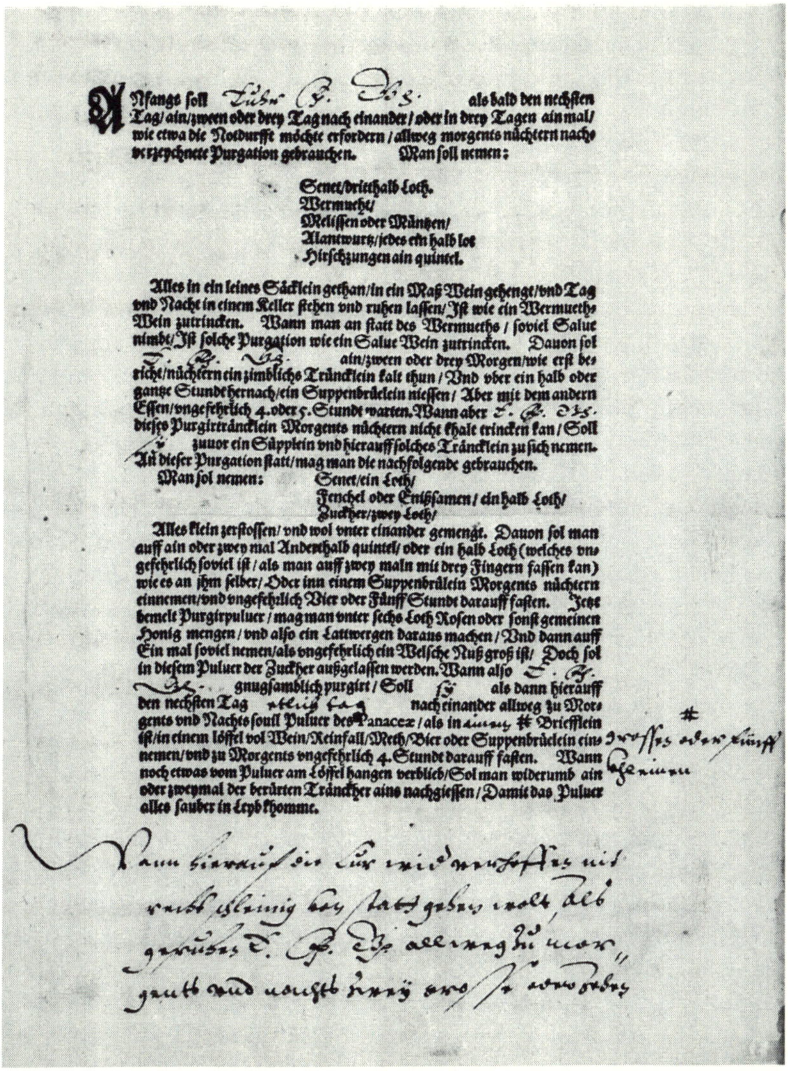

ANfangs soll als bald den nechsten Tag/ ain/zween oder drey Tag nach einander / oder in drey Tagen ain mal/ wie etwa die Notdurfft möchte erfordern / allweg morgents nüchtern nachverzeychnete Purgation gebrauchen. Man soll nemen:

Senet/dritthalb Loth.
Wermuetz/
Melissen oder Müntzen/
Alantwurtz/jedes ein halb lot
Hirschzungen ain quintel.

Alles in ein leines Säcklein gethan/ in ein Maß Wein gehengt/ vnd Tag vnd Nacht in einem Keller stehen vnd ruhen lassen/ Jst wie ein Wermueth-Wein zutrincken. Wann man an statt des Wermueths / soviel Salue nimbt/ Jst solche Purgation wie ein Salue Wein zutrincken. Dauon sol ain/zween oder drey Morgen/wie erst bericht/ nüchtern ein zimbliches Träncklein kalt thun / Vnd vber ein halb oder gantze Stunde hernach/ ein Suppenbrülein niessen/ Aber mit dem andern Essen/ vngefehrlich 4. oder 5. Stundt warten. Wann aber dieses Purgirträncklein Morgents nüchtern nicht khalt trincken kan / Soll zuuor ein Süpplein vnd hierauff solches Träncklein zu sich nemen. An dieser Purgation statt/ mag man die nachfolgende gebrauchen.

Man sol nemen: Senet/ein Loth/
Fenchel oder Enißsamen / ein halb Loth/
Zuckher/zwey Loth/

Alles klein zerstossen/ vnd wol vnter einander gemengt. Dauon sol man auff ain oder zwey mal Anderthalb quintel/ oder ein halb Loth (welches vngefehrlich soviel ist / als man auff zwey maln mit drey Fingern fassen kan) wie es an jhm selber/ Oder inn einem Suppenbrülein Morgents nüchtern einnemen/ vnd vngefehrlich Vier oder Fünff Stundt darauff fasten. Jetz bemelt Purgirpuluer / mag man vnter sechs Loth Rosen oder sonst gemeinen Honig mengen / vnd also ein Lattwergen daraus machen / Vnd dann auff Ein mal soviel nemen/ als vngefehrlich ein Welsche Nuß groß ist/ Doch sol in diesem Puluer der Zuckher außgelassen werden. Wann also gnugsamblich purgirt / Soll als dann hierauff den nechsten Tag nach einander allweg zu Morgents vnd Nachts souil Puluer des Panaceæ / als in Brieflein ist/ in einem löffel vol Wein/ Reinfall/ Meth/ Bier oder Suppenbrülein einnemen/ vnd zu Morgents vngefehrlich 4. Stundt darauff fasten. Wann noch etwas vom Puluer am Löffel hangen verblieb/ Sol man widerumb ain oder zweymal der berürten Tränckher ains nachgiessen/ Damit das Puluer alles sauber in Leyb khomme.

6.9 Form letter from Georg am Wald for patients taking his panacea (here for Landgrave Georg I of Hesse-Darmstadt). With kind permission of the Hessischen Staatsarchiv Darmstadt.

included a series of quotations from classical and contemporary authorities that, he felt, demonstrated "to my enemies, before they mock and chide my panacea," either the imperfection of Galenic medicine or the virtues of metallic or alchemical cures.[118] Crucially, am Wald also added a series of letters from happy patients and practitioners testifying to the successful use of his panacea. While the treatise itself is not quite 90 pages, the appendix, which he called *Testimonials of Various Doctors*, stretched to over 150 pages of evidence in praise of his panacea. Am Wald intended these testimonials as irrefutable evidence that his panacea worked on varying cases of illness. The criticism of his "enemies," he claimed, had forced him to include "letters and missives," in order to rebut the falsehoods that had been spread about his cure.[119]

Am Wald was far from the first healer to use patient testimonials as evidence. The Italian empiric Leonardo Fioravanti had used testimonial letters in a very similar way in his 1570 *Treasury of Human Life* (*Tesoro della vita humana*), although it is questionable whether am Wald would have known this work.[120] His letters also functioned similarly to Berthold's official poison trial testimonials and panegyrics. For all of these authors, the inspiration likely came from existing practices outside of the purview of print. As David Gentilcore has noted, itinerant charlatans and other drug-selling empirics needed some means to vouch for their cures when they traveled outside of their home networks. They thus carried testimonial letters with them as evidence of the worth of their wares. These letters could come from patients, practitioners, hospitals, patrons—any source that was willing to provide a voucher.[121] Fioravanti's innovation, soon followed by Berthold and am Wald, was to display these testimonies to a virtual audience in print. As in the case of charlatans, it was up to the reader to decide whether the letters appeared trustworthy.

To judge by the response to Berthold's poison trials, his letters did inspire trust in his readers. They were documents that had been officially sanctioned for use as evidence, and they described, in detail, contrived experiments conducted by physicians and princes, never Berthold himself. Am Wald was again a different story. His testimonials represented anecdotal accounts drawn from first-person experience. Most of the letters were from the early 1590s, following his invention of the Panacea Amwaldina, but there were a few from the 1570s and 1580s that referred to his terra sigillata instead. It is unclear how many, if any, of the authors had given am Wald permission to publish their testimony, but it appears that most had not. Am Wald essentially acknowledged that lack of consent by pleading, "I hope no one is miffed that I included their letters in print."[122]

"Mineral dragon's blood"

Unsurprisingly, some letter writers were indeed miffed. The physician Andreas Libavius, in particular, was horrified to see his letter to am Wald used as an endorsement of his panacea. Libavius did not deny writing a friendly letter of inquiry in 1591, but he was incensed at its appearance in print, a story that Bruce Moran has told in detail.[123] Just months after his letter first appeared in the expanded 1594 edition of am Wald's *Report*, Libavius published a Latin treatise, *Neoparacelsica*, which excoriated am Wald.[124] The following year, he published an expanded attack on am Wald in German, called *Counter-report on the Panacea Amwaldina*.[125] In both treatises, Libavius pushed the argument that the panacea was a fraud, through and through. The German *Counter-report*, in particular, represented a page-by-page and point-by-point takedown of am Wald's book, starting with the title page. It stretched to 144 pages, almost as long as am Wald's book with all of its testimonials.

Libavius went through the many ailments am Wald claimed to cure one by one and explained why one medicine could not do all he promised. He noted that some of the herbs that am Wald prescribed with the panacea would have worked for the condition anyway, and in one place he literally laughed at am Wald, *"ha, ha, he,"* for prescribing a drink in a "roughly Galenic" manner.[126] He protested am Wald's "false" comparison of the panacea with the philosopher's stone, which, he noted slyly, had not actually been found. He cast aspersions on am Wald's qualifications and nobility, referring to him derisively as "Sir Amwald" (*Junckher Amwaldt*) throughout the book. Particularly concerned by am Wald's use of the word "panacea" to describe his cure, he complained that am Wald had twisted the long-understood meaning of a panacea from an especially effective medicine to a cure-all. In his view, am Wald had created a new classification of drug that was completely invalid.[127]

Toward the end of the book, Libavius addressed the testimonial letters head-on. It was true that some of the letters came from happy patients who might have thought that they felt better after taking the panacea, but patients did not always understand why they improved. Anyway, most of the letters did not support the Panacea Amwaldina in the way am Wald claimed. Many authors (including himself) had changed their minds about the value of the panacea and now found it worthless; many letters were taken out of context; several letters lauded the terra sigillata rather than the panacea; and some of the physicians who wrote were just deluded.[128] Far from proof of the panacea, Libavius claimed, the letters showed that it was "nothing other than mineral dragon's blood."[129] Importantly, Libavius did not invalidate the testimonial as a source on principle. To do would have been rather hypocritical, as his *Neoparacelsica* began with a series of panegyric poems. Instead, he weighed each letter or poem in its own context and found nearly every one of them wanting.

Libavius did not need to pay any attention to am Wald's boasts and testimonials anyway, because he had a different method of proof: chemical analysis. Although he was a staunch Galenist, Libavius was interested in alchemical techniques and wrote what is often considered the first chemistry textbook (called, ironically, *Alchemia*). Johannes Posthius had already analyzed the panacea in 1593 and found it contained sulfur and mercury, contrary to am Wald's statement that it was far too light in weight to contain "common, coarse purges [such as] *Antimonium, Mercurius Præcipitatus*." Posthius recommended that the Nuremburg city council warn its citizens against am Wald's "fraud."[130] Libavius similarly analyzed the substance chemically and determined that it consisted of nothing but sublimated sulfur and mercury (much like cinnabar, or red mercury sulfide), saffron, and powdered mother of pearl. It was not even concocted particularly subtly.[131] At the end of his book, he added a section called "On the trial of the Panacea," in which he instructed his readers how to conduct their own chemical analysis of the powder. He then gave two recipes on how (in his estimation) to make the panacea. He concluded, "I have mostly included this for am Wald's sake, so that he sees that the Galenists have also tried something."[132] The force and length with which Libavius countered am Wald only highlight how much of a threat he perceived in the Panacea Amwaldina. His choice to publish his book first in Latin and then in German points to a recognition that the panacea might prove very appealing to both learned and lay audiences.

Epistolary Authority

Libavius's fear was justified. Whether he liked it or not, testimonials, both written and oral, enticed patients of all walks of life. They figured strongly into am Wald's interaction with his most prominent patient, Landgrave Georg of Hesse-Darmstadt, the same prince who had tested Berthold's terra sigillata on a prisoner condemned to die by the wheel. By 1593, Landgrave Georg had been debilitated by a series of strokes that had caused "a throbbing and screaming in the body" as well as lameness, and he could no longer carry out the duties of governing.[133] In his desperate state, am Wald's panacea promised a chance at regaining the life he had lost.

On the seventh of March, 1594, the ailing Landgrave Georg penned a letter to the "worthy and highly learned" am Wald, who he understood to be a "proper *medicus*." He had become interested in am Wald's panacea in late 1593, when he had heard talk of it at court, and he had asked a courtier to inquire around about it.[134] In his first letter to am Wald, using the royal "we," he noted that "we have heard much about the *Panacea* that you discovered and have also heard it praised by many." Georg felt certain that his medical problems came from the "slime and excess moisture in us, which then resolved itself and fell down into the arm and leg." He asked am Wald to come visit him, so that he could "first take note of our complexion" and then explain how he should take the new medicine.[135] The very next day, he asked another courtier to chase a rumor that an agent of am Wald was selling the Panacea Amwaldina in Frankfurt am Main.[136] All of these references point to a steady level of rumor about the panacea at the small Darmstadt court, all of it seemingly complimentary.

Am Wald had no intention of traveling to Darmstadt to read Georg's complexion, but he also did not want to lose such an illustrious client. On March 13, he replied regretfully that he would not be able to visit His Princely Grace in person, as he had a number of existing obligations. However, he explained, "it is unnecessary for me to be there in person, for the special reason that the *Panacea,* shown to me by God, is not the same as other usual medicines for which one has to regard a number of *Circumstantias*. As long as I know the age, complexion, if a person is fat or thin by nature, the illness, and its length, that is enough. And I have already received enough of a report from the letter Y[our] P[rincely] G[race] sent to me and from his courtly functionary." He promised that the Panacea was completely safe, despite what some "seedy people" might say about it, but that "it depends entirely on the premise that one uses enough of it and *continues* to use it." To further entice Georg to try the Panacea, he sent along a copy of his book, complete with the testimonial letters.[137]

Georg apparently found the explanation convincing, for on March 18 he sent a servant to Thurnhöfen with a letter stating that he would have preferred to have am Wald read his complexion in person, but that he was inclined to try the cure given that "the Panacea you discovered has been praised by so many." The testimonials were quite literally worth their weight in gold. Georg sent along a servant to purchase forty gulden worth of the panacea, which bought him one hundred small packets of the panacea and sixty large ones. On March 23, am Wald sent the medicine along with one of his printed form letters, which listed a purgation that Landgrave Georg should take before he tried the medicine (see fig. 6.9). At the bottom of the printed document, he wrote, "If afterward this cure, against hopes, does not bring proper relief, then YPG should use two large or ten small envelopes of Panacea in the aforementioned manner. And this dose should cause no hesitation, for 20, 30, 40, and even more *grana* can be used at one time without any harm or detriment, which is an advantage it has over common medicines, with which one cannot easily give one single *grana* too much." He suggested that Georg abstain from "crass, difficult to digest, very hot and fragrant foods, and also from too strong wine, from all excess in eating and drinking, colds, anger, displeasure and sadness," trying his best to sound like a proper Galenic physician.[138]

The servant must have ridden back to Darmstadt quickly, as Georg began taking the panacea just three days later, on March 26. On April 2, he wrote that he had felt "no particular improvement" in his symptoms. How long, he asked, should he double the dose without concern?[139] Am Wald responded that he should have no fears about the lack of improvement, for "the cure happens faster for one person than for another." Indeed, am Wald stated, Landgrave Georg could read "in the people's testimonials . . . that have appeared in print" that some people were cured in two or three days, some people in eight to fourteen, and for some people it took three or four weeks.[140] The printed letters, he argued, showed Georg that he was not alone. Georg's response, dated two weeks later on April 18, noted that indeed, his illness had been with him a long time, so it was not surprising that it might take a long time for his condition to change. He still felt no improvement, but since he also felt no worse, he would continue with the Panacea for the moment. He thus asked for 20 gulden more of the panacea, and he said that if it did end up working on him, he intended to buy 200 gulden worth and distribute it to "our poor people, who cannot pay for it themselves."[141] The testimonials continued to convince Georg to keep on with the unusual cure. Although no further letters have survived, someone recorded all of the doses Georg took, beginning "when my dear lord began to use the panaziha," from March 26 to May 30 (see fig. 6.10). For the vast majority of that time period, he took at least one dose a day, usually two, and sometimes three. Only toward the end did his doses begin to peter out.

It is clear from this correspondence that am Wald felt the patient testimonial letters would help support his case, and he appears to have been correct. While we cannot know for sure what convinced Landgrave Georg to continue with the Panacea Amwaldina for such a long period of time, he expressly gave credence to the testimonials of others for whom the cure had taken some time to take effect. As in the case of online medical forums today, the many examples of patients experiencing similar complaints—and feeling better after taking the Panacea—appears to have been persuasive.

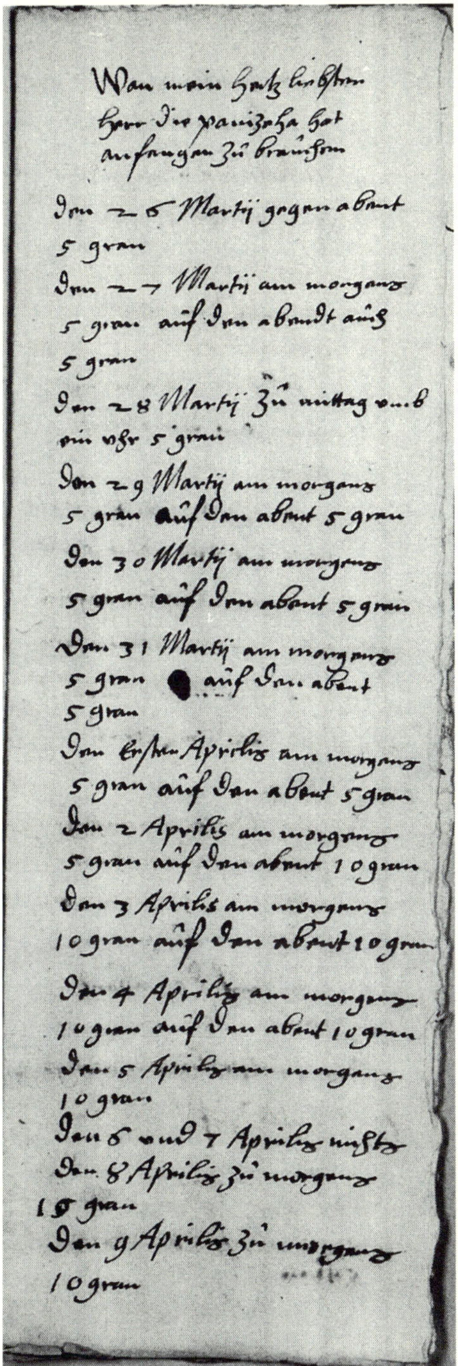

Wan mein hertz liebster
herr die panizeha hat
anfangen zu brauchen

den 26 Martij gegen abent
5 gran
den 27 Martij am morgens
5 gran auf den abendt auch
5 gran
den 28 Martij zu mittag umb
ein uhr 5 gran
den 29 Martij am morgens
5 gran auf den abent 5 gran
den 30 Martij am morgens
5 gran auf den abent 5 gran
den 31 Martij am morgens
5 gran auf den abent
5 gran
den ersten Aprilis am morgens
5 gran auf den abent 5 gran
den 2 Aprilis am morgens
5 gran auf den abent 10 gran
den 3 Aprilis am morgens
10 gran auf den abent 10 gran
den 4 Aprilis am morgens
10 gran auf den abent 10 gran
den 5 Aprilis am morgens
10 gran
den 6 und 7 Aprilis nichts
den 8 Aprilis zu morgens
10 gran
den 9 Aprilis zu morgens
10 gran

6.10 List of Landgrave Georg I's panacea doses. With kind permission of the Hessischen Staatsarchiv Darmstadt.

The Problem of Proof

The cases of Berthold and am Wald highlight the utter confusion that everyone faced in making sense of both drugs and medical authority in the sixteenth century. Berthold, who never pretended he was a physician, gained a certain level of respect around Germany with his Latin texts and his seemingly irrefutable poison trials. Yet in the end, he was a victim of his own success. He could never escape the connection with Montanus's Silesian terra sigillata, even as tensions with Montanus mounted. By 1590, he had been kicked out of the Strigan business altogether. He was thus left with the unenviable task of trying to undo all of the evidence he had amassed in favor of the drug. In a short, German-language book dedicated to Emperor Rudolf II in 1594, he attempted to recast the Silesian terra sigillata as his drug alone. *He* had found it outside of Striga, and *he* had spent years and enormous sums developing the proper "preparation, separation, and extraction" of the raw material. No one would have known about the Silesian terra sigillata without his arduous travels through western Germany in the 1580s, he complained, but the ungrateful Strigans were nevertheless "taking the bread out of my mouth" by kicking him out of the business.[142]

Berthold attempted to use the concept of the alchemical arcanum to claim the Silesian terra sigillata as his own. Even though the seal may look the same as always, he stated, the terra sigillata from Striga was no longer valid, as the people who had forced him out could muster only a "clumsy, futile, and careless" preparation.[143] He once again mentioned the poison trials, but he insisted that they proved the efficacy not of the terra sigillata clay, but rather of his own process of preparation—his Arcanum. In the meantime, he had found a new earth in the Vogelsberg mountains in Hesse with even greater powers. *That* terra sigillata was what the poison trials now confirmed.[144] It was a difficult argument to make, and Berthold's short book suggests he had already fallen on hard times. The copy of his book in the Saxony State Library includes a prefacing letter to Elector Christian I of Saxony, in which Berthold advertised drinking vessels he had made with terra sigillata, which suggests he was searching for new lines of income. There is scant evidence of his whereabouts after 1594, but he eventually found his way to the town of Freiburg in his native Saxony. On April 14, 1610, he died at the age of eighty in the Freiburg hospital. A chronicle of the city noted that he was a "diligent seeker of natural things" but did not mention the terra sigillata. The city council sold his books and "mathematical instruments" after his death.[145] The fact that he died in the public hospital and left his effects to the city suggests he had no heirs or close friends to manage his affairs. His name was mostly forgotten.

In contrast, Montanus, who had died in 1604, continued to be celebrated for his Silesian terra sigillata. A legend later spread that Montanus's hair had gone from gray to black in his old age, owing to his success at finding a youth-giving elixir.[146] The poison trials that Berthold had conducted so diligently benefited Montanus in the end. The Silesian terra sigillata remained a widely respected remedy, eventually becoming ensconced in the European pharmacopoeia. It appeared in the 1618 *Pharmacopoeia Londonensis,* among various other kinds of earth. In Nuremberg in 1704, you could buy it for ten kreutzer to the ounce, among the pricier drugs but far from a luxury.[147] It continued to be a major industry for the town of Striga, until an earthquake collapsed the mine in 1883.[148] In 2001, the city of Strzegom dedicated a plaque in Montanus's honor, complete with images of the Silesian terra sigillata (see fig. 6.11). Berthold's poison trials were enormously influential in helping establish the Silesian terra sigillata as a valid drug among learned physicians, but once accepted and tied to Montanus, the drug lived on without him.

Am Wald, for his part, continued to enjoy great success with his Panacea. The dispute with Libavius served only to call greater attention to his cure, and it continued for several years.[149] Despite Libavius's chemical proof that (in Libavius's opinion) the Panacea Amwaldina was a fraud, enough people still bought it to keep am Wald reasonably wealthy for the rest of his life. He died in 1616 in his Thürnhofen estate.[150] Although the interest in the Panacea Amwaldina slowly petered out after his death, his model proved enticing. From the early seventeenth century, a host of alchemical "panaceas" began appearing around Europe, most prominently the *Panacea Aurea* by the English chemical physician Francis Anthony (1618) and German polymath Johan Rudolf Glauber's "Panacea or Universal Mineral Medicine" (1651).[151] The panacea became a model alchemical cure, and a thorn in the side of many physicians, who continued to fight the idea of a universal cure.[152]

This chapter could end with a point about the increasing division between physicians and laypeople in their understanding of trustworthy evidence. I could point to Libavius's chemical analysis of the Panacea Amwaldina as a sign that physicians had begun to go in a more "scientific" direction, while the pseudo-physician am Wald and his followers clung

6.11 Plaque in Strzegom honoring Johannes Montanus. Photo by Bonio. Wikimedia Commons.

to their panaceas and their dubious letters of support. But that would misrepresent the situation. Despite Libavius's public denunciation of am Wald, privately he remained intrigued by the Panacea Amwaldina. As Joel Klein has noted, Libavius believed that am Wald had initially found an effective medicine, which he had received from an Italian nobleman during his time in Padua. When that ran out, in Libavius's estimation, he did not know how to make more. Libavius spent considerable effort trying to reconstruct this original panacea.[153] In these efforts he was typical of the physicians we have seen throughout this book. Despite their professed rejection of empirics and their methods, Galenic physicians remained enticed by the promise of wonder cures.

A derivative of the poison antidote terra sigillata, the Panacea Amwaldina demonstrated what could happen once poison no longer remained the focus. As such, it brought questions of evidence and proof to a head. As we have seen, physicians conducted poison trials to test the efficacy of antidote-cure-alls and circulated evidence of these tests. They also, as we saw in chapter 5, provided their own testimonials of the success of wonder cures in cases of illness. Am Wald's patient testimonials resembled these forms of evidence enough to muddy the waters considerably. Although Libavius strongly denounced am Wald's testimonials as false evidence, other physicians were less resistant, and even Libavius remained enticed by the Panacea. As in the case of wonder drugs from the Levant and the New World, physicians did not reject wonder cures out of hand, even ones that sounded far-fetched. They, too, struggled with what counted as proof.

Testing and Testimony: Orviétan

In 1680, an empiric named Giuseppe Toscano applied for a license to sell a poison antidote and cure-all known as orviétan throughout the Holy Roman Empire.[1] To prove its worth, he planned to hold a public poison trial in Prague on April 8 of that year. A printed advertisement for the trial, addressed to "honored lords," promised a "wondrously rare experience" in which Toscano would drink various poisons and then take the orviétan. He invited everyone to bring the deadliest poison they could think of for him to try. Additionally, if anyone else had a better remedy for poison and could come prove it on two animals, he would pay them one hundred gold duplones.[2] Toscano submitted the small, rectangular advertisement for this trial as part of the documentation for his license approval (see fig. 7.1).

One might assume that seventeenth-century physicians would dismiss this sort of evidence as outdated quackery, but poison trials formed an important part of Toscano's overall case. His application consisted mainly of testimonials from princes and physicians certifying the drug's worth. A long, handwritten letter from physician Nicolas Franchemont on behalf of the college of physicians of Frankenfelt specifically listed a public poison trial (*proba publice*) as part of its evidence.[3] A printed certificate from King Charles II of England, which listed the many places where Toscano had received a license since 1668, mentioned that he had tested orviétan on pigeons.[4] Along with this evidence drawn from poison trials, several witnesses described its effective use on

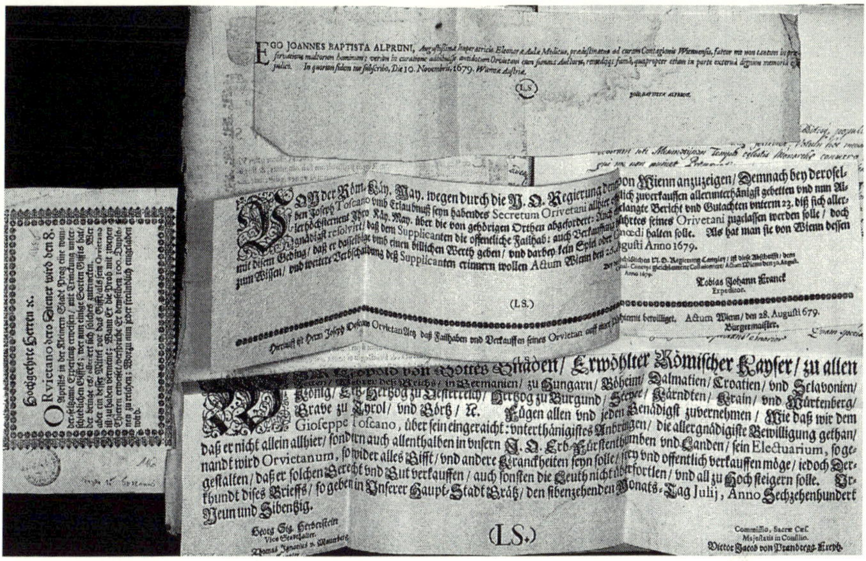

7.1 Testimonials from Jacopo Toscano's orviétan license application, 1681. The leftmost document advertises his poison trial. The three other documents are testimonials from respected individuals. With kind permission of the Austrian State Archive, Vienna.

patients in practice. Gianbattista Alpruni, personal physician to the Empress Eleonora Magdalena, claimed that he had cured "many people" with the orviétan. On April 17, 1680, Emperor Leopold I agreed to an official license for the drug. His letter approving the license noted both that it had been proven publicly in a poison trial and that it had been shown to be useful on patients in practice.[5]

Poison trials lived on for centuries in these special drug licenses, an idiosyncratic convention across Europe that stood outside of the usual process for approving drugs.[6] Public tests were particularly common in the case of orviétan, a derivative of theriac that originally appeared in Italy in the late sixteenth century and quickly gained renown as a cheaper alternative. As a number of historians have noted, its popularity prompted fierce battles for licenses among empirical practitioners across Europe, particularly in France.[7] Orviétan became associated with dramatic marketplace demonstrations that included self-poisoning and, often, the poisoning of various animals. In the 1680s, after Toscano had returned to continental Europe, a German named Cornelius Tilburg famously impressed King Charles II with a self-poisoning demonstration. He not only received the orviétan license in England but also was appointed as a royal physician.[8]

These marketplace shows remind us of Pietro Andrea Mattioli's long diatribe against charlatans hawking their fraudulent poison antidotes, and like Mattioli, seventeenth-century physicians derided orviétan and the showmanship around it. Yet a number of physicians, particularly those working at princely courts or on town licensing boards, participated in the approval process, and they relied on both poison trials and testimonials as markers of validation.[9] In 1735, the application of Joseph Heinrich Thelenius to sell his orviétan in the Holy Roman Empire included a printed notification of a poison trial, to which he specifically invited all of the city's physicians and apothecaries. He also presented official printed certification from the medical faculty of Württemberg, as well as dozens of patient testimonials, all sealed with wax and notarized.[10] A number of applications, as well, promised alchemical universal medicines. A Viennese merchant named Johann Michael Dietmann applied for and received permission to sell a "panacea and bezoardic tincture" in the Holy Roman Empire, particularly Vienna. His file contained a number of "attestations" from physicians and patients.[11] A "distiller and chymicus" from Nuremberg named Nikolaus Bäumler included not only patient testimonials (sealed with wax) but also a one-page printed advertisement for his "universal balsam."[12] The varying forms of evidence we have seen throughout this book—poison trials, anecdotes from clinical practice, and patient testimonials—all became codified as part of these special license applications, with the input of city and princely physicians (despite their continued rhetoric deriding empirical practitioners).[13]

The connection between poison antidotes and alchemical panaceas discussed in chapter 6 also continued in publications by chymical physicians. Angelo Sala (1576–1637), an Italian autodidact raised in Geneva, published a book in 1616 advertising three alchemical antidote-cure-alls, called *Ternarius bezoarticorum, or Three Sovereign Bezoardic Medicines*.[14] Sala, who became known for his alchemical cures and his theories about chemical composition, invoked the long tradition of poison antidotes on his title page, which included engravings of "King Mithridates" and "the physician Andromachus." He juxtaposed these images with an engraving depicting a man at his still, merging the old traditions of poison antidotes with the new chymical medicine (see fig. 7.2).[15] He addressed this theme even more explicitly in his introduction. Mithridates, he noted, had developed his marvelous antidote by testing it on "malefactors who had deserved death for their crimes." Meanwhile, Andromachus, Galen, and many other ancient princes and physicians had developed many useful antidotes through experience. "In our time," he continued, new poison

7.2 The engraved title page to Angelo Sala's *Ternarius bezoarticorum*, 1618. Mithridates and Andromachus frame the scene, with an alchemical laboratory below. Courtesy of the Harvard Medical Library in the Francis A. Countway Library of Medicine. Photo by Jessica Murphy.

antidotes had been discovered by "men learned in medicine, who traveled the world . . . and also, today, [through] the great princes of Europe, who delight in the noble, speculative, and natural art of chymistry."[16] Sala thus hearkened back to Mithridates' ancient poison trials, while also invoking the early modern drive for both exotic antidotes like bezoar and new alchemical cures.

Sala's own bezoardics, he claimed, were the culmination of these varying traditions. He placed poison, plague, and poisonous illnesses at the top of the list of ailments they cured, but the full catalogue stretched to twenty-seven different complaints.[17] His first and most powerful bezoardic, called Diasolis, drew on the power of the sun and moon. It could comfort the heart, fortify the natural heat, restore vital forces, open obstructions, resist corruption, and expel the malignancy of poisons.[18] Unlike Georg am Wald, however, he claimed he did not intend to suggest that his drugs had an "absolute property" that could cure "every single malady without the help of other remedies and medical interventions." He contrasted his own position with that of "many alchemists" searching in vain "for their unknown and invisible philosopher's stone."[19] Sala thus positioned himself at the crossroads of Galenic and alchemical medicine, stopping short of a universal cure while also highlighting the wide range of diseases that his antidotes could cure. His approach diverged from that of the influential Dutch chymist Jan Baptist van Helmont, his near contemporary, who explicitly advocated a universal medicine with an underlying theory and who also drew on the concept of poisons.[20] In a range of ways, the story of poison antidotes, wonder drugs, and alchemical cures remained intertwined.

And what of learned physicians? Did their interest in poison extend beyond approvals of special drug licenses? Extensive accounts of poison trials on condemned criminals disappeared by the late sixteenth century, and the focus of medical experimentation overall shifted from humans to animals. William Harvey's discovery of the circulation of the blood was based on extensive experiments on fish and frogs, and the excitement surrounding Harvey's new physiology encouraged further anatomical experiments on animals.[21] Intensive (and often gruesome) animal experimentation became a hallmark of seventeenth-century medical investigation, with dogs as the most common research animal.[22] Sometimes the interest in poison and physiology overlapped. In the 1650s, members of the Royal Society of London, including John Wilkins, Christopher Wren, and Robert Boyle, injected a solution of opium into the veins of a large dog in order to test the intravenous transfer of poisonous substances. Boyle reported that the "tortured dogs violent strugglings" ceased once the opium took effect, and it immediately collapsed. Eventually, it survived and "grew fat," and Boyle noted with regret that it was later stolen from him.[23]

Boyle viewed this experiment (and others like it) in the context of research on poison, including the question of whether experimental results on animals also pertained to humans—a problem that continued to vex researchers. In a discussion of poisonous diseases, he noted that most poisons affected "both Man and brutes" in similar ways. However, he stated, there were some differences, and "the liberty of exhibiting" poisons on animals, "when, and in what manner we please . . . (which we dare not do to [humans]), allows us great opportunities of observing their manner and investigating their Nature." For example, he noted, dogs could withstand a far larger quantity of opium than humans, "as we have more than once given to a Dog, without much harming him, such a quantity of Opium, as would probably have suffic'd to have kill'd severall Men." When they tried the same experiment on a cat, it was driven mad, "so her Keeper was fain to kill her."[24] Boyle did not, however, explain how this knowledge would help solve the problem of poisonous diseases. His interests centered on the action of poison, not the search for a cure.

Boyle's inquiries took place within a Europe-wide surge of interest in poison as a subject of investigation in its own right, focused on animals. As mentioned in the introduction, the Medici physician Francesco Redi conducted a series of prominent experiments on viper venom and snakestone (an exotic antidote) in the 1660s. Redi's disputes with other researchers about proper experimental methodologies, particularly with French royal apothecary Moyse Charas and German Jesuit Athanasius Kircher, led to a more extensive discussion of how one should conduct an experiment than we saw in the case of sixteenth-century poison trials. Redi advocated repetition far beyond the scale we have seen previously, and he also disproved many long-standing ideas about vipers with his experiments.[25] Yet the overall context of his experimental endeavors contained echoes of the past. Redi was a court physician, and the snakestone he tested (and rejected) was an exotic import that represented a commercial threat to local theriac. He described his experiments as a spectacle, much as his sixteenth-century predecessors had done, and he too provided extensive description of his experiments using poison on animals.[26] Although he criticized medical authorities from Galen to Mattioli, he also drew on that precedent when it suited him—including, as it happened, the rooster experiments in *Theriac to Piso*.[27] The ancient precedent of poison trials, introduced in chapter 1, cast long shadows.

A contemporary of Redi's, the Swiss physician Johann Jakob Wepfer (1620–95), conducted expansive investigations into the effects of various poisons in the 1670s, in ways that invoked past precedents. Wepfer began a study of toxicity after eight children were accidentally poisoned with water hemlock in the village of Almanshofen in Baden. They all suffered terrible pains in the stomach and intestines, five of them came close to death, and the two who had eaten the most hemlock died. Wepfer visited the village and carefully recorded each child's case history based on the testimony of the parents, siblings, and surviving children. His eventual study on water hemlock included an extensive description of the symptoms and progression of the poisoning in each case, much like the careful records of scholarly poison trials that we saw in chapters 3 and 5. Wepfer added an additional step, however, by describing the autopsies of the children who had died.[28]

Wepfer also conducted his own contrived experiments with poisons. As in the sixteenth-century poison trials, he received ample support from his aristocratic patrons, who remained eager to uncover the secrets of poison and its cures. In lieu of criminals, they sent him animals to use as test subjects, particularly dogs, but also cats, birds, and, in one case, a young wolf.[29] Like several of the authors described in chapter 3, Wepfer recounted each experiment in excruciating detail, including each incidence of the animal's vomiting, defecation, or cramping, as well as the elapsed time at which every symptom took place. Some of these experiments were downright grisly: he gave varying doses of poisons and then, as the animals were dying, vivisected them to view the action of poison. On July 14, 1676, at two in the afternoon, for example, he gave a half dram of aconite, in milk, to a two-week-old puppy. I will spare readers the gruesome description of the symptoms and vivisection that followed, but suffice it to say, Wepfer's account was comprehensive.[30] As in the case of the poisoned children, he added a focus on anatomy and physiology that did not appear in sixteenth-century poison trial accounts.

Were the similarities between mid-sixteenth- and late seventeenth-century poison experiments coincidental, or were there direct ties? Drawing firm links over such a long period of time is always difficult, especially given the vastly different experimental landscape by the later seventeenth century.[31] Mattioli's desire to create a legitimate category of learned experiment was shared by many others, and it became mainstream in the 1600s. Redi explicitly criticized Mattioli in several places, suggesting that his status as an authority had faded. At the same time, the conventions of sixteenth-century poison experiments were known to their seventeenth-century successors. Both Wepfer and Redi read and cited Mattioli, even if they disagreed with him. In one striking section of Wepfer's book on poisons, moreover, he related a detailed description of Mattioli's Prague poison trial and used it as evidence of the symptoms of aconite poisoning in humans.[32] His focus on this particular example from Mattioli demonstrates an awareness both of the existence of sixteenth-century poison trials and of the detailed method of reporting them.

Wepfer did not, however, mention that Mattioli's aconite trial was a contrived test on a criminal condemned to death. Instead, he noted only that Mattioli observed the effects of aconite on a "soldier," leaving the reader to assume that the poisoning was accidental.[33] This omission highlights that quite a good deal had changed in many respects. In particular, conventions had shifted away from using condemned criminals for fatal tests. The hesitations that we saw in chapter 4 had only increased, and the rise in animal experimentation had rendered the use of humans even less palatable. Wepfer specifically addressed this issue in a response to criticism of his torturous use of animals as test subjects.[34] Some people objected to his experimentation on animals as cruel or barbarous, he complained. However, he had read that imperial physicians of the past, like Mattioli, had used condemned criminals to test poison. He, in contrast, preferred to study "poisons on animals, in order to help humans" rather than harm them.[35] While questions about the transferability between humans and animals remained, poison trials on condemned criminals could now be used as a foil. At the same time, Mattioli's poison trials remained compelling enough that Wepfer used a sanitized version as evidence.

By the seventeenth century, contrived experiments had become a standard and unquestioned method to study natural processes. Nevertheless, the association between physicians' poison experiments and charlatans' marketplace shows continued to lurk in the background. One series of Redi's experiments demonstrated that viper venom was poisonous when applied externally but harmless when taken internally. This division, Redi argued, explained why charlatans could ingest scorpions or viper heads and drink buckets of their venom without harm; they would have been fine even without their so-called antidotes. To demonstrate this point, Redi used a charlatan, a man named Jacopo Sozzi, in a dramatic show in front of the Accademia del Cimento.[36] Wepfer, for his part, told a story of a Neapolitan empiric named Agyrta, who had invited him to act as a witness in a self-poisoning demonstration to prove his orviétan—presumably to obtain a testimonial from Wepfer for licensing purposes. Knowing the fraudulent practices of orviétan hawkers, Wepfer declined, and he added a long story about the man's deceitful practices.[37] The fact that both Redi and Wepfer addressed charlatans' marketplace tests alongside their own poison experiments suggests that the comparison remained a concern.

Despite Wepfer's characterization of human testing as a thing of the past, moreover, isolated incidences of giving poison to humans continued. An Italian surgeon named Johannes Puntaeus had one of his servants poisoned when demonstrating orviétan to the physicians of Oxford in the 1650s (the servant became very ill but survived).[38] By his own account, Boyle once paid a man to be bitten by vipers in order to test his theory that a hot iron held near the wound would draw out the poison. While the iron did not seem to reduce the swollen viper bite, the swelling also did not spread, and the man, "glad of his mony, without further Ceremony, went about his affairs." The man later reported that he had "got mony by repeating the Experiment" several times.[39] In French Saint-Domingue in the 1720s, royal physicians received permission to test poisons and antidotes on three slaves sentenced to death for poisoning.[40] Despite these isolated examples, poison trials on humans were rare enough that most authors viewed them as a thing of the past.

In contrast, tests of promising cures in cases of illness, as well as other, more invasive medical procedures, expanded in the seventeenth and (especially) the eighteenth centuries. Increasingly, tests using human subjects focused especially on large, vulnerable populations like hospital patients, orphans, soldiers, and slaves.[41] Prisoners, as well, continued to represent attractive test subjects. Famously, six prisoners in London's Newgate Prison were used to test inoculation against smallpox in 1722. The respected physician Hans Sloane, who had experience testing poisons on dogs, oversaw the experiments, which James Delbourgo has described as "rare trials on human subjects." Derivative inoculation experiments later took place on orphans in the Foundling Hospital and on African slaves.[42] Although these tests were dangerous and invasive, they had a potential health benefit to the test subjects. Sloane's poison experiments took place on dogs.

Although human poison trials all but disappeared, concern about poison did not wane, and neither did the drive to find effective antidotes. The special drug licenses for fantastic-sounding wonder drugs, many of them antidotes, reached their apogee in the eighteenth century, suggesting an ongoing demand.[43] Similarly, the fascination with exotic antidotes from afar continued. Sloane's famous collection included many antidotes, including various bezoars, from both India and the Americas.

Sloane drew on the testimony of trusted connections across the world to assess their efficacy.[44] Poison also continued to be a focus in scientific circles. Andreas Holger Maehle has shown that poison testing of one kind or another represented the majority of medical experiments in eighteenth-century scientific journals.[45] The questions of experiment, authority, authenticity, and ethics that this book has raised began long before the days of Clement VII, Mattioli, Berthold, and am Wald, and they continued long past them. Yet in the Renaissance revival of poison trials, these questions came into focus in a particularly potent way. The promise of wonder drugs drove investigations into how to test them.

I began this book with the story of Caravita's oil, the marvelous antidote against poison and plague that sparked the Renaissance revival of ancient poison trials. In the end, human poison trials were a limited phenomenon, but they nevertheless reveal a depth of thought about experimental procedures and medical authority that has not previously been recognized. Poison's peculiar ability to harm allowed for the possibility of a contrived trial, an opportunity that had been recognized since ancient times. In the sixteenth century, the long-standing tradition of verifying theriac by testing it on poisoned animals expanded into a drive to test new antidotes on human subjects, to carefully record the results, and to share those records with other learned individuals. Evidence taken from "experience" and "experiment" became ever more integrated into many areas of scholarly science and medicine in the sixteenth century. Poison trials represent only one aspect of that shift. Nevertheless, they provided the first in-depth attempts to develop protocols for contrived trials, and they did so as an explicit contrast to the well-known marketplace poison experiments by mountebanks and other empirical practitioners. The resulting protocols set up a model for how to conduct a scholarly contrived experiment, with a focus on extensive descriptions of the symptoms of poisoning and the response to the antidote, often with specific time markers.

Alongside experimental protocol, Renaissance poison trials addressed questions of transferability between humans and animals, and in so doing confronted challenges to the use of humans, even those already condemned to die, for fatal tests. Poison trials on condemned criminals remained limited in part because it was not easy to remove a condemned criminal from the usual cultural and religious rituals that surrounded the process of execution. Only powerful monarchs could manage such a feat, and human poison trials thus were tied up in projections of princely authority. It was no accident that all of the trials on criminals tested valuable, exotic imports or promising new, local drugs. Outside of the princely courts, the effects of these wonder drugs on humans had to be observed in the course of regular clinical experience in cases of disease or (more rarely) accidental poisoning. The historic conception of antidotes as near cure-alls intensified throughout the sixteenth century, making it easier to use the remarkable recoveries of sick patients as evidence. Quick-and-dirty marketplace poison trials on animals remained a common device for demonstrating authenticity and thereby allaying the fears of fraud connected to exotic imports, but the time-honored idea still held that drugs were best tested on humans. Georg am Wald introduced both his poison antidote and his panacea by denigrating the notion of testing poison on "dumb animals." His patient testimonials, he argued, gave a superior sense of the remedy's worth.

In the seventeenth century, the advantages of using animals to test toxic substances overcame the concerns about transferability between animals and humans. Landgrave Wilhelm IV's extensive dog trials presaged Redi's and Wepfer's later poison experiments, which focused more on questions about toxicity than on whether or not an antidote worked. Even so, seventeenth- and eighteenth-century poison experiments and special drug licenses, both led by learned physicians, engaged with the same questions of experimental protocol, the use of humans and animals in experiment, and the yearning for effective drugs that could cure both poison and disease. Renaissance poison trials intensified a set of conversations about experiment, authority, and ethics that had begun in ancient times and continued into the eighteenth century and beyond.[46] For the most part, physicians solidified their authority over other medical practitioners during that time period.[47] Yet the optimistic hope for a wonder cure collapsed professional boundaries and forced knotty questions about the parameters of drug testing. As substances imbued with deeper medical and cultural meanings, poison antidotes and their related panaceas proved the catalyst for an enduring debate about the meaning of testimony, evidence, and proof.

Acknowledgments

This book began in a castle. While researching my first book in January 2006, I spent a week at the Hohenlohe Zentralarchiv, housed in a glorious Renaissance palace in the tiny southwest German town of Neuenstein. In the midst of helping me with my research on noblewomen's medicinal recipes, the archivist, Wilifried Beutner, suggested that I take a look at "another very interesting file" on sixteenth-century medical history. In the file was a treasure trove of fascinating documents describing the poison trial now at the center of chapter 4 of this book. I immediately knew that I had the basis for a new research project. I owe my heartfelt thanks first and foremost to Herr Beutner, without whom this book would not have existed, and to Katharine Park and Bruce Moran, who helped me realize that poison trials were a phenomenon well beyond Germany. I also warmly thank my editor at the University of Chicago Press, Karen Darling, for her encouragement and interest from a very early stage of this project and for cajoling me to finish it.

I could not have completed the book without significant research travel, and I am grateful for the funding I received from Tufts University; Trinity College, Cambridge; an ACLS Ryskamp Fellowship; and two Visiting Researcher fellowships at the Max Planck Institute for the History of Science in Berlin.

As always, conversations with and suggestions from colleagues helped this book take shape. Special thanks to Elaine Leong, Elly Truitt, Debby Levine, and Jeremy Greene, who provided enthusiastic support (and friendship) throughout this project and stepped in at the last minute to read drafts. Brad Bouley generously sent me scans of the poison trial in the Strozzi archive, and William Eamon sent me his notes from the Venetian archives in addition to giving me advice on panaceas. Tara Nummedal was the source of many helpful conversations about poison in the German-speaking world, and Hannah Marcus helped me with Italian paleography and gave many excellent suggestions. A huge thanks also to the many scholars who gave me feedback on portions of this material, including Lauren Kassell, Larry Principe, Michael McVaugh, Emma Spary, Justin Rivest, Joel Klein, Erik Heinrichs, Montserrat Cabré, Andrew Sparling, and Heidi Hausse. I was fortunate to have had the opportunity to present talks and chapter drafts from various stages of this book to numerous conferences, seminars, and colloquia in the United States, Canada, the United Kingdom, and continental Europe, and I thank the audiences and organizers for their sage input and perceptive questions.

I am extraordinarily lucky in my colleagues at Tufts, both within the History Department and in the wider university. The History Department's Works in Progress writing group commented on early drafts of most chapters. I am grateful to Elizabeth Foster for organizing the group and hosting us in her home, and to the regular attendees, including Beatrice Manz, Kendra Field, Rachel Appelbaum, James Rice, Hugh Roberts, and Gary Leupp. I also shared chapters with my medical humanities writing group, Sarah Pinto and Rosemary Taylor, and I thank them for the very helpful perspective of nonhistorians. Riccardo Strobino spent an afternoon helping me decipher the Guido da Vigevano manuscript, a very generous gift of time and enthusiasm. Thanks to Jeanne Penvenne and Bárbara Brizuela for their warm support and advice along the way, and a special thanks to Elizabeth Foster and David Ekbladh, whose friendship helped keep me sane over the long path to publication. Finally, the students I taught in many Tufts classes over the past decade helped me think through many of the ideas in this book.

Like most books, mine depended heavily on the help of archivists, librarians, and museum directors. The wonderful support staff at the Hohenlohe Zentralarchiv Neuenstein answered my many questions and provided afternoon Kaffee and Kuchen on a return visit to the archive. Special thanks to Jan Weichert and Helmut Wörner, who spent a day helping me understand the world of Wendel Tümler and Zacharius Hyso. Martin Kluge at the Basel Pharmacy Museum generously gave me a tour of the museum's terra sigillata and other antidote specimens, helpfully answered my questions, and provided some bibliography. I am also grateful to the staff at the Österreichische Nationalbibliothek, the

Österreichische Staatsarchiv, the Hessische Staatsarchiv Marburg, the Hessische Staatsarchiv Darmstadt, the Bibliothèque nationale de France, the Bibliothèque de l'Arsenal, Houghton Library, the Harvard Botany Libraries, and the Countway Library of Medicine. Special thanks to Jessica Murphy at Countway, who came to the rescue with last-minute images.

The University of Chicago Press has been an absolute dream to work with. Warm thanks to Tristan Bates for answering my many questions so patiently and helpfully and to Susan Tarcov, who skillfully edited the manuscript. Thanks also to the two anonymous readers for engaging so constructively with my book manuscript and giving such useful suggestions.

Friends and family outside academia helped make this book happen as well. Leah Ekbladh and Shira Rosan stepped in to help with childcare more times than I can count. My parents, stepparents, and in-laws always had welcome encouragement about the book, and I deeply regret that my mother-in-law, Linda Kuczwara, did not live to see its completion. Thanks especially to my mother, Marjorie Gormley, who helped with childcare during my archival trips; to my talented sister, Sarah Rankin, who made the beautiful map in chapter 4; and to my other talented sister, Kathleen Rankin, who provided friendship and support.

Last but not least, I could not have written this book without the help of John Kuczwara, who shared my fascination with the topic from the beginning of the project and gave me the amazing trifecta of editorial help, moral support, and childcare throughout. This book has accompanied my children, Nico and Eli, through their entire lives, and they endured my travel (more or less) cheerfully, asked questions about my research, and even created their own "poisons" using food coloring. My family was an antidote to my many years of hard work, and this book is dedicated to the three of them with deepest love and gratitude.

Notes

INTRODUCTION

1. Giovanni Battista degli Abbati ("l'Abbatino") to Federico Gonzaga, August 4, 1524, in Giovio, *Lettere inedite*, 47–48.
2. L'Abbatino to Gonzaga, in Giovio, *Lettere inedite*, 48.
3. *Testimonium*, A1r–A4r. Unless otherwise noted, all translations are my own.
4. *Testimonium*, A1v–4r.
5. See, for example, T. C. P. Zimmermann, *Paolo Giovio*, 16.
6. Mattioli, *Di Pedacio Dioscoride Anazarbeo libri cinque*, bk. 4, chap. 80.
7. On materiality and the Renaissance, see Findlen, "Possessing the Past."
8. "In questa nostra età floridissima." Mattioli, *Il Dioscoride*, bk. 6, 3v.
9. On medical pluralism in early modern Europe, see especially Lindemann, *Medicine and Society*; Gentilcore, *Healers and Healing*; Park, *Doctors and Medicine*; Pelling, *Medical Conflicts in Early Modern London*; Marland and Pelling, *Task of Healing*.
10. Stolberg, "Learning from the Common Folks."
11. Some of these ideas have been developed in Rankin, "On Anecdote and Antidotes"; Leong and Rankin, "Testing Drugs and Trying Cures"; Rankin and Rivest, "Medicine, Monopoly, and the Premodern State."
12. Findlen, *Possessing Nature*; Ragland, "Making Trials"; Cook, *Decline of the Old Medical Regime*; Cook, "Sir John Colbatch and Augustan Medicine"; Cook, "Victories for Empiricism"; McVaugh, "Experience-Based Medicine."
13. For more on this point, see Leong and Rankin, "Testing Drugs and Trying Cures," 163–65. On mathematics and experiment, see Dear, *Discipline and Experience*; Dear,

"Meanings of Experience"; Schmitt, "Experience and Experiment." For the efforts to demonstrate the influence of alchemy, see Newman and Principe, *Alchemy Tried in the Fire*; Principe, *Aspiring Adept*; Principe, *Secrets of Alchemy*; Newman, *Promethean Ambitions*; Newman, *Newton the Alchemist*; Nummedal, *Alchemy and Authority in the Holy Roman Empire*.

14. Park, "Country Medicine"; Gentilcore, *Medical Charlatanism*, 174–80.
15. Smith, *Body of the Artisan*; Newman and Principe, *Alchemy Tried in the Fire*; Nummedal, *Alchemy and Authority in the Holy Roman Empire*; Smith, Meyers, and Cook, *Ways of Making and Knowing*; Gentilcore, *Medical Charlatanism;* Eamon, *Professor of Secrets;* Rankin, *Panaceia's Daughters*; Leong, *Recipes and Everyday Knowledge*; Werrett, *Thrifty Science*.
16. Dyck and Stewart, *Uses of Humans in Experiment*; Guerrini, *Experimenting with Humans and Animals*. An exception is McVaugh, "Determining a Drug's Properties."
17. Foucault, *Order of Things*; Foucault, *Birth of the Clinic*; Foucault, *Discipline and Punish*. See also Dyck and Stewart, *Uses of Humans in Experiment*, 1–3.
18. Carlino, *Books of the Body*; Park, "Life of the Corpse"; Park, "Criminal and the Saintly Body."
19. The nineteenth-century Italian historian Alfonso Corradi compiled accounts of several poison trials in an attempt to provide context for a case in which the revered physician Gabriele Falloppio was granted a condemned criminal to use for an experiment. He gave the man fatal doses of opium. Corradi, "Degli esperimenti tossocologici," *Annali universali*, 73–74. Pastore has particularly examined the legal response to poisoning cases. Pastore, *Veleno*; Pastore, *Il medico in tribunale*; Pastore, "Médicine Légale et Investigation Judiciaire." Sheila Barker discussed poison trials in an article about poison at the Florentine ducal court. Barker, "Poisons and the Prince."
20. M. Baldwin, "Snakestone Experiments"; Findlen, "Controlling the Experiment"; Schickore, "Trying Again and Again"; Schickore, "Significance of Re-doing."
21. Schickore, *About Method*, 30–60.
22. Tribby, "Cooking (with) Clio and Cleo."
23. M. Baldwin, "Snakestone Experiments," 394.
24. Gibbs, *Poison, Medicine, and Disease*, chap. 4.
25. Gibbs, *Poison, Medicine, and Disease*, chap. 6.
26. See letters between Cosimo I de' Medici and Francesco di Paolo Vinta, October 3 and December 19, 1547, Archivio di Stato Florence, vol. 3101a, fol. 807ff., and vol. 9, fol. 259. Accessed via Medici Archive Project, bia.medici.org.
27. Collard, *Crime of Poison*, 172–73.
28. For excellent overviews of this system, see Siraisi, *Medieval and Early Renaissance Medicine*; Lindemann, *Medicine and Society*, esp. chaps. 1–4.

29. McVaugh, "Quantified Medical Theory and Practice at Fourteenth-Century Montpellier"; McVaugh, "Experience-Based Medicine," 109–11; Levey, *Early Arabic Pharmacology*.
30. Avicenna, *Liber canonis*, 2.2, 82r. See also McVaugh, "Determining a Drug's Properties," 183–209; Nasser, Tibi, and Savage-Smith, "Ibn Sina's Canon of Medicine"; Crombie, *Science, Optics, and Music*, 97–99; Crombie, "Avicenna's Influence," 89–90; Brater and Daly, "Clinical Pharmacology"; Daly and Brater, "Medieval Contributions."
31. McVaugh, "Determining a Drug's Properties," 183–209.
32. Marks, *Progress of Experiment*.
33. Gibbs, *Poison, Medicine, and Disease*, esp. 156; Gibbs, "Poisonous Properties."
34. Gibbs, *Poison, Medicine, and Disease*, 174–78, 194.
35. Rankin, "On Anecdote and Antidotes," 274–302.
36. McVaugh, "Experience-Based Medicine," 109–13.
37. Pomata and Siraisi, *Historia*; Grafton and Siraisi, *Natural Particulars*; Smith, *Body of the Artisan*; Ogilvie, *Science of Describing*; Siraisi, *Communities of Learned Experience*.
38. Smith, *Body of the Artisan*; Long, *Artisan/Practitioners*; Taape, "Hieronymus Brunschwig."
39. Pomata and Siraisi, *Historia*; Ogilvie, *Science of Describing*; Rankin, *Panaceia's Daughters*, 43–54; Leong and Rankin, "Testing Drugs and Trying Cures," 167–72; Heinrichs, *Plague, Print, and the Reformation*; Klestinec, *Theaters of Anatomy*.
40. Pomata notes that there were two separate Latin genres, the *Curationes* and the *Observationes*. Pomata, "Observation Rising"; Pomata, "Praxis Historialis."
41. The document by Claudius Richardus was printed in Jordan, *Pestis phaenomena*, 600–618. See also Rankin, "On Anecdote and Antidotes," 274–302.
42. Rankin, "On Anecdote and Antidotes," 281. See also Ragland, "Making Trials," 503–28. This outlook fits the general interest in making knowledge that Pamela Smith and others have noted. Smith, Meyers, and Cook, *Ways of Making and Knowing*; Smith and Schmidt, *Making Knowledge*; Shaw and Welch, *Making and Marketing Medicine*.
43. Smith and Findlen, *Merchants and Marvels*; Wallis, "Exotic Drugs and English Medicine"; Bleichmar, "Books, Bodies, and Fields"; Cagle, *Assembling the Tropics*; Crawford, *Andean Wonder Drug*; Crawford and Gabriel, *Drugs on the Page*; Breen, *Age of Intoxication*; Barrera-Osorio, *Experiencing Nature*.
44. Everaerts, *De herba panacea;* Norton, *Sacred Gifts, Profane Pleasures*, 117; Dickson, *Panacea or Precious Bane*, 88–89; Ziser, "Sovereign Remedies."
45. See especially Rankin, *Panaceia's Daughters*; Rankin, "Empirics, Physicians, and Wonder Drugs;" Leong, *Recipes and Everyday Knowledge*; Taape, "Hieronymus Brunschwig"; Heinrichs, "Plague Cures of Caspar Kegler."

46. Moran, *Paracelsus*, chap. 3; Gantenbein, "Poison and Its Dose"; Gibbs, *Poison, Medicine, and Disease*, 200–204; Hedesan, "Alchemy, Potency, Imagination"; Cunningham, Introduction.
47. Wittich, *Bericht*, A3r.
48. Collard, *Crime of Poison*, 97–100; Rankin, "Gender, Poison, and Antidotes in Early Modern Europe"; Bednarski, *Poisoned Past*; Mollenauer, *Strange Revelations*.
49. Scot, *Discouerie of Witchcraft*, 116; Codronchi, *De morbis veneficis ac veneficiis libri quatuor*.
50. Naphy, *Plagues, Poisons, and Potions*.
51. Walker, *Crime, Gender and Social Order*, 143–48.
52. R. Martin, *Women, Murder, and Equity*, 123.
53. Collard, *Crime of Poison*, 101. This otherness is also mentioned in Wilson, *Poison's Dark Works*, introduction.
54. Walker, *Crime, Gender and Social Order*, 144–145.
55. Behringer, *Witchcraft Persecutions in Bavaria*, 189.
56. R. Martin, *Women, Murder, and Equity*, 123–29. Studies of crime in early modern Germany mention poison only as an aside, although Behringer's work on Bavaria suggests that accusations of poisoning against women increased in the late seventeenth century. Rublack, *Crimes of Women*; Behringer, *Witchcraft Persecutions in Bavaria*, 189; Wiltenburg, *Crime and Culture*.
57. On the trade in exotica, see Borschberg, "Euro-Asian Trade in Bezoar Stones"; M. P. Stark, "Mounted Bezoar Stones"; Novoa, "Unicorns and Bezoars in a Portuguese House in Rome"; Bleichmar, "Seeing the World in a Room"; Koeppe, "Exotica and the Kunstkammer"; Matthee, "Exotic Substances"; Wallis, "Exotic Drugs and English Medicine."
58. On drugs and the apothecary's trade, see Shaw and Welch, *Making and Marketing Medicine*; Pugliano, "Pharmacy, Testing, and the Language of Truth"; Mez-Mangold, *History of Drugs*, 94–110; Stille, *Krankheit und Arznei*.
59. Gentilcore, *Medical Charlatanism*, esp. 203–5.
60. Mayor, *Poison King*; Totelin, "Mithridates' Antidote."
61. Jo Wheeler, *Renaissance Secrets, Recipes and Formulas*, 76–77. Additional information can be found at http://renaissancesecrets.blogspot.com/2013/06/medici-anti-poison-oil.html.
62. Rankin, "Gender, Poison, and Antidotes in Early Modern Europe"; Barker, "Poisons and the Prince."
63. Pastore, *Veleno*, 223.
64. Codices Palatini germanici 177 and 287. Rankin, "Gender, Poison, and Antidotes in Early Modern Europe," 140–42; Rankin, *Panaceia's Daughters*, chap. 2.
65. See especially Shapin, *Social History of Truth*; Shapin and Schaffer, *Leviathan and the Air-Pump*; Biagioli, "Knowledge, Freedom, and Brotherly Love"; Gooding, Pinch, and Schaffer, *Uses of Experiment*. On women's roles in sci-

ence and medicine, see Schiebinger, *Mind Has No Sex?*; Rankin, *Panaceia's Daughters*; Leong, *Recipes and Everyday Knowledge.*

CHAPTER ONE

1. Mayor, *Poison King*, 348–52; Mayor, "Mithradates of Pontus," 21–33; Totelin, "Mithradates' Antidote," 1–19.
2. Touwaide, "Nicander, Thêriaka, and Alexipharmaka"; Gibbs, *Poison, Medicine, and Disease*, 9–12.
3. Totelin, "Mithridates' Antidote," 3–6; Mayor, *Poison King*, 214; Mayor, "Mithridates of Pontus," 21–28.
4. Galen, *On Theriac to Piso*, 71.
5. Gibbs, *Poison, Medicine, and Disease*, 4–7.
6. Gibbs, *Poison, Medicine, and Disease*, 9–10; Touwaide, "Nicander, Thêriaka, and Alexipharmaka," 48–49; Watson, *Theriac and Mithridatium*, 1–93.
7. Watson, *Theriac and Mithridatium*, 1–93; Dousset, *Histoire des médicaments*, 392–96; Dann, *Einführung in die Pharmaziegeschichte*, 22.
8. On the likely spurious nature of *Theriac to Piso*, see Leigh, Introduction, 61–63; Nutton, "Galen on Theriac"; Mattern, *Prince of Medicine*, 214.
9. Galen, *De theriaca ad Pisonem*, 110; Galen, *On Theriac to Piso*, 71. Avicenna, using Galen's account, specified that the "wood roosters" were pheasants; al-Shirazi, on the other hand, used "wood cock." See, e.g., Levey, *Early Arabic Pharmacology*, 87–89.
10. Galen, *De theriaca ad Pisonem*, 110; Galen, *On Theriac to Piso*, 71.
11. On the history of the control, see Schickore, *About Method*, 151–55.
12. Ptolemy I Soter (387–283 BCE) and Ptolemy II Philadelphus (283–246 BCE) were the kings who potentially allowed human dissection. On the likelihood that Herophilus performed vivisections, see Staden, *Herophilus*, 138–56. See also Guerrini, *Experimenting with Humans and Animals*, 11–12.
13. The argument for using animals as a stand-in for humans came from Aristotle's notion of a "ladder of life" (*scala naturae*), a hierarchical system of the world often related to the concept of a Great Chain of Being that put humans on top, followed by animals, plants, and minerals. Spaces between the different species were slight, which meant that many aspects of animal anatomy and physiology also pertained to humans. Lovejoy, *Great Chain of Being*; Guerrini, *Experimenting with Humans and Animals*, 9.
14. Galen, *De theriaca ad Pisonem*, 110–12; Galen, *On Theriac to Piso*, 71.
15. Galen, *On Theriac to Piso*, 65. See also Leong and Rankin, "Testing Drugs and Trying Cures," 167–68.
16. Pomata, "Word of the Empirics," 1–25.
17. Galen, *On Theriac to Piso*, 67.
18. Levey, *Early Arabic Pharmacology*, 87.
19. Crombie, "Avicenna's Influence."
20. Avicenna, *Liber canonis*, 5.1; Avicenna, *Zusammengesetzte Heilmittel*, 6–7.

21. The Persian physician al-Sharazi (1236–1311) later recounted these same poison trials as a means to test whether theriac was "good and active," without noting the Galenic precedent. Levey, *Early Arabic Pharmacology*, 87.
22. Levey, *Early Arabic Pharmacology*, 132.
23. Jābir ibn Ḥayyān, *Buch der Gifte*; Ardestani et al., "Golden Age."
24. Levey, "Medieval Arabic Toxicology," 48.
25. Avenzoar, *Liber theysir*, 1r–v, 10v; Azar, *Sage of Seville*, 57.
26. Maimonides, *On Poisons*, 54.
27. Maimonides, *On Poisons*, 51.
28. Maimonides, *On Poisons*, 51–54.
29. Razi, *Liber nonus ad Almansorem*, 35r–40r.
30. The famous ibn Rushd, or Averroes (1126–98), a contemporary of ibn Zuhr, composed a freestanding treatise on theriac. However, he argued that experience on patients was inferior to "reason" in testing most drugs. Ricordel, "Le traité sur la thériaque d'Ibn Rushd," 86–87.
31. Gil drew strongly from Avicenna, but only books 2 and 4. For his knowledge of Galen, he seems to have relied on several medieval encyclopedists, especially Vincent of Beauvais and Gilbertus Anglicus. See Zamora and Ferrero Hernández, "Liber contra venena"; Gibbs, *Poison, Medicine, and Disease*, 46–50.
32. McVaugh, "Theriac at Montpellier," 113–14. On the manuscript tradition of *Theriac to Piso*, see Leigh, Introduction, 1–8.
33. This interest followed the translation of Averroes' treatise on theriac into Latin. See McVaugh, "Introduction to 'De Dosi Tyriacalium,'" 60.
34. McVaugh, 63; Gibbs, *Poison, Medicine, and Disease*, 83.
35. Arnald, *Speculum medicinae*, quoted in McVaugh, "Introduction to the 'Epistola,'" 64n22.
36. Gibbs, *Poison, Medicine, and Disease*, 86–87; McVaugh, "Experimenta of Arnald of Villanova."
37. Villanova, "De arte cognisciendi venenis."
38. Michael McVaugh builds a convincing case attributing authorship to Peter Cellararius, a Spanish physician practicing at the court of Aragon. McVaugh, "Introduction to the 'Epistola,'" 66.
39. Villanova, "De venenis," 203.
40. Galen's own writings on theriac were not available to medieval Europeans and instead filtered in through other sources. McVaugh, "Introduction to the 'Epistola,'" 57.
41. Gordon, "Ueber den Theriak," 6. Bernard cites Avicenna's *Canon*, book 5, and Galen's *On Simples*, chap. 2.
42. Gordon, "Ueber den Theriak," 8; McVaugh, "Experience-Based Medicine," 111; Demaitre, *Doctor Bernard de Gordon*, 72–77.
43. McVaugh, "Experience-Based Medicine," 112–13.
44. McVaugh, "Experience-Based Medicine," 111; Gordon, "Ueber den Theriak," 8.

45. Abano, *Tractatus de venenis*, chaps. 2–3; Gibbs, *Poison, Medicine, and Disease*, 87–94.
46. Abano, *Tractatus de venenis*, chaps. 2–3; Gibbs, *Poison, Medicine, and Disease*, 22–23, 150–80; Gibbs, "Poisonous Properties."
47. Pietro called prassius "the matrix and palace of emerald." Abano, *De venenis*, 26–27.
48. Gibbs, *Poison, Medicine, and Disease*, 90–94. On Pietro, see also Siraisi, "Reflections on Italian Medical Writings of the Fourteenth and Fifteenth Centuries"; Siraisi, "Pietro d'Abano and Taddeo Alderotti."
49. "Ego quandoque sum expertus." Abano, *De venenis*, chap. 5.
50. Avicenna, *Liber canonis*, 2.2, 82r. See also Nasser, Tibi, and Savage-Smith, "Ibn Sina's Canon of Medicine"; Crombie, *Science, Optics, and Music*, 97–99; Crombie, "Avicenna's Influence," 89–90; Brater and Daly, "Clinical Pharmacology"; Daly and Brater, "Medical Contributions."
51. Avicenna, *Liber canonis*, 2.2, 82r–83v.
52. McVaugh, "Determining a Drug's Properties."
53. Quoted in McVaugh, "Experience-Based Medicine," 111.
54. McVaugh, "Determining a Drug's Properties," 7–12; McVaugh, "Experience-Based Medicine," 112, 119.
55. McVaugh, "Quantified Medical Theory and Practice at Fourteenth-Century Montpellier," 399.
56. Vigevano, *Texarus Regis Franciae*. The entire *Texarus* runs from fol. 32r to 54v; the medical section runs from 32r to 41v. See also Hall, "Guido Da Vigevano's Texaurus Regis Franciae"; Thorndike, *History of Magic*, 3: 26–27.
57. Vigevano, *Texarus Regis Franciae*, 40r.
58. Vigevano, *Texarus Regis Franciae*, 40r; the story of the mouse can be found in book 4 of Avicenna's *Canon*.
59. He added the worms to a recipe for a terra sigillata theriac that he had already recommended. Vigevano, *Texarus Regis Franciae*, 40r; Hall, "Guido Da Vigevano's Texaurus Regis Franciae," 40–41.
60. Vigevano, *Texarus Regis Franciae*, 40v.
61. Vigevano, *Texarus Regis Franciae*, 40r. On manicules, see Sherman, Used Books, chap. 2.
62. Collard, *Crime of Poison*, 127–28.
63. Collard, *Crime of Poison*, 192–93; Collard, "Ouvrir pour découvrir."
64. Commynes, *Preuves des Mémoires de Philippe de Comines*, 199–200; Collard, *Crime of Poison*, 192.
65. Stein, *Charles de France*, 464; Collard, *Crime of Poison*, 192.
66. J. W. Baldwin, "Intellectual Preparation for the Canon of 1215 against Ordeals"; Langbein, *Prosecuting Crimes in the Renaissance*, 134–39. They did return to use to a limited extent during the witch craze of the sixteenth century. See Pihlajamaki, "'Swimming the Witch.'"
67. Collard, *Crime of Poison*, 184.

68. Michel Foucault argued that the medieval French word for proof (*épreuve*) could cover a number of acts or tests aimed at obtaining evidence, such as gathering witnesses or submitting to a trial by ordeal. See Foucault, "La vérité et les formes juridiques"; Hildebrandt, "Trial of the Expert."
69. Scully, *Art of Cookery*, 249.
70. Lentsch, "La proba," 155.
71. Lentsch, "La proba," 155; Collard, *Pouvoir et poison*, 143–50; Hoberg, *Die Inventare des päpstlichen Schatzes in Avignon*; Guiffrey, *Inventaires*, cxxxix and inventories. On the tradition of wonders in medieval Europe, see Daston and Park, *Wonders and the Order of Nature*, esp. chaps. 1–4; Truitt, *Medieval Robots*, 17–18; and Truitt, "Virtues of Balm."
72. Marra, *Opusculum quod dicitur Sertum Papale de Venenis*, 8.
73. Levey, "Medieval Arabic Toxicology," 21–22.
74. Rosner, *Medical Legacy of Moses Maimonides*, 35; Maimonides, *On Poisons*, introduction; McVaugh, "Introduction to the 'Epistola'"; Demaitre, *Doctor Bernard de Gordon*, 32.
75. Collard, *Crime of Poison*, 238–48; Collard, *Pouvoir et poison*, 167–210.
76. Lentsch, "La proba," 155; Hoberg, *Die Inventare des päpstlichen Schatzes in Avignon*; Guiffrey, *Inventaires*, cxxxix and inventories; Collard, *Pouvoir et poison*, 143–65; Daston and Park, *Wonders and the Order of Nature*, 74–75.
77. Lentsch, "La proba," 155–56.
78. The documents on this case were printed in Königshofen, *Die Alteste Teutsche so wol Allgemeine Als insonderheit Elsassische und Straßburgische Chronicke*, 1031–36.
79. Königshofen, *Die Alteste Teutsche so wol Allgemeine Als insonderheit Elsassische und Straßburgische Chronicke*, 1031–36. See also Horrox, *Black Death*, 210–20; Arrizabalaga, "*Pestis Manufacta*," 67–69.
80. For more on the Black Death, see especially Green, *Pandemic Disease in the Medieval World*; Herlihy, *Black Death and the Transformation of the West*; Lehfeldt, *Black Death*; Ziegler, *Black Death*; Dols, *Black Death in the Middle East*. Rosemary Horrox's excellent collection of primary sources gives a good sense of the impact. Horrox, *Black Death*.
81. The city council of Strasbourg sent out a call to localities in Savoy, the Rhineland, and Cologne to provide hard evidence of Jewish poison plots. Of the sixteen towns that responded, only one (Cologne) expressed skepticism at Jewish malfeasance. The other fifteen provided "proof." Cohn, "Black Death and the Burning of Jews," 19. On the Black Death and the Jews, see also Bergdolt, "Die Pest und die Juden"; Barkai, "Jewish Treatises," 6; Jankrift, "Judenpogrome"; Schabel and Pedersen, "Miraculous, Natural, or Jewish Conspiracy?"
82. Cohn points out that most pogroms were led by noblemen and other elites, not by popular protest. Cohn, "Black Death and the Burning of Jews," 21–26.

83. Thorndike, *History of Magic,* 3: 234; Wickersheimer, *Les accusations d'empoisonnement,* 1–8.
84. Gibbs, *Poison, Medicine, and Disease,* 116–41; Arrizabalaga, *"Pestis Manufacta,"* 61–74.
85. Collard, *Crime of Poison,* 104–6.
86. Gibbs, *Poison, Medicine, and Disease,* 118–19; Siraisi, *Medieval and Early Renaissance Medicine,* 128–29; Arrizabalaga, *"Pestis Manufacta,"* 61–74.
87. Foligno, *Gentilis Fulginatis medici illustris contra pestilentiam consilium,* 2; Gibbs, *Poison, Medicine, and Disease,* 122–23.
88. On Penna, see Gibbs, *Poison, Medicine, and Disease,* 123. On the Paris report, see Michon, *Documents inédits.*
89. Arrizabalaga, *"Pestis Manufacta,"* 248; Gibbs, *Poison, Medicine, and Disease,* 123.
90. Fabbri, "Treating Medieval Plague."
91. The question of whether theriac should be taken by the healthy was a matter of heated disagreement. McVaugh, "Theriac at Montpellier."
92. Foligno, *Gentilis Fulginatis medici illustris contra pestilentiam consilium,* b3v. He specifically mentioned Pietro d'Abano in conjunction with the gems.
93. Report of the Paris medical faculty in Michon, *Documents inédits,* 62. "Smaragdus est insignis medicina contra omne venenum."
94. Sies, *Das "Pariser Pestgutachten"*; Holste, *Der Theriakkrämer,* 5: 36.
95. Haage, *Das gereimte Pestregimen,* 71. I have taken a few liberties with the translation to preserve some sense of the rhyming. The original reads as follows:

    Och tiriaca alß ain aerbis genossen
    Under dem win ist guet usß der maussen.
    Es sterkt dz hertz on maesse fast,
    Gift mag nit sin by im ain gast.
    Bolusarmenus ist nuechter mit essich guet.
    Von terra sigillata erfroewt sich dz blut.

96. Marra, *Opusculum,* 29–30.
97. Marra, *Opusculum,* 37; Thorndike, *History of Magic,* 3: 529–34.
98. Foligno, *Gentilis Fulginatis medici illustris contra pestilentiam consilium,* b1v–2r.
99. Marra, *Opusculum,* 29–30; Thorndike, *History of Magic,* 3:529–34.
100. Guaineri, *De venenis,* in *Opera,* fols. 112r–121v. Lynn Thorndike worked out the dating. Thorndike, *History of Magic,* :215–17.
101. Gibbs, *Poison, Medicine, and Disease,* 152–54, 24–26.
102. He merely described Guido as a physician from Pavia. Guido was also known as Guido da Pavia. Guaineri, *De venenis,* fol. 117r.
103. Guaineri, *De pestis, Opera,* 109v–110r.
104. Guaineri, *De venenis, Opera,* 116; *De pestis, Opera,* 110r.

105. Gibbs, *Poison, Medicine, and Disease*, 155–56.
106. Arduino, *Opus de venenis*, 15–16. On the use of "probatum est," see Jones, "Formula and Formulation"; Pugliano, "Pharmacy, Testing, and the Language of Truth," 233.
107. Arduino, *Opus de venenis*, 66.
108. Gibbs, *Poison, Medicine, and Disease*, 178–99.
109. Cabanès, *Poisons et sortilèges*, appendix C, 295–99.

CHAPTER TWO

1. Pastore, *Veleno*; Pastore, "Médecine légale"; Pastore, *Il medico in tribunale*.
2. *Testimonium*, A2r: "ad publicam pertinerent utilitaté, in corporibus damnatorum integra fide & diligenter experiremur."
3. For example, a letter from Cosimo I de' Medici to Francesco di Paolo Vinta, written December 19, 1547, casually notes that two soldiers had offered to kill Piero Strozzi on behalf of Ferrante Gonzaga. Archivio di Stato Florence, vol. 9, fol. 259, Medici Archive Project, doc. 4601.
4. *Testimonium*, A1r.
5. On the use of condemned criminals in medicine, see especially Park, "Criminal and the Saintly Body"; Carlino, *Books of the Body*. On dissection in Clement VII's Rome, see Andretta, *Roma medica*, 500–501.
6. Andretta, *Roma medica*, 500.
7. Pastore, *Veleno*; Pastore, "Médecine légale"; Pastore, *Il medico in tribunale*; Conforti, "Visible Signs, Invisible Processes"; Andretta, "Pre- and Post-Mortem Inquiries."
8. Abano, *De venenis*, 10–11.
9. Park, *Secrets of Women*, 122–26; Park, "Criminal and the Saintly Body."
10. Florentine physician Antonio Benivieni's private notes recorded autopsies in eighteen different patients, a selection of which was published after his death. Benivieni, *De abditis nonnullis*; Park, *Secrets of Women*, 124–26. On the expansion of post-mortems in the sixteenth and seventeenth centuries, see especially the essays in De Renzi et al., *Pathology in Practice*.
11. According to Faith Wallis, medieval readers had not realized that Galen never dissected humans: "because he wrote so much about anatomy, and recommended dissection so fervently, they just assumed he had." Wallis, *Medieval Medicine*, 231. On the influence of humanism on Renaissance anatomists, see especially Park, *Secrets of Women*, 165–68.
12. Anatomy and dissection have been the topic of much historical research over the past few decades. See especially Carlino, *Books of the Body*; Park, *Secrets of Women*; Park, "Life of the Corpse"; Park, "Criminal and the Saintly Body"; Klestinec, *Theaters of Anatomy*; Cunningham, *Anatomical Renaissance*; Kusukawa, *Picturing the Book of Nature*; Guerrini, *Courtiers' Anatomists*; Nutton, "Wittenberg Anatomy."

13. My students are always surprised to learn that dissection was not taboo in the Catholic Church. Park, "Life of the Corpse," 113. See also Park, "Criminal and the Saintly Body," 3; Park, *Secrets of Women*, 47–48.
14. On apprehensions around dissection, see especially Carlino, *Books of the Body*, 92–119; Park, "Criminal and the Saintly Body," 8–12; Park, "Life of the Corpse," 129–30. Renaissance Italians did not, however, view dissection itself as a penal event, as would later become common in England. See Sawday, *Body Emblazoned*, chap. 4.
15. Carlino, *Books of the Body*, 98.
16. Benedetti, "History of the Human Body [1502]," 82–83.
17. Benedetti, "History of the Human Body [1502]," 82–83. See also Ferrari, *L'esperienza del passato*, esp. chap. 2.
18. Benedetti, "History of the Human Body [1502]," 82–83.
19. Benedetti, "History of the Human Body [1502]," 83; Carlino, *Books of the Body*, 92–94; Park, "Criminal and the Saintly Body."
20. On dissection and Carnival, see Ferrari, "Public Anatomy Lessons and the Carnival"; Carlino, *Books of the Body*, 80–81. On the suspension of usual rules during Carnival, see especially Bakhtin, "Carnival and the Carnivalesque," 50–54; Bakhtin, *Rabelais and His World*, introduction.
21. Quoted in Carlino, *Books of the Body*, 84.
22. Carlino, *Books of the Body*, 104–8.
23. Gentilcore, *Food and Health*, chap. 3.
24. Carlino, *Books of the Body*, 93.
25. On using hospital patients and soldiers as test subjects, see Rivest, "Testing Drugs and Attesting Cures"; Rankin and Rivest, "Medicine, Monopoly, and the Premodern State."
26. Schiebinger, "Medical Experimentation"; Schiebinger, *Secret Cures of Slaves*.
27. Park, "Criminal and the Saintly Body," 19.
28. An abbreviated version of Celsus was available in the Middle Ages, but more complete manuscripts were rediscovered by humanists in the 1420s. Granados, "New Light on Celsus' 'De Medicina'"; Lo Presti, "Anatomy as Epistemology."
29. Celsus, *De Medicina*, proemium. See also Carlino, *Books of the Body*, 156–70.
30. Celsus, *De Medicina*, proemium. See also Carlino, *Books of the Body*, 156–70.
31. Park, "Criminal and the Saintly Body," 19.
32. Galen, *Claudii Galeni opera omnia*, 14: 1–3.
33. Benedetti, "History of the Human Body [1502]," 82.
34. Falloppio, *Gabrielis Falloppii . . . Libelli duo*, chap. 14; Corradi, "Degli esperimenti tossocologici" (1886), 73–74; Park, "Criminal and the Saintly Body," 20.
35. *Testimonium*, A1r.
36. Foucault, *Discipline and Punish*, 55.
37. Terpstra, "Theory into Practice."

38. Fosi, *Papal Justice*, chap. 2; Rebecchini, "Rituals of Justice and the Construction of Space in Sixteenth-Century Rome."
39. Carlino, *Books of the Body*, chap. 2; Andretta, *Roma medica*, 499–500.
40. Gaisser, "Seeking Patronage," esp. 304–5; Noel, *Renaissance Popes*, 266.
41. Nutton, "Humanist Surgery," 78–79.
42. Andretta, *Roma medica*, 500–502.
43. Statuta 1584, quoted in Carlino, *Books of the Body*, 78. See also Fosi, *Papal Justice*, 23–39.
44. T. C. P. Zimmermann, *Paolo Giovio*, chap. 2. Nancy Siriasi notes that the Renaissance ties between history and medicine were closer than has previously been acknowledged. Siraisi, *History, Medicine, and the Traditions of Renaissance Learning*, 195.
45. T. C. P. Zimmermann, *Paolo Giovio*, esp. chaps. 2 and 6.
46. It is possible that Tiraboschi's statement was a misrepresentation of the 1524 *Testimonium*, as the author does not appear to have seen the publication himself, but the details do not match exactly (the initials D.P.P do not appear in the *Testimonium*). Tiraboschi, *Storia della Letteratura Italiana*, vol. 7, part 3, p. 902.
47. Clayton, "Leonardo's Anatomy Years," 316; Clayton and Philo, *Leonardo da Vinci*, 19–23; Wilson, *Collaborations in Art and Medicine*, 183–84; Schutz, *Art and Anatomy*, 96–100.
48. Giovio, *Elogia doctorum virorum ingenii monumentis illustrium*, 128. On Giovio's relationship with della Torre, see T. C. P. Zimmermann, *Paolo Giovio*, 9.
49. On autopsies and the Catholic Church, see Bouley, *Pious Postmortems*; Park, *Secrets of Women*.
50. Grassi, *Il diario di Leone X*, 88–89. My thanks to Bradford J. Bouley for this reference.
51. Bijloos, *Adrianus VI*, 104.
52. Visceglia, *Morte e elezione*, 41.
53. Ponzetti himself received one vote in the initial scrutiny. Pastor, *History of the Popes*, 9: 17n1; Thorndike, *History of Magic*, 5: 472–74.
54. Ponzetti, *Libellus de venenis*, 2.1.5; Park, "Country Medicine," 111.
55. Park, "Country Medicine." See also Gentilcore, *Medical Charlatanism*, 174–80.
56. Castiglione, *Lettere inedite*; Stinger, *Renaissance in Rome*, 38–39.
57. Pastor, *History of the Popes*, 9:104.
58. Giovio, *Pauli Iovii Opera*, 1:105. Letters from 1522 indicate that he was in Florence and Genoa, and he mentioned the need for "comfort at not being in Rome with the *patrone*."
59. Castiglione, *Lettere*, 127 (July 8), 129 (July 17).
60. Peter Partner notes that Giovio's 1516–17 figures are generally distrusted, but he argues that they should be given some credence owing to the twin scourges of plague and the Sack of Rome. Partner, *Renaissance Rome*, 81–82.

61. Brackett, *Criminal Justice and Crime*, 2. See also Fosi, *Papal Justice.*
62. *Testimonium*, A2r.
63. Fosi, *Papal Justice*, 23–29.
64. Borghese appears to have died in the Sack of Rome, so it is not unreasonable to place him in Rome in 1524. Olivieri, *Il senato Romano*, 312–14; Vicchi, *Villa Borghese nella storia e nella tradizione del popolo romano*, 155–57.
65. *Testimonium*, A4r. "Ita est & attestor Petrus borgesius senator Vrbis."
66. *Testimonium*, A2r.
67. Partner, *Renaissance Rome*, 76.
68. *Testimonium*, A2v, A3v.
69. *Testimonium*, A2v–3r.
70. Carlino, *Books of the Body*; Edgerton, *Pictures and Punishment.*
71. Although the good thief became a particularly strong theme in Protestant regions, Feinberg notes that he appeared on confraternities' images in Florence, Bologna, and Rome. Merback, *Thief, the Cross, and the Wheel*; Feinberg, "Imagination All Compact"; Timmermann, "'Locus Calvariae.'"
72. *Testimonium*, A4r.
73. *Testimonium*, A3v.
74. Collard, *Crime of Poison.*
75. Chiaramonte and Tozzi, "Un medico umanista," 62.
76. Giovio, *Lettere inedite*, 47. Federico later became duke of Mantua.
77. "Mando qua alligata in stampa La regula de operare L'olio del veneno / et Cosa il Testimonio de la prova depso." Houghton Library, MS Typ 739.
78. Giovio, *Lettere inedite*, 48.
79. Cortese, *I secreti*, 1; Olivieri, *Il senato romano*, 314. Olivieri, however, lists Borghese as being senator only in 1515.
80. Cellini, *My Life*; Holman, "For 'Honor and Profit,'" 522.
81. Corti was also known as Corte or Curtius. He became known primarily for his anatomical work on veins. Gibbs, *Poison, Medicine, and Disease*, 243–44; Ferrari, "Matteo Corti."
82. Falloppio, *Gabrielis Falloppii . . . Libelli duo*, fol. 68v.
83. Noel, *Renaissance Popes*, 278.
84. Falloppio, *Gabrielis Falloppii . . . Libelli duo*, 68v. "ego saepe usus felicissimo cum successu in vermibus."
85. The missive is described in Corradi, "Degli esperimenti tossocologici" (1886), 80–81.
86. Corradi, "Degli esperimenti tossocologici" (1886), 80–81.
87. Corradi, "Degli esperimenti tossocologici" (1886), 79–80; Terpstra, "Confraternal Prison Charity and Political Consolidation in Sixteenth-Century Bologna."
88. Bologna was a Papal State. Fosi, *Papal Justice*, 14–19.
89. Corti came off looking like a stodgy old Galenist in the face of a brash young renegade. Nutton, "'Qui magni Galeni.'"
90. Corradi, "Degli esperimenti tossocologici" (1886), 80–81.

91. Brasavola, *De medicamentis*, 354; Corradi, "Degli esperimenti tossocologici" (1886), 80.

CHAPTER THREE

1. Mattioli, *Commentarii*, 1096.
2. Mattioli, *Commentarii*, 1096.
3. Mattioli, *Commentarii*, 1096.
4. Mattioli, *Commentarii*, 1096.
5. On Mattioli, see Fabiani, *La Vita di Pietro Andrea Mattioli*; Ferri, *Pietro Andrea Mattioli*; Findlen, "Formation of a Scientific Community."
6. Mattioli, *New Kreüterbuch*, 472v–473r.
7. Mattioli, *Commentarii*, 1096.
8. On contrived experiences see Rankin, "On Anecdote and Antidotes." On the increase of observation among naturalists, see Ogilvie, *Science of Describing*.
9. Mattioli, *Il Dioscoride*, bk. 6, 15r.
10. Mattioli, *Il Dioscoride*, bk. 6, 15r–16r. See also Gentilcore, *Medical Charlatanism*, 175–76; Bayle, "Thériaque et triacleurs chez Pierre-André Mathiole."
11. *Testimonium*, 3v.
12. Mattioli, *Il Dioscoride*, 16r–v. See also Gentilcore, *Medical Charlatanism*, 175–76; Bayle, "Thériaque et triacleurs chez Pierre-André Mathiole."
13. Gentilcore, *Medical Charlatanism*; Holste, *Der Theriakkrämer*; Klairmont Lingo, "Empirics and Charlatans in Early Modern France"; Eamon, *Science and the Secrets of Nature*; Eamon, *Professor of Secrets*.
14. Schmitt, "Experience and Experiment"; Dear, *Discipline and Experience*; Dear, "Meanings of Experience."
15. Park, "Natural Particulars"; Pomata and Siraisi, *Historia*.
16. Drug testing had always represented something of an exception to that rule, as in the example of Arnau of Villanova and his poisoned cheese in chapter 1, but there were virtually no instances in which experimental results received widespread attention. McVaugh, "Determining a Drug's Properties"; McVaugh, "Experience-Based Medicine"; Leong and Rankin, "Testing Drugs and Trying Cures."
17. Scholars had previously seen the term *periculum facere* as emerging in the later sixteenth-century mathematical sciences, especially in the context of Galileo's experiments. Ragland, "Making Trials"; Findlen, *Possessing Nature*; Schmitt, "Experience and Experiment"; Dear, "Narratives, Anecdotes, and Experiments"; Dear, *Discipline and Experience*.
18. See especially Shapin and Schaffer, *Leviathan and the Air-Pump*; Collins, *Golem*; Collins and Pinch, *Dr. Golem*; Gooding, Pinch, and Schaffer, *Uses of Experiment*.
19. Klairmont Lingo, "Empirics and Charlatans in Early Modern France," 583–89; Holste, *Der Theriakkrämer*, 5: 81–84; Gentilcore, *Medical Charlatanism*, pp.

20. Mattioli, *De i Discorsi*, 1467–68.
21. Park, "Country Medicine"; Gentilcore, *Medical Charlatanism*, 174–80; Klairmont Lingo, "Empirics and Charlatans in Early Modern France."
22. Cordus, *Von der vielfaltigen tugent vnnd waren bereitung*, title page. On Cordus, see Ogilvie, *Science of Describing*, 133–38; Ogilvie, "Many Books of Nature," 30–32.
23. Cordus, *Von der vielfaltigen tugent vnnd waren bereitung*, A2v.
24. Cordus, *Von der vielfaltigen tugent vnnd waren bereitung*, A2v–3r.
25. For example, the ingredient called squillam, required for the second kugel, was a species of large onion, while the measurement quintin was "an apothecary's pound." The third kugel required the reader to "take 4 or 5 vipers, cut off three fingers' length of their heads and tails, skin and gut them well, and simmer them in a new kettle, well washed" along with a good handful of salt and dill. The cooked flesh should then be pounded to a pulp and mixed with bread crumbs. Cordus, *Von der vielfaltigen tugent vnnd waren bereitung*, C2v–3r.
26. Cordus, *Von der vielfaltigen tugent vnnd waren bereitung*, B3v–4r.
27. Cordus, *Von der vielfaltigen tugent vnnd waren bereitung*, B4r.
28. Ogilvie, *Science of Describing*, 132–37; Ogilvie, "Many Books of Nature," 30–31.
29. Mattioli translated Dioscorides from an early sixteenth-century Latin commentary by Jean Ruel. On the general importance of Dioscorides in the Renaissance, see Ogilvie, *Science of Describing*, esp. 137–38; Stannard, "Dioscorides and Renaissance Materia Medica"; Kusukawa, *Picturing the Book of Nature*, chaps. 5 and 6.
30. Stannard, "Dioscorides and Renaissance Materia Medica," 8–10.
31. Ferri, "Il 'Dioscoride,' i 'Discorsi,' i 'Commentarii'"; Findlen, "Formation of a Scientific Community," 373–75.
32. Mattioli, *Il Dioscoride*, bk. 4, chap. 80, p. 305.
33. Fabiani, *La Vita di Pietro Andrea Mattioli*, 6; Chiaramonte and Tozzi, "Un medico umanista," 62.
34. Lenzi, "Dal 'regno di Iatria,'" 84.
35. Mattioli, *Epistolarum medicinalium libri quinque.*
36. Listed as Christofano Macruccio in Mattioli's dedication. Mattioli, *Il Dioscoride*, bk. 6, 2r.
37. "In questa nostra età floridissima." Mattioli, *Il Dioscoride*, bk. 6, 3v.
38. David Gentilcore explains that the term *cerretani* linked charlatans to the ultimate deceiver, Satan. See Gentilcore, *Medical Charlatanism*, 12–15.
39. Mattioli, *Il Dioscoride*, bk. 6, 15v. See also Gambaccini, *Mountebanks and Medicasters*, 68; Gentilcore, *Medical Charlatanism*, 120, 175–76; Findlen, *Possessing Nature*, 269.
40. Mattioli specifically mentioned the *pauliani* in this diatribe. Mattioli, *Il Dioscoride*, book 6, 16r.
41. Mattioli, *Il Dioscoride*, bk. 6, 16v.

42. Mattioli, *Il Dioscoride*, bk. 6, 16v–17r.
43. Mattioli, *Il Dioscoride*, bk. 6, 16r–v.
44. Gambaccini, *Mountebanks and Medicasters*, 68–69.
45. Gentilcore has noted that charlatans sold antidote oils at a higher price than apothecaries. Gentilcore, *Medical Charlatanism*, 218. On the Medici oil, see Barker, "Poisons and the Prince," 79; Wheeler, *Renaissance Secrets, Recipes and Formulas*, 76–77.
46. Mattioli, *Il Dioscoride*, bk. 6, 3r.
47. Mattioli, *Il Dioscoride*, bk. 6, 15r.
48. Mattioli, *I Discorsi* (1568), **3r–v.
49. Mattioli, *I Discorsi* (1568), **3v–**4r. On Calzolari, see Palmer, "Pharmacy in the Republic of Venice," 110–12; Findlen, *Possessing Nature*, chap. 4; Pugliano, "Pharmacy, Testing, and the Language of Truth."
50. Mattioli, *Commentarii*, 1096.
51. On medical signs, see Maclean, *Logic, Signs, and Nature in the Renaissance*.
52. Mattioli, *I Discorsi* (1568), 1153.
53. Mattioli, *I Discorsi* (1568), 1153. On Mattioli and Fuchs, see Kusukawa, *Picturing the Book of Nature*, 130–31.
54. Stolberg, "Empiricism in Sixteenth-Century Medical Practice"; Stolberg, "Learning from the Common Folks"; Stolberg, "Many Uses of Writing."
55. Österreichische Nationalbibliothek (ÖNB), esp. cod. 11183, 11205–7.
56. Mattioli, *New Kreüterbuch*, sig. 1:6v–2:1r.
57. Mattioli, *Herbář*. The Czech edition, more of an excerpt than a translation, was prepared by Tadeáš Hájek von Hájek, with woodcuts by Giorgio Liberale and Wolfgang Meyerpeck. See Moran, "Preserving the Cutting Edge," 403–9. On Mattioli's distillation section, see Pereira, "Utili segreti."
58. Mattioli, *New Kreüterbuch*, 472v–73.
59. Mattioli, *New Kreüterbuch*, 473r.
60. ÖNB 11183, fols. 126v–127r.
61. ÖNB 11183, fols. 126v–136v and 243r–v.
62. ÖNB 11183, fol. 127v.
63. Mattioli, *New Kreüterbuch*, 473r.
64. ÖNB 11183, fol. 126r.
65. Jordan, *Pestis phaenomena*, 621–30; Wittich, *Bericht*, 17–22. The two versions are almost identical, but the Latin original reads rather clinically while Wittich's translation adds more drama.
66. Jordan spent the 1560s mainly in Vienna, both as a student and as a professor of medicine. His book is dedicated to Johannes Crato von Krafftheim, one of Ferdinand's personal physicians at the time of the poison trial. Hirsch, "Thomas Jordan," 520.
67. Mattioli listed Richardus among the physicians and surgeons present for the dissection of a beaver. Mattioli, *Commentarii*, bk. 2, chap. 23, p. 330. The coronation document listed four medical personnel under the general

title of "Medici" (doctors)—two physicians, Richardus, and an apothecary. Strotschius, *De Electione et Inauguratione Maximiliani Austrii II*, D4v.

68. An unnamed surgeon was listed in the courtly household (*Hofstaat*) of Emperor Maximilian II, who succeeded Ferdinand in 1564—possibly a reference to Richardus, but it is impossible to be sure. It is possible that Richardus continued on under Maximilian, as many of Ferdinand's physicians did. The 1564 record simply lists "surgeon" (*Wundarzt*); from 1565 to 1567 the man is listed as *Leibwundarzt*. He was paid 33 gulden a month, significantly less than the physicians (at 50–60 gulden/mo), but still a hefty fee. Crystal, "Medicine in Vienna in the Sixteenth and Seventeenth Centuries," 354.
69. Richardus mentioned conversation with the prisoners, and he included a German phrase in his Latin document.
70. A higher level of medical learning was common among surgeons trained in Italy, and it is possible that Richardus trained there, which may explain his apparently high position at court. On surgery in Germany and Italy, see Nutton, "Humanist Surgery," esp. 92–96; Siraisi, *History, Medicine, and the Traditions of Renaissance Learning*, 195–96; Klestinec, "Translating Learned Surgery." The German humanist Walter Hermann Ryff called himself a Medicus and Chirurgicus (doctor and surgeon) in his works, although he appears to have trained as an apothecary. See Rankin, "How to Cure the Golden Vein," 120–21.
71. Wittich, *Bericht*, 17.
72. Wittich, *Bericht*, 17. The Latin version is less evocative but makes the same point: "Cum vero infirmus assereret ni statum subueniretur, se penitus deficere." Jordan, *Pestis phaenomena*, 626.
73. Jordan, *Pestis phaenomena*, 627; Wittich, *Bericht*, 18.
74. Jordan, *Pestis phaenomena*, 628; Wittich, *Bericht*, 19.
75. Jordan, *Pestis phaenomena*, 628–29; Wittich, *Bericht*, 19–21.
76. Jordan, *Pestis phaenomena*, 630; Wittich, *Bericht*, 21. The insertion of the German phrase "gantz wol" into the Latin suggests that Latin was the original language of composition.
77. The first edition was based on the original Latin edition and contained only the trial of Caravita's oil; the second edition included the anecdotes about trials at Emperor Ferdinand's court. Mattioli, *Les commentaires . . . sur les six livres des simples de Pedacius Dioscoride Anazarbeen*; Mattioli, *Commentaires* (1572).
78. Paré, *Les Oeuvres d'Ambroise Paré* (1579), bk. 20, chap. 45.
79. The significance of this assertion will be discussed further in chapter 4. Mattioli, *Commentaires* (1572): "qu'il valoit mieus mourir de venin dans la prison que d'estre publiquement pendu"; Paré, *Les Oeuvres d'Ambroise Paré* (1579), bk. 20, chap. 45. "qu'il aimeroit trop mieux encore mourir de ladicte poison en la prison, que d'estre estranglé à la veuë du peuple."

80. Paré, *Les Oeuvres d'Ambroise Paré* (1579), bk. 20, chap. 45.
81. Wheeler, *Renaissance Secrets, Recipes and Formulas*, 76–77; Barker, "Poisons and the Prince," 78–81.
82. Duke Cosimo to Francesco di Paolo Vinta, December 19, 1547, Medici Archive Project doc. ID 4601, vol. 9, fol. 259; January 11, 1548, doc. ID 7450, vol. 9, fol. 329. See also Barker, "Poisons and the Prince," 79.
83. Archivio di Stato di Firenze (hereafter ASF), Carte Strozziane, ser. 1, 97, fol. 2r.
84. This information is listed in a later note attached to the record. ASF, Carte Strozzi, 1, 97, fol. 11
85. ASF, Carte Strozzi, 1, 97, fol. 2r–3r.
86. ASF, Carte Strozzi, 1, 97, fols. 3v–7r.
87. ASF, Carte Strozzi, 1, 97, fol. 9r.
88. Pomata, "Observation Rising"; Siraisi, *History, Medicine, and the Traditions of Renaissance Learning*; Ogilvie, *Science of Describing*, 174–82; Rankin, "On Anecdote and Antidotes."
89. From 1567, the College of Physicians had to sanction the drug before the Magistrate would approve it; in this case, the noblemen appear to have sought physicians' opinions. Gentilcore, *Medical Charlatanism*, 104.
90. Archivio di Stato Venezia (ASV), Provveditori alla Sanità, inv. 300, reg. 731, fol. 1r. Warmest thanks to William Eamon for his archival notes.
91. Gentilcore, *Medical Charlatanism*, 104–6 and 140–44. A recipient of one of these special drug licenses was Leonardo Fioravanti, an alchemist who we shall meet again in chapter 6.
92. ASV, Provveditori alla Sanità, inv. 300, reg. 731, fol. 1r.
93. Berthold, *Terrae sigillatae*, C4r–v. On Berthold, see also Dannenfeldt, "Introduction of a New Sixteenth-Century Drug"; Rankin, "Empirics, Physicians, and Wonder Drugs."
94. Hessisches Staatsarchiv (HStA) Marburg, Bestand 17, 4750.
95. Berthold, *Terrae sigillatae*, B4v.
96. Berthold, *Terrae sigilatae*, C1v.
97. Berthold, *Terrae sigilatae*, C1v–2r.
98. Berthold, *Terrae sigillatae*, C2r.
99. Berthold, *Terrae sigillatae*, C2r.
100. As Gianna Pomata and Katharine Park have noted, the term *observatio* had once been used to refer to careful observations of the heavens. Park, "Observation in the Margins"; Pomata, "Observation Rising," 49–54.
101. Pomata, "Observation Rising," 48.
102. Codices Palatini germanici 177, fols. 16r–17r. See, for example, a letter from Archbishop Gebhard of Cologne to Landgrave Wilhem's brother Ludwig from November 1580. HStA Marburg, Bestand 4f Köln 84.
103. Berthold, *Terrae sigilatae*, C1v–C2r.
104. Findlen, *Possessing Nature*; Ragland, "Making Trials"; Pugliano, "Pharmacy, Testing, and the Language of Truth."
105. Findlen, *Possessing Nature*, 279.

106. Palmer, "Pharmacy in the Republic of Venice," 110.
107. Schickore, *About Method.*
108. Dear, "Narratives, Anecdotes, and Experiments," 135.
109. Pomata, "Observation Rising."
110. See especially Guerrini, *Courtiers' Anatomists*; Ragland, "Experimental Clinical Medicine."

CHAPTER FOUR

1. Hohenlohe Zentralarchiv (HZA) Neuenstein, La 5 Bü 400, no. 1–4.
2. Weyer, *Graf Wolfgang II. von Hohenlohe und die Alchemie.*
3. Berthold, *Terrae sigillatae*, B4r–C3v.
4. Fischer, *Geschichte des Hauses Hohenlohe*, 2: 18–35.
5. Wolfgang's original letter is lost, but the reply from Hyso puts the date at January 5, 1581. HZA Neuenstein, La 5 Bü 400, no. 7.
6. HZA Neuenstein, La 5 Bü 400, no. 4, 7.
7. HZA Neuenstein, La 5 Bü 400, no. 8.
8. *Trials of War Criminals*, 2: 181–82.
9. See especially Gere, *Pain, Pleasure, and the Greater Good*; Lederer, *Subjected to Science.*
10. As a side note, Percival expressly noted the public good but not patient consent. Percival, *Medical Ethics*, 14–15, 162–68.
11. Foucault, *Order of Things*; Dyck and Stewart, *Uses of Humans in Experiment*, 2–3; Schiebinger, *Secret Cures of Slaves*, 4–8.
12. See especially the essays in Geyer-Kordesch, Wear, and French, *Doctors and Ethics.*
13. Hippocrates, *Opera Omnia*, 1. Roger French notes that the prominent Italian physician Gabriele de Zerbi (1445–1505) used the oath as a central source in contructing his description of the good doctor. French, "Medical Ethics of Gabriele de Zerbi," 76, 84–86.
14. Collard, *Crime of Poison*, 141–46.
15. The Ottomans had taken control of Lemnos in the mid-fifteenth century. Hasluck, *Christianity and Islam under the Sultans*, 2:672–88. See also Dannenfeldt, "Introduction of a New Sixteenth-Century Drug"; Mattern, *Prince of Medicine*, 103–4; Mez-Mangold, *History of Drugs*, 107; Stille, *Krankheit und Arznei*, 136–39.
16. Montanus, *Terrae Sigillatae Vtriusque*, D1r–4r.
17. Berthold, *Terrae sigillatae*, B4v–C3v.
18. Berthold, *Vvonderful and Strange Effect*, 1–2.
19. HStA Marburg, Best. 17I.
20. As evidenced in the aforementioned letters from Philipp of Hessen-Marburg to Gebhard of Cologne, November 16, 1580, and April 6, 1581, HStA Marburg, 4f Köln 84.
21. Dinkel was convicted of fornicating with his stepdaughter and fathering a child by her, for which he was thrown into prison. He then could not pay

the prison fee, went into great debt, and was accordingly *justificiert*, an ambiguous term that may not necessarily indicate an execution. No method of execution is listed, in any case. HZA Hohenlohe, La 30 Bü 46, fol. 37.

22. HZA Neuenstein, La 5 Bü 400, no. 1.
23. HZA Neuenstein, La 5 Bü 400, no. 3.
24. Rublack, *Astronomer and the Witch*.
25. His letter to Hyso does not survive but can be ascertained from Hyso's response, which exists both in draft and in the fair copy. HZA Neuenstein, La 5 Bü 400, nos. 4, 4 ½.
26. Dülmen, *Theatre of Horror*; Harrington, *Faithful Executioner*.
27. HZA Neuenstein, La 5 Bü 400, no. 4.
28. This questioning appears to have taken place on December 27, as planned, although the document is undated. HZA Neuenstein, La 5 Bü 400, no. 2.
29. HZA Neuenstein, La 5 Bü 400, nos. 5–6.
30. HZA Neuenstein, La 5 Bü 400, no. 6.
31. Dülmen, *Theatre of Horror*, chap. 1.
32. Zacharius Hyso to Count Wolfgang of Hohenlohe, January 6,1581, HZA Neuenstein, La 5 Bü 400, no. 7.
33. Guerrini, *Experimenting with Humans and Animals*; Guerrini, *Courtiers' Anatomists*; Dyck and Stewart, *Uses of Humans in Experiment*; Schiebinger, "Medical Experimentation"; Ragland, "Experimental Clinical Medicine"; Maehle, "Ethical Discourse on Animal Experimentation."
34. Paracelsus, "Diseases That Deprive Man," 205; Gessner, *Historiae animalium*, A2v.
35. Avicenna, *Liber canonis*, bk. 2, chap. 1; McVaugh, "Determining a Drug's Properties."
36. SLUB Dresden, Ms.ger. K350.
37. Gessner, *Historiae animalium*, A2r.
38. McVaugh, "Experience-Based Medicine," 111.
39. Brasavola, *De medicamentis*, 123v.
40. *Testimonium*, A2v–3v.
41. Corradi, "Degli esperimenti tossocologici" (1886), 79–80.
42. Falloppio, *Libelli duo*, chap. 14. See also Corradi, "Degli esperimenti tossocologici" (1886), 73–74; Park, "Criminal and the Saintly Body," 20.
43. *Testimonium*, A3r.
44. Mattioli, *New Kreüterbuch*, 473. Handsch was also present at this poison trial.
45. Paré, *Les Oeuvres d'Ambroise Paré* (1579), bk. 20, chap. 45. As we saw in chap. 3, Paré referred specifically to Mattioli in the printed marginalia to this edition.
46. Dülmen, *Theatre of Horror*, 97; Edgerton, *Pictures and Punishment*, 135. As Edgerton notes, beheading originally signified the removal of the crown from a disgraced monarch, known as *poena capatum*, "capital punishment," or literally "punishment of the head." Edgerton, 128–9.

47. Dülmen, *Theatre of Horror,* 97–102. See also Stuart, *Defiled Trades and Social Outcasts,* introduction and chap. 3.
48. Benedetti, "History of the Human Body [1502]," 82.
49. *Testimonium,* A4r.
50. Falloppio, *Gabrielis Falloppii . . . Libelli duo,* chap. 14.
51. This evidence comes from Georg Handsch's private notebook, ÖNB 11183, fol. 128r–v. See also Stolberg, "Empiricism," 512.
52. From a manuscript note by the Compagnia del Tempio in Florence written on June 3, 1567. Corradi, "Degli esperimenti tossocologici" (1886), 84.
53. Collard argues that this can be seen in contrast to a murder under fisticuffs, which was a comparatively honorable crime. Collard, *Crime of Poison,* 128–32.
54. Dülman, *Theatre of Horror,* 120–124.
55. Dülmen, *Theatre of Horror;* Edgerton, *Pictures and Punishment,* esp. chaps. 4–5
56. Dülmen, *Theatre of Horror,* 2–23. On the Carolina and the procedures it introduced, see also Langbein, *Prosecuting Crimes in the Renaissance,* 167–209.
57. Dülmen, *Theatre of Horror,* 74–118.
58. See for example O'Conner, *Art of Dying Well;* Beaty, *The Craft of Dying;* Wagner, *Ars Moriendi;* Boyden, "Worst Death becomes a Good Death," 240–65; Geary, *Living with the Dead.*
59. Dülman, *Theatre of Horror,* 119–32. Edgerton, *Pictures and Punishment,* chap. 5.
60. Reinis, *Reforming the Art of Dying;* Buchrucker, "Wer so stirbt, der stirbt wohl," 125–36; Dülmen, *Theatre of Horror,* 119–32.
61. Dülmen, *Theatre of Horror,* 58–79.
62. Collard, *Crime of Poison,* 127.
63. *Testimonium,* A2r–3v.
64. Mattioli, *I Discorsi* (1585), 1153.
65. Paré, *Les Oeuvres d'Ambroise Paré* (1579), bk. 20, chap. 45.
66. Paré, "Résponse." See also Packard, *Life and Times of Ambroise Paré,* 108–9.
67. Paré, "Résponse," 242–43.
68. *Testimonium,* A2r.
69. Corradi, "Degli esperimenti tossocologici" (1886), 84.
70. Archivio di Stato Venezia, Provveditori alla Sanità, inv. 300, reg. 731, fol. 1r.
71. HStA Marburg, Best. 17I.
72. Collard, *Crime of Poison,* 141–46.
73. Meyer, *Borgias;* Noel, *Renaissance Popes;* Cobb, "Case against the Borgias."
74. Moryson, *Shakespeare's Europe,* 406.
75. In thanks for Hyso's loyal service, Wolfgang's father, Count Ludwig Casimir (d. 1568), awarded him a letter of privileges (*Freiheitsbrief*) freeing him from taxes and yearly rent in 1564 and gave him a house and a meadow in 1566. The Hohenlohe Zentral Archiv Neuenstein still has the official

documentation of these gifts. HZA Neuenstein, Oe1 Bü 17124. Hyso also lent a deft hand in helping to develop the first Lutheran Church Order in Hohenlohe in 1581. Wibel, *Hohenlohische Kyrchen- und Reformationshistorie*, 260. Fischer, "Die älteste evangelische Kirchenordnung und die frühesten Kirchenvisitationen in Hohenlohe," 46.

76. HZA Neuenstein, We Bü 2120.
77. HZA Neuenstein, La 5 Bü 400, no. 7.
78. Robisheaux, *Rural Society and the Search for Order*, 48–67 and 175–94. Robisheaux also documented these complexities in a seventeenth-century witchcraft trial in Langenburg. Robisheaux, *Last Witch of Langenburg*.
79. HZA Neuenstein, La 5 Bü 400, no. 7.
80. "Process und Form, welcher gestalt mit dem verhafften zu Langenburg von wegen eingebung der Gifft und Prob der Terra Sigillatae zuhandlen." HZA Neuenstein, LA 5 Bü 400, no. 8. The four advisers were Christoffel Rauchhaupt, Hyso, Johann Bomenstock, and Albrecht Wilhelm Heber. All four men received *Freiheitsbriefe* for their service. See HZA Neuenstein, La 30 Bü 1115.
81. HZA Neuenstein, LA Bü 400, no. 8.
82. Ibid.
83. Dülmen, *Theatre of Horrors*, 74–118.
84. *Testimonium*, A2v–4r.
85. As Thomas V. Cohen has shown, scripts could help temper transgressive events in early modern Europe. Cohen describes a case in which a man was acquitted after murdering his adulterous wife by following a known script. Cohen, *Love and Death*, chap. 1.
86. HZA Neuenstein, La 5 Bü 400, no. 7–8.
87. HZA Neuenstein, La 5 Bü 400, no. 8.
88. Berthold, *Vvonderful and Strange Effect*, 33.
89. Berthold, *Vvonderful and Strange Effect*, 35.
90. Berthold, *Vvonderful and Strange Effect*, 35.
91. ÖNB 11183, fols. 126r–128v, 243r.
92. Berthold, *Epistola Panegyrica*, A3v–4r.
93. On the situation in Germany, see Dülmen, *Theatre of Horror*, 29–30. For France, see Langbein, *Prosecuting Crimes in the Renaissance*, 242–43. Clemency was rarer in the Papal States. See Fosi, *Papal Justice*, 224–36.
94. *Trials of War Criminals*, 2: 181.
95. Gostin, Vanchieri, and Pope, *Ethical Considerations for Research Involving Prisoners*, esp. 79–80.
96. Rankin and Rivest, "Medicine, Monopoly, and the Premodern State."
97. The last record I have found is a "poison proof" overseen by Duke Vincenzo I of Mantua in 1597. Corradi, "Degli esperimenti tossocologici" (1886), 81.
98. Dodart, *Mémoires*, 10. Schiebinger, "Human Experimentation," 394.
99. As Susan Lederer has noted, the long list of humans used for testing new therapies included hospital patients, orphans, the mentally disabled, physicians' family members, and physicians themselves, while Londa Schiebin-

ger has discussed medical testing on slaves. Lederer, *Subjected to Science*, 1–26; Schiebinger, "Medical Experimentation"; L. Stark, "Work, Welfare, and the Values of Volunteerism"; L. Stark, *Behind Closed Doors*.

CHAPTER FIVE

1. Wittich, *Bericht*, A3r.
2. Wittich, *Bericht*, A3r.
3. Orta, *Coloquios*; Monardes, *Dos libros*; Orta, *Aromatum*; Monardes, *Simplicium medicamentorum*.
4. On Wittich's life, see Hafemann, *Magister Johann Wittich*.
5. In contrast, poison itself did not enter into discussion as a wondrous substance, as Frederick Gibbs has argued. Gibbs, *Poison, Medicine, and Disease*, 176–77. Surekha Davies has shown that narratives of wonder also helped shape narratives of indigenous peoples, not just their drugs. Davies, *Renaissance Ethnography and the Invention of the Human*, esp. chap. 6.
6. Jenner and Wallis, *Medicine and the Market*; Costa, *Medicine, Trade and Empire*; P. Wallis, "Exotic Drugs and English Medicine"; Matthew, "Exotic Substances"; Schiebinger, "Prospecting for Drugs."
7. On bezoar see, e.g., Borschberg, "Euro-Asian Trade in Bezoar Stones"; Stephenson, "From Marvelous Antidote"; Novoa, "Unicorns and Bezoars in a Portuguese House in Rome"; Figueroa, "Bezoar Stone"; Sameiro Barroso, "Bezoar Stone."
8. *Testimonium*, A4v.
9. Wittich, *Bericht*, A3r.
10. Monardes, *Dos libros*, k2r–r3r.
11. Brévart, "Between Medicine, Magic, and Religion," 2.
12. Brévart, "Between Medicine, Magic, and Religion," 7–18 and 37–47.
13. Bock, *Kreüter Buch*, 82r–v.
14. Lonicer, *Kreuterbuch*, 167v. Similarly, rosemary, an important symbol of love and fertility in antiquity, became seen as a near cure-all. V. Zimmermann, "Rosmarin als Heilpflanze," 361–75.
15. Rowling, *Harry Potter and the Sorcerer's Stone*, 137–38. Prof. Snape mentions its efficacy against wolfsbane/monkhead (napellus).
16. Sameiro Barroso, "Bezoar Stone," 79.
17. Avenzoar, *Liber theysir*, fol. 20v.
18. Daston and Park, *Wonders and the Order of Nature*, 75; Guiffrey, *Inventaires*; Hoberg, *Die Inventare des päpstlichen Schatzes in Avignon*.
19. Marra, *Opusculum quod dicitur Sertum Papale de Venenis*, 30.
20. Arduino's text contains dozens of references to bezoar—see, e.g., *Opus de venenis*, 20–25, 40–57.
21. Borschberg, "Euro-Asian Trade in Bezoar Stones," 29–36. On the flood of exotic substances coming from Portugal, see Cagle, *Assembling the Tropics*, esp. 210–211.

22. Monte's commentary on the Hippocratic Epidemics was published posthumously in 1554, while Lusitano's commentary on Dioscorides first appeared in 1553 but was particularly widespread in 1554 and 1558 editions. Monte, *In tertium primi epidemiorum*, 155r–65; Lusitanus, *In Dioscoridis Anazarbei De medica materia libros quinque*, 269–71; Mattioli, *Di Pedacio Dioscoride Anazarbeo libri cinque*, bk. 5, chap. 74.
23. Acosta, *Tractado*, 155–57.
24. See, e.g., Ctesias, *Ctesias: On India*; Pliny, *Natural History*, bk. 8, chap. 31; Isidore, *Etymologies*, bk. 12. Hildegard of Bingen's *Physica*, composed in the second half of the twelfth century, mentioned some medicinal uses of unicorn hoof, hide, and liver, but nothing on the horn itself. Lavers, *Natural History of Unicorns*, chap. 5.
25. Lavers, *Natural History of Unicorns*, chap. 5.
26. Rhinoceros horn was also seen as an antidote unto itself, albeit somewhat less valuable. Christen and Christen, "Unicorn and the Narwhal."
27. The tradition continued to be mentioned by sixteenth-century authors such as Pierre Belon and Felix Platter. See Daston and Park, *Wonders and the Order of Nature*, 74–75; Fischer and Cossu Ferra Fischer, "La licorne et la corne de licorne chez les apothicaires et les médecins," 269.
28. Gessner, *Historiae animalium*, 689–95. See also Pinon, "Gessner and the Historical Depth of Renaissance Natural History"; Egmond and Kusukawa, "Circulation of Images and Graphic Practices in Renaissance Natural History."
29. Bacci, *L'alicorno*; Paré, *Discours*, 15r–37v. Paré noted that descriptions of the unicorn varied widely, that claims of success were confused, and that no ancient authors had discussed the horn's healing powers.
30. Daston and Park, *Wonders and the Order of Nature*, 161.
31. Daniela Bleichmar emphasizes the slipperiness and fluidity of exotic objects, while Anna Winterbottom discusses the notion of hybrid knowledge. Bleichmar, "Seeing the World in a Room," 20; Winterbottom, *Hybrid Knowledge*, 2–3.
32. M. P. Stark, "Mounted Bezoar Stones."
33. Findlen, *Possessing Nature*, esp. chaps. 5–6; Daston and Park, *Wonders and the Order of Nature*, esp. chaps. 4, 6, and 7; Pugliano, "Natural History"; Impey and MacGregor, *Origins of Museums*.
34. Olivi, *De reconditis et praecipuis Collectaneis ab honestissimo, et solertiss. Francisco Calceolario Veronensi in Musaeo adservatis*, index and 33–38.
35. Olivi, *De reconditis et praecipuis Collectaneis . . . Francisco Calceolario Veronensi*, unpaginated front matter.
36. Pugliano, "Natural History," 47–49; Cook, *Matters of Exchange*, 29–31; Meadow, "Merchants and Marvels."
37. M. P. Stark, "Mounted Bezoar Stones."
38. ". .tres piedras beçares de las de la Yndia de Portugal metidas las dos en su guarniçion de oro de feligran y la otra sin guarniçion metida en una cajilla

de madera; mas van en la dicha cajilla çinco piedras beçares de las de la Yndia de Castilla grandes y quatro o cinco pequeñas . . . un coco de Maldivar guarniçido de plata." Inventory, 1580. Archivio di Stato, Florence, vol. 1212, fol. 821, Medici Archive Project doc. ID 4284.

39. Fuiková, "Collection of Rudolf II at Prague."
40. Menzhausen, "Elector August's Kunstkammer," 70; Seelig, "Munich Kunstkammer," 80.
41. Daston and Park, *Wonders and the Order of Nature*, 75.
42. Koeppe, "Exotica and the Kunstkammer"; Kappel, "Turned Serpentine Works."
43. Meadow, "Merchants and Marvels," 193.
44. Gessner, *Historiae animalium*, 689–95.
45. Schultze and Fillitz, *Prag um 1600*, 1: 528; Distelberger, "Hapsburg Collection"; Fuiková, "Collection of Rudolf II at Prague"; Seelig, "Munich Kunstkammer," 81–82; Kappel, "Turned Serpentine Works."
46. Gessner, *Epistolarium medicinalium*, 33v.
47. Wittich, *Bericht*, A3r.
48. Jordan, *Pestis phaenomena*, 621–30.
49. Olahus acted as Regent of Hungary in 1562, and he came up with a plan to halt the spread of Reformed denominations in Hungary through education reform. Pop, "Religiones and Nationes."
50. Jordan, *Pestis phaenomena*, 622–30.
51. Jordan, *Pestis phaenomena*, 630.
52. Lusitanus, *In Dioscoridis Anazarbei De medica materia libros quinque*, 283.
53. Marra, *Opusculum quod dicitur Sertum Papale de Venenis*, 30.
54. Rankin, *Panaceia's Daughters*; Rankin, "Gender, Poison, and Antidotes in Early Modern Europe."
55. Mattioli, *Discorsi* (1585), bk. 6.
56. Paré, *Les Oeuvres d'Ambroise Paré* (1579), bk. 20, chap. 45.
57. Winnerling, "Der Fürsten-Theriak."
58. Albert Hyperius to Landgrave Wilhelm IV of Hesse-Kassel, November 27, 1588, HStA Marburg, Best. 17a, Hyperius 5.
59. Wittich, *Bericht*, a4r.
60. Mattern, *Prince of Medicine*, 99–104.
61. Gibbs, *Poison, Medicine, and Disease*, 13–14; Watson, *Theriac and Mithridatium*.
62. Galen, *On Theriac to Piso*, chap. 2.
63. See, e.g., Gordon, "Ueber den Theriak," 6; McVaugh, "Experience-Based Medicine," 111.
64. On Aldrovandi's tests, see Findlen, *Possessing Nature*, 279.
65. On theriac in general, see Watson, *Theriac and Mithridatium*. On theriac in Italy, see Pugliano, "Pharmacy, Testing, and the Language of Truth," 233–73. On regulating apothecaries in Nüremberg, see Murphy, *New Order of Medicine*, 23–29.

66. Pugliano, "Pharmacy, Testing, and the Language of Truth," 267–68.
67. Cooper, *Inventing the Indigenous*, 29; Pugliano, "Pharmacy, Testing and the Language of Truth," 249–50.
68. Gentilcore, *Medical Charlatanism*, 225.
69. Fries, *Spiegel der Artzny*, 9v–10r.
70. On Turkish attempts to market terra sigillata in the Renaissance, see Hasluck, *Christianity and Islam under the Sultans*, 672–88; Mez-Mangold, *History of Drugs*, 107; Dannenfeldt, "Introduction of a New Sixteenth-Century Drug," 174–88; Stille, *Krankheit und Arznei*,136–39.
71. Park, "Country Medicine," 104–20.
72. Antonio Brasavola, *De hominum aequalitate*, 789.
73. Dannenfeldt, "Introduction of a New Sixteenth-Century Drug," 175.
74. Belon, *Observations* (1553), 52–53.
75. Sameiro Barroso, "Bezoar Stone," 77–98.
76. Borschberg, "Euro-Asian Trade in Bezoar Stones," 33.
77. Gessner, *Historiae animalium*, 691.
78. Pugliano, "Pharmacy, Testing, and the Language of Truth," 255.
79. Calzolari to Ulisse Aldrovandi, January 18, 1568, in Calzolari, *Francesco Calzolari da Verona*, 40; Pugliano, "Pharmacy, Testing, and the Language of Truth," 255.
80. Cooper, *Inventing the Indigenous*, esp. 21–50; Johnson, *German Discovery of the World*, chap. 5; Ogilvie, *Science of Describing*; Harkness, "'Strange' Ideas and 'English' Knowledge."
81. Bock, *Kreüter Buch*, 154v.
82. Bock, *Kreüter Buch*, 56r.
83. Cooper, *Inventing the Indigenous*, 21–30. Johnson, *German Discovery of the World*, chap. 5.
84. "*Trans marina* oder *peregrina*." Schnellenberg, *Experimenta von zwentzig pestilentz wurtzeln*, fol. 2v. See also Heinrichs, *Plague, Print, and the Reformation*, 140–45.
85. Lusitano settled in the city of Salonica. Mattioli's attacks came on the heels of violent campaigns against Portuguese Jews in Ancona by papal forces. Lopes Andrade, "Garcia de Orta and Amato Lusitano's Views on Materia Medica"; Pinto, "Preconceito e ciência"; Findlen, "Formation of a Scientific Community."
86. Even Mattioli's friend and colleague Georg Handsch cited Lusitano on bezoar and unicorn horn. Österreichische Nationalbibliothek, 11205, fol. 186v–187v.
87. Lusitanus, *In Dioscoridis Anazarbei De medica materia libros quinque*, 270–71. Beatrix de Luna is the Christianized name of Gracia Mendes Nasi, who later became a powerful figure in the community of Jewish exiles in the Ottoman Empire. Birnbaum, *Long Journey of Gracia Mendes*.
88. Lusitanus, *In Dioscoridis Anazarbei De medica materia libros quinque*, 282.

89. Lusitanus, *In Dioscoridis Anazarbei De medica materia libros quinque*, 283. See also Lopes Andrade, "Garcia de Orta and Amato Lusitano's Views on Materia Medica," 161.
90. Dioscorides Pedanius of Anazarbos, *Des Pedanios Dioskurides*, 37.
91. Orta, *Coloquios*, 271. Antonio Manuel Lopes Andrade suggests that Orta was referencing Lusitano directly here. Lopes Andrade, "Garcia de Orta and Amato Lusitano's Views on Materia Medica," 160–61.
92. Monardes, with typical optimism, speculated that it would work better on a human. Monardes, *Dos libros*, sig. l8r.
93. Hieblin, *Tractat*, 24v–25r.
94. Wittich, *Bericht*, 5. This point was an embellished version of a comment by Clusius in his translation of Orta. Orta, *Aromatum*, 173.
95. Pimentel and Soler, "Painting Naked Truth."
96. Orta, *Coloquios*. There is a decent (if quite literal) translation by Clements Markham: Orta, *Colloquies*. An excellent volume of essays edited by Palmira Fontes da Costa examines all aspects of Orta's career. Costa, *Medicine, Trade and Empire*.
97. Hugh Cagle argues that Orta's organization represented an attempt to speak to the diverse audiences in colonial Goa, including drug merchants. Cagle, "Cultures of Inquiry"; Cagle, *Assembling the Tropics*, 85–103.
98. Orta, *Colloquies*, 30.
99. Orta, *Colloquies*, 160.
100. Orta, *Colloquies*, 363–66.
101. Orta, *Colloquies*, 270. The rulers involved in this exchange were Sultan Muzafan II of Cambay, King Manuel II of Portugal, and Pope Leo X. Sadly, the rhinoceros never made it to Rome, as it perished in a shipwreck off the coast of Italy in 1516. Dürer's sketch was likely based on the drawing of an Indian artist. Dackerman, "Dürer's Indexical Fantasy," 164; Bedini, *Pope's Elephant*, 119–21.
102. Orta, *Colloquies*, 271. Lopes Andrade argues that Orta did not cite Lusitano by name owing to Mattioli's campaign against Lusitano and Orta's own precarious situation as a *converso*. Lopes Andrade, "Garcia de Orta and Amato Lusitano's Views on Materia Medica," 160–61.
103. Orta, *Colloquies*, 470–71; Duffin, "Porcupine Stones."
104. Norton, *Sacred Gifts, Profane Pleasures*, 111–13.
105. Stephenson, "From Marvelous Antidote," 9; Barrera-Osorio, *Experiencing Nature*; Bleichmar, "Books, Bodies, and Fields."
106. Monardes, *Dos Libros*.
107. The translation history is complicated. Dutch naturalist Carolus Clusius translated the *Dos Libros* into Latin in 1574, and John Frampton produced an English translation in 1577 with the evocative title *Joyfull Newes out of the Newe Founde World*. Both men then produced updated editions, in 1579 and 1580 respectively, once the *Historia medicinal* was available to them.

As José Pardo-Tomás and Antonio Barrera-Osorio have pointed out, the Clusius translation was highly modified. Pardo-Tomás, "Two Glimpses"; Barrera-Osorio, "Translating Facts."

108. Monardes, *Dos Libros*, a1r.
109. Monardes, *Dos Libros*, k2v.
110. José Pardo-Tomás has shown Orta's likely influence on Monardes's *Historia medicinal*, but the direct influence on the bezoar treatise is less clear. Pardo-Tomás, "East Indies, West Indies," 200–201.
111. Monardes, *Dos Libros*, n8r.
112. Monardes, *Dos Libros*, o3r–p2r. Quotation at o3r.
113. Monardes, *Primera y segunda y tercera partes de la historia medicinal*, 72r.
114. Monardes, *Primera y segunda y tercera partes de la historia medicinal*, 110v–112v.
115. One seventeenth-century priest, Father Bernabé Cobo, noted that only one or two bezoars would be found for every hundred animals killed. Stephenson, "From Marvelous Antidote," 9.
116. In contrast, much less mercantile attention was devoted to the other antidote Monardes highlighted, the herb escuerçonera.
117. Monardes, *Segunda parte*, 11v.
118. Norton, *Sacred Gifts, Profane Pleasures*, 117; Dickson, *Panacea or Precious Bane*, 88–89; Ziser, "Sovereign Remedies."
119. Monardes, *Segunda parte*, 12r–13r.
120. The doctor's name was written Bernaldo in the 1571 edition but changed to Bernardo in the 1574 *Historia medicinal*. A Dr. Bernardo appears elsewhere as one of Philip II's *protomédicos*. Michele L. Clouse, *Medicine, Government and Public Health*, 84.
121. Monardes, *Segunda parte*, 13v.
122. Norton points out that Mesoamerican healing practices classified tobacco as a hot substance, whereas the mostly closely analogous European drugs all were classified as cold. Norton, *Sacred Gifts, Profane Pleasures*, 117.
123. The edition had 1,250 copies. Egmond, "Figuring Exotic Nature," 168.
124. On the many Clusius editions, see Egmond, "Figuring Exotic Nature," 172–80; Pardo-Tomás, "East Indies, West Indies," 210–11.
125. Pardo-Tomás, "East Indies, West Indies"; Pardo-Tomás, "Two Glimpses"; Barrera-Osorio, "Translating Facts"; Egmond, "Figuring Exotic Nature."
126. Egmond, "Figuring Exotic Nature," 181; Egmond, *World of Carolus Clusius*, chap. 1.
127. Clusius's 1582 edition of Acosta is dedicated to Landgrave Wilhelm IV. Acosta, *Aromatum*, 3–6.
128. Egmond, "Figuring Exotic Nature," 170–71.
129. Orta, *Aromatum*.
130. Lonicer, *Kreuterbuch*, 381r–382.
131. Lonicer, *Kreuterbuch*, 381v.

132. Wittich, *Vade mecum*, 39r–40r.
133. Wittich, *Bericht*, 25.
134. Wittich, *Bericht*, 5.
135. Wittich, *Bericht*, 132–46.
136. Bauhin, *De lapidis bezaaris*, 223–44.
137. Bauhin, *De lapidis bezaaris*, 244–54.
138. Duffin, "Bezoar Stones and Their Mounts," 3.
139. Bauhin, *De lapidis bezaaris*, 281–87.

CHAPTER SIX

1. Am Wald, *Bericht und Erklerung*, A1r–v.
2. Am Wald, *Bericht und Erklerung*, A2r.
3. Am Wald, *Bericht und Erklerung*, C4r.
4. The second 1582 edition was the last one. Müller-Jahncke, "Georg am Wald," 241–42.
5. Am Wald, *Kurtzer Bericht* (1591), 17v–46v.
6. For an extensive biography of am Wald and his works, see esp. Müller-Jahncke, "Georg am Wald"; Eichhorn, "Der Paracelsist Georg Am Wald"; Moran, *Andreas Libavius and the Transformation of Alchemy*, 125–32; Rankin, "Empirics, Physicians, and Wonder Drugs"; Klein, "Chymical Medicine," 199–210; Zedler, "Georg Amwald"; Hirsch, "Georg Amwald."
7. Everaerts, *De herba panacea*; Dickson, *Panacea or Precious Bane*, 77.
8. Moran, *Andreas Libavius and the Transformation of Alchemy*, 125–32.
9. Am Wald, *Kurtzer und zum andernmal gemehrter Bericht*, 6r.
10. Gentilcore, *Medical Charlatanism*, 119; Fioravanti, *Il tesoro*, 97–213; Eamon, *Professor of Secrets*, 220–26.
11. Müller-Jahncke, "Georg am Wald," 214–18. Gertaud Eichhorn has noted that the family name was usually given as Amwald in Passau, although also occasionally as "am Wald." There is, however, no indication that the family belonged to the lower gentry. The surname Baldinus, under which Georg and his brothers matriculated in the university, was likely a derivative of Baldersheim, where Jacob Amwald had attended the famous bookbinding school. Eichhorn, "Der Paracelsist Georg Am Wald," 45n25.
12. This information is cited in his early publications. See, e.g., Berthold, *Newe Zeitung und Beschreibung*.
13. Müller-Jahncke, "Georg am Wald," 215–16.
14. Berthold, *Newe Zeitung und Beschreibung*; Berthold, *Lob vnd Danckpsalmen*; Am Wald, *Gerichtsteuffel*.
15. Berthold, *Terrae sigillatae*, B4r.
16. Müller-Jahncke, "Georg am Wald," 217–19.
17. In a small book of laudatory poems about the Silesian terra sigillata published in 1586, the poems' authors consistently referred to him as "D. Andreas Berthold" (the D. signified "Doctor"). Berthold, *Epistola Panegyrica*.

18. He bought the estate from Jorg von Ellrichshausen. The deed of sale mentioned a "Ringmauer, das Schloßtor, und ein Tanzhaus erwähnt." Ramisch, *Landkreis Feuchtwangen*, 121.
19. Müller-Jahncke, "Georg am Wald," 214–18; Eichhorn, "Der Paracelsist Georg Am Wald," 36–37 and 45. Eichhorn notes a discrepancy with the date of am Wald's death (45n25).
20. Paula Carusi notes the strong relationship between medical and alchemical traditions in medieval Islam. Carusi, "Il filosofo e il marinaio," 22–23. See also Kopp, *Geschichte der Chemie*, 1: 47–52.
21. On Roger Bacon, see Paravicini Bagliani, "Ruggero Bacone e l'alchimia"; Pereira, "*Mater Medicinarum*," 28–29; DeVun, *Prophecy, Alchemy, and the End of Time*, 81–89; Matus, *Franciscans and the Elixir of Life*, 41–49; Principe, *Secrets of Alchemy*, 69–73. It is unclear whether Arnald actually wrote this work, called the *Rosarius philosophorum*, or, indeed, any of the alchemical works attributed to him, but readers assumed he did. DeVun, *Prophecy, Alchemy, and the End of Time*, 92.
22. Michaela Pereira has done definitive work on this topic. Pereira, "*Mater Medicinarum*," 26–52; Pereira, "Heavens on Earth"; Pereira, *Arcana sapienza*; Pereira, *L'oro dei filosofi*. See also DeVun, *Prophecy, Alchemy, and the End of Time*, 96–99.
23. Pereira, "Medicina," 1–9; DeVun, *Prophecy, Alchemy, and the End of Time*; Matus, *Franciscans and the Elixir of Life*, 41–49. On promoting fertility, see Nummedal, *Anna Zieglerin and the Lion's Blood*, chaps. 5–6.
24. Pereira, "Medicina," 9; Pereira, "*Mater Medicinarum*," 33–40.
25. Brunschwig, *Liber de arte distillandi de compositis*, esp. bk. I, chaps. 2–19; Taape, "Hieronymus Brunschwig," 109–17; Rankin, "How to Cure the Golden Vein."
26. DeVun, *Prophecy, Alchemy, and the End of Time*, 81–82; Paravicini Bagliani, *Pope's Body*, 199–211.
27. This tale has been brought to life by Michaela Pereira. Pereira, "Medicina," 1; Pereira, "*Mater Medicinarum*," 39–40.
28. The recipe for another alchemical oil attributed to a Gregorio Mezzo Capo, allegedly also tested in Rome in November 1524, appeared in the *Secrets of Isabella Cortese*. Cortese, *I secreti*, 1.
29. Heinrichs, *Plague, Print, and the Reformation*, 107–31; Heinrichs, "Plague Cures of Caspar Kegler."
30. Paracelsus, "Seven Defensiones," 20–22. See also Moran, *Paracelsus*, chap. 3; Gantenbein, "Poison and Its Dose"; Gibbs, *Poison, Medicine, and Disease*, 200–204; Hedesan, "Alchemy, Potency, Imagination"; Cunningham, Introduction.
31. Paracelsus, *Archidoxorum*.
32. Moran, *Paracelsus*, chap. 3.
33. *Wunder Artzney, vonn allerley Leibs Gebrechen, unnd zu fallenden Kranckheiten.* The book was first printed in 1555.

34. Rankin, *Panaceia's Daughters.*
35. Eamon, "Masters of Fire," 141–42.
36. Eamon, *Professor of Secrets.*
37. Eamon, *Professor of Secrets,* 77–85 and 205–7.
38. See the helpful appendix in Eamon's *Professor of Secrets,* 322–23.
39. Several editions of Fioravanti were published in early seventeenth-century German cities. Fioravanti, *Compendium;* Fioravanti, *Corona oder Kron der Artzney;* Fioravanti, *M. Leonardi Fioravanti . . . Allgemeiner Kunst vnd Weltspiegels drey Bücher.*
40. Möhsen, *Beiträge,* 55.
41. These were his *Archidoxa* (1569) and *Pison* (1572).
42. Möhsen, *Beiträge,* 120–30. See also Spitzer, *—und die Spree führt Gold.*
43. Thurneisser's Arzneibücher can be found in the Staatsbibliothek Berlin, MSS germ. 101, 103, and 104. See especially MS germ. 104, fols. 226r–256v. See also Moran, "Art and Artisanship."
44. He became particularly known for his urinalysis. See Stolberg, *Experiencing Illness;* Stolberg, *Homo patiens.*
45. Möhsen, *Beiträge,* 120–30; Spitzer, *—und die Spree führt Gold.*
46. Thurneisser zum Thurn, *Historia,* 69.
47. Paracelsus, *Der Grosse Wundarzney,* 177r-v.
48. One of his pupils, Johannes Huser, later produced an edition of Paracelsus based on Montanus's library. Dannenfeldt, "Introduction of a New Sixteenth-Century Drug," 176.
49. Montanus, *Breve, sed exquisitum, vereque philosophicum Iudicium Doctrinis,* A2r–3v; Dannenfeldt, "Introduction of a New Sixteenth-Century Drug," 179.
50. Montanus, *Breve, sed exquisitum, vereque philosophicum Iudicium Doctrinis,* D2v.
51. The letter introduced Franz Krettschwer as "ein Throphrastischer amator Medicine vnd Chimiae." Johannes Trimontanus to Leonhard Thurneisser, May 15, 1576, Staatsbibliothek Berlin, MS germ. 421a, fol. 86r.
52. Berthold, *Terrae sigillatae,* D1r–4v.
53. In his definitive (and excellent) study of the Silesian terra sigillata, Karl H. Dannenfeldt assumed that there was only one terra sigillata and described Berthold as "undoubtedly one of Montanus' trusted friends." Dannenfeldt, "Introduction of a New Sixteenth-Century Drug," 178. I followed this assumption in my initial foray into the terra sigillata. Rankin, "Empirics, Physicians, and Wonder Drugs," 680–710. However, Hermann Schelenz posited long ago that Berthold's terra sigillata was different from Montanus's. Schelenz, *Geschichte der Pharmazie,* 455.
54. Berthold specified that Kupferberg was near the river Bober, "about seven miles from the head of the river called the Elbe." Berthold, *Terrae Sigillatae,* E1r.
55. Berthold, *Terrae sigillatae,* D1r–4v.

56. Berthold, *VVonderfull and Strange Effect*, 17.
57. Moran, "'Herbarius' of Paracelsus," 104–5; Cooper, *Inventing the Indigenous*, 22–41.
58. Berthold, *Vvonderful and strange effect*, 9–15.
59. Am Wald, *Bericht und Erklerung*, A1r.
60. Am Wald, *Bericht und Erklerung*, A1v. See also Müller-Jahncke, "Georg am Wald."
61. Am Wald, *Bericht und Erklerung*, C4v.
62. Am Wald, *Bericht und Erklerung*, A2r.
63. A handful of other new terra sigillatas appeared around the Holy Roman Empire in the 1580s and 1590s, but they mostly postdated Berthold's book. Karl H. Dannenfeldt has noted that treatises concerned with a single drug were rare in sixteenth-century Germany. Dannenfeldt, "Introduction of a New Sixteenth-Century Drug," 177–78.
64. The Batzen was a southern German coin (different from the Swiss Batzen). Its worth varied significantly with place and time, but it tended to be between 15–17 batzen to a gulden. Although it was technically outlawed in the Imperial Coin Law (Reichsmünzordnung) of 1559, it continued to be used in many places. Four Kreuzer were also sometimes called a Batzen. Döring, *Handbuch der Münz-, Wechsel-, Mass- und Gewichtskunde*, 186, 252.
65. Am Wald's friend Johann Muselius later asked him for a recipe. Müller-Jahncke, "Georg am Wald," 226.
66. Am Wald, *Bericht und Erklerung*, A2r.
67. Am Wald, A3r–B2v.
68. Am Wald, C4v.
69. Am Wald, C4r.
70. Wittich, *Bericht*, 132–46.
71. Marius, *Terra sigillata*, D2r.
72. Berthold, *Terrae sigillatae*, A2r–v.
73. Berthold, *Terrae sigillatae*, A2r.
74. Johannes Posthius to Joachim Camerarius, June 15, 1582, in Muller-Jahnke, "Georg am Wald," 269.
75. Johannes Posthius to Joachim Camerarius, July 17, 1582, in Müller-Jahncke, "Georg am Wald," 269–70.
76. Johannes Posthius to Joachim Camerarius, July 17, 1582, in Müller-Jahncke, "Georg am Wald," 269–70.
77. His correspondence with Theodor Zwinger and Peter Monau mentions the Terra Sigillata Amwaldina several times. The letters were published in Crato von Krafftheim, *Consiliorum et epistolarum medicinalium*, 2: 277, 281, 337, 348, 358. See also Muller-Jahnke, "Georg am Wald," 271–72.
78. Monau to Weidner, December 5, 1583, in Crato von Krafftheim, *Consiliorum*, 2: 337. See also Muller-Jahnke, "Georg am Wald," 271.
79. Montanus, *Breve, sed exquisitum, vereque philosophicum Iudicium Doctrinis*, A1r, C1v.

80. Montanus, *Breve, sed exquisitum, vereque philosophicum Iudicium Doctrinis*, A2v. Terra sigillata has sometimes been called the first trademarked drug owing to the fixation on the seal. Dannenfeldt, "Introduction of a New Sixteenth-Century Drug," 174–76. On later trademark developments, see Gabriel, *Medical Monopoly*, esp. chap. 1.
81. Berthold, *Terrae sigillatae* (1583), A1r. "Andreas Bertholdus, der Natur in mineralischen Krefften Erforscher und Liebhaber."
82. Berthold, *Terrae sigillatae* (1583), A2r–v.
83. Berthold, *Terrae sigillatae* (1583), B2r–3r.
84. Berthold, *Epistola Panegyrica*, A2r–3r.
85. Berthold, *Epistola Panegyrica*, A3v–4v.
86. Berthold, *Epistola Panegyrica*, C2r.
87. The copy is now in the Bibliothèque nationale de France in Paris. The handwriting matches letters from Berthold in SLUB Dresden, introductory letter to Berthold, *Außführliche Beweysung*, and Codices Palatini germanici, 177. The verse, a highly altered version of Job 22:22–24, reads, " Wirst du dich bekehren zu dem Almechtigen so wird er dir gemalen gold geben/ Und reiche seiffen / von den felssen / wie im Ophir" (If you return to the Almighty/ so will He give you powdered gold /and rich soaps from the fells, as in Ophir). Berthold used the golden ink to letter the line about powdered gold. Berthold, *Epistola Panegyrica*, E4r.
88. Am Wald, *Kurtzer Bericht* (1591).
89. Am Wald, *Kurtzer Bericht* (1591), 6v.
90. Am Wald, *Kurtzer Bericht* (1591), 9r–11v.
91. Am Wald, *Kurtzer Bericht* (1591), 3r–9r, 50v.
92. Am Wald, *Kurtzer Bericht* (1591), 47r.
93. Am Wald, *Kurtzer Bericht* (1591), 11v.
94. Am Wald, *Kurtzer Bericht* (1591), 16v.
95. Am Wald, Kurtzer Bericht (1591), 46v–51r.
96. Am Wald, *Kurtzer Bericht* (1591), 48r–49v.
97. Müller-Jahncke, "Georg am Wald," 227–28; Eichhorn, "Der Paracelsist Georg am Wald," 37.
98. Johann Wittich complained about their openness. See Wittich, *Consilium apoplecticum*, 55; Hafemann, *Magister Johann Wittich*, 35–36.
99. Müller-Jahncke has transcribed the entire text of this letter, dated August 25, 1590. The original can be found among Count Gottfried's papers in Schloß Harburg. Müller-Jahncke, "Georg am Wald," 257–58.
100. Am Wald, *Kurtzer Bericht* (1591), 16v–46v.
101. Am Wald, *Kurtzer Bericht* (1591), 11r, 17v.
102. Am Wald, *Kurtzer Bericht* (1591), C1v.
103. Bock, *Kreüter Buch*, fols. 154v–155r.
104. Erasmus, *Correspondence of Erasmus*, 64–65.
105. Drabitius, *Panacea*.

106. See, e.g., Herring, *Panacea Christiana*; G. R., *Panacea*. In German-speaking regions, it became a religious motif in a handful of printed eulogies. Lindius, *Iustorum Panacea*; Schönfeldt, *Panacea contra vim mortis*.
107. Novenianus, *Von den bösen vmbflechtenden Bauchflüssen*, 3; Rankin, *Panaceia's Daughters*, 2–3.
108. Everaerts, *De herba panacea*.
109. Dickson, *Panacea or Precious Bane*, 77, 94.
110. Am Wald, *Kurtzer Bericht* (1591), 12r–v.
111. Am Wald, *Kurtzer Bericht* (1591), 4r–v.
112. Müller-Jahncke, "Georg am Wald," 220–22.
113. J. Zimmermann, *Carmina*; Lauterbach, *Lauterbachius, pro Panacea Am Waldina*; Müller-Jahncke, "Georg am Wald," 252.
114. Am Wald, *Kurtzer und zum andernmal gemehrter Bericht*, 70v; Müller-Jahncke, "Georg am Wald," 268, 278.
115. HStA Darmstadt, 41/3 (unpaginated).
116. Am Wald spoke up against Donauwörth's religious tolerance, particularly the practice of letting Catholics attend the weddings of Protestant relatives and vice versa. Müller-Jahnke, "Georg am Wald," 220–21.
117. See the title pages of am Wald, *Kurtzer Bericht* (1592); am Wald, *Kurtzer und zum andernmal gemehrter Bericht*.
118. Am Wald, *Kurtzer Bericht* (1592), 50v
119. Am Wald, *Kurtzer Bericht* (1592), 91v.
120. Fioravanti, *Il tesoro*, 97–213; Eamon, *Professor of Secrets*, 220–26.
121. Gentilcore, *Medical Charlatanism*, 119.
122. Am Wald, *Kurtzer und zum andernmal gemehrter Bericht*, 91v.
123. Moran, *Andreas Libavius and the Transformation of Alchemy*, 125–32.
124. That treatise also attacked another Paracelsian, Johann Graman. Libavius, *Neoparacelsica*; Moran, *Andreas Libavius and the Transformation of Alchemy*, 125–42.
125. Libavius, *Gegenbericht*; Moran, *Andreas Libavius and the Transformation of Alchemy*, 127; Müller-Jahncke, "Georg am Wald," 259–60.
126. The laughter is in a marginal callout and is therefore especially visible. Libavius, *Gegenbericht*, 60.
127. Moran, *Andreas Libavius and the Transformation of Alchemy*, 127–32.
128. Libavius used this explanation particularly in the case of the former imperial physician Johaness Crato von Crafftheim. Libavius, *Gegenbericht*.
129. Libavius, *Gegenbericht*, 112–28.
130. Am Wald, *Kurtzer Bericht* (1592), 10r; Johannes Posthius to Joachim Camararius the Younger, December 12, 1593, in Müller-Jahncke, "Georg am Wald," 273–74.
131. Libavius, *Gegenbericht*, b2r–4v.
132. Libavius, *Gegenbericht*, 144.
133. Georg I to Georg am Wald, March 7, 1594, HStA Darmstadt, D4, 31/3. "Nuhn ist vns von der Panacea so ihr erfunden viel gesagt, auch von etzlichen vns sehr gerumbt worden."

134. Georg of Hesse-Darmstadt to Georg Nigrinius, November 13, 1593, HStA Darmstadt, D4, 31/3. Unpaginated. For an account of Landgrave Georg's health problems near the end of his life, see Noack, *Landgraf Georg I,* 99–103.
135. Georg I to Georg am Wald, March 7, 1594, HStA Darmstadt, D4, 31/3. "Nuhn ist vns von der Panacea so ihr erfunden viel gesagt, auch von etzlichen vns sehr gerumbt worden."
136. Georg I to Uffsteiner, March 8, 1594, HStA Darmstadt, D4, 31/3.
137. Georg am Wald to Landgrave Georg I, March 13, 1594, HstA Darmstadt, 41/3. Emphasis in the original.
138. Georg am Wald to Landgrave Georg I, March 23, 1594, HstA Darmstadt, 41/3.
139. Landgrave Georg I to Georg am Wald, April 2, 1594, HstA Darmstadt, 41/3.
140. Georg am Wald to Landgrave Georg I, April 7, 1594, HstA Darmstadt, 41/3.
141. Landgrave Georg I to Georg am Wald, April 18, 1594, HstA Darmstadt, 41/3.
142. Berthold, *Außführliche Beweysung,* A4v.
143. Berthold, *Außführliche Beweysung,* A4v.
144. Berthold, *Außführliche Beweysung,* A2v.
145. Möller, *Theatrum Freibergense Chronicum,* 395.
146. See, for example, Zedler, "Johann Montanus."
147. Royal College of Physicians of London, *Pharmacopœia Londinensis,* 6–7; Collegio Medico Nuremberg, *Verneuerte Gesetz, Ordnung, Und Tax . . . Deß Hey[Ligen] Reichs Statt Nürnberg,* H2v.
148. Wiese, *Biedermeierreise durch Schlesien,* 192; Partsch, *Schlesien,* 341.
149. Am Wald, *Vortrab*; Libavius, *Panacea Ambaldina victa et prostrata.*
150. Müller-Jahncke, "Georg am Wald," 239.
151. It is unclear whether am Wald influenced Anthony's panacea or not. Anthony's original work appears to have been published in Hamburg in 1598, but is very rare today (listed in only two Slovenian editions in WorldCat). Much more common was the 1618 London edition. Glauber discussed his panacea in his *Operis Mineralis* (1651). Anthony, *Panacea aurea* (1598); Anthony, *Panacea aurea* (1618); Glauber, *Operis mineralis.*
152. On the increased interest in a universal cure, see Hedesan, *Alchemical Quest for Universal Knowledge.*
153. Klein, "Chymical Medicine," 198–206.

CONCLUSION

1. Toscano had previously received permission to sell orviétan in Vienna if he offered it for a cheap price, and he wanted to expand this privilege. Österreichisches Staatsarchiv Vienna (ÖStA), HHStA RHR Grat et Feud Ärzte und Arzneiprivilegien 15–8, Toscano Johann, fol. 116, docs. 3 and 4.
2. ÖStA, HHStA RHR Grat et Feud Ärzte und Arzneiprivilegien 15–8, Toscano Johann, fol. 116, doc. 1.

3. ÖStA, HHStA RHR Grat et Feud Ärzte und Arzneiprivilegien 15–8, Toscano Johann, fol. 119.
4. ÖStA, HHStA RHR Grat et Feud Ärzte und Arzneiprivilegien 15–8, Toscano Johann, fol. 115.
5. There is a handwritten draft of the license in the archival file. ÖStA Vienna, HHStA RHR Grat et Feud Ärzte und Arzneiprivilegien 15–8, Toscano Johann, fol. 113r–114v.
6. On this topic, see especially Rivest, "Beyond the Pharmacopoeia?"; Gentilcore, *Medical Charlatanism*; Rivest, "Testing Drugs and Attesting Cures."
7. On the history of orviétan, see Gentilcore, *Healers and Healing*, chap. 4; Catellani, *L'orvietano*; Le Paulmier, *L'orviétan*; Rivest, "Beyond the Pharmacopoeia?," 81–100.
8. Tilburg, *By His Majesties License*; Mullini, "I Cornelius à Tilbourn."
9. Gentilcore notes that orviétan had become firmly entrenched in "established medical traditions" by the late seventeenth century. Gentilcore, *Healers and Healing*, 118.
10. ÖStA Vienna, HHStA RHR Grat et Feud Ärzte und Arzneiprivilegien, 15–3.
11. ÖStA Vienna, HHStA RHR Grat et Feud Ärzte und Arzneiprivilegien, 2–18.
12. ÖStA Vienna, HHStA RHR Grat et Feud Ärzte und Arzneiprivilegien 1, fols. 176r–192v.
13. For more on these special licenses, especially in the eighteenth century, see Gentilcore, *Medical Charlatanism*, esp. chap. 4; Lindemann, *Health and Healing in Eighteenth-Century Germany*, chap. 3.
14. The book appeared first in French and later in Latin. Sala, *Tenarius bezoarticorum*; Sala, *Ternarius ternariorum*.
15. Sala, *Tenarius bezoarticorum*, title page.
16. Sala, *Tenarius bezoarticorum*, A2r–3v.
17. Sala, *Tenarius bezoarticorum*, table of contents.
18. Sala, *Tenarius bezoarticorum*, 1.
19. Sala, *Tenarius bezoarticorum*, 8.
20. Hedesan, *Alchemical Quest for Universal Knowledge*; Hedesan, "Jan Baptist Van Helmont"; Pagel, *Joan Baptista van Helmont*.
21. Guerrini, *Experimenting with Humans and Animals*, chap. 2.
22. On animal experimentation in the seventeenth century, see especially Guerrini, *Courtiers' Anatomists*; Ragland, "Experimental Clinical Medicine"; Maehle, "Zur wissenschaftlichen und moralischen Rechtfertigung toxikologischer Tierversuche."
23. Boyle, *Some Considerations Touching the Usefulnesse of Experimental Naturall Philosophy*, 53–54.
24. Boyle, *Some Considerations Touching the Usefulnesse of Experimental Naturall Philosophy*, 48.
25. Schickore, *About Method*, 30–60; Schickore, "Trying Again and Again"; M. Baldwin, "Snakestone Experiments"; Findlen, "Controlling the Experiment."

26. On this point, see especially Findlen, "Controlling the Experiment," 39–45.
27. Redi did not cite Galen as the author. Redi, *Opere di Francesco Redi*, 4: 180–81. See also Schickore, *About Method*, 31.
28. Wepfer, *Cicutae*, 107–13; Maehle, *Johann Jakob Wepfer als Toxikologe*, 23–33.
29. Maehle, *Johann Jakob Wepfer*, 124–25.
30. Wepfer, *Cicutae*, 176–77.
31. On this point, see Dear, "Narratives, Anecdotes, and Experiments."
32. Wepfer, *Cicutae*, 49–50.
33. Wepfer, *Cicutae*, 49–50.
34. Anita Guerrini has noted that Robert Hooke, for example, expressed distaste with experiments that hurt animals. Guerrini, *Experimenting with Humans and Animals*, chap 2.
35. Wepfer, *Cicutae*, 132–34.
36. Redi, *Opere di Francesco Redi*, 4: 20. See also Gentilcore, *Medical Charlatanism*, 42.
37. Wepfer, *Cicutae*, 150.
38. Gentilcore, *Healers and Healing*, 99.
39. Boyle, *Some Considerations Touching the Usefulnesse of Experimental Naturall Philosophy*, 50.
40. Schiebinger, *Secret Cures of Slaves*, 127–28.
41. Schiebinger, *Secret Cures of Slaves*; Schiebinger, "Medical Experimentation"; Rivest, "Testing Drugs and Attesting Cures"; Gentilcore, *Medical Charlatanism*, 147–49; Rankin and Rivest, "Medicine, Monopoly, and the Premodern State," 106–9.
42. Delbourgo, *Collecting the World*, 175–77.
43. Gentilcore, *Medical Charlatanism*, 146–49.
44. Sloane, "Letter from Sir Hans Sloane"; Delbourgo, *Collecting them World*, 118–19, 211–13.
45. See Maehle's table of fields of study. Maehle, *Drugs on Trial*, 11.
46. On the continued relevance of these questions today, see Greene, "Therapeutic Proofs and Medical Truths," 420–29.
47. See especially Murphy, *New Order of Medicine*; Lindemann, *Health and Healing in Eighteenth-Century Germany*.

Bibliography

Archival Sources

Archivio di Stato, Florence

Carte Strozziane, Prima Serie, 97

Medici Papers. Medici Archive Project, bia.medici.org

Archivio di Stato, Venice

Provveditori alla Sanità, Inv. 300, Reg. 731

Berliner Staatsbibliothek

Papers of Leonhard Thurneisser. MS germ. 97, 99, 101, 103, 104, 419, 421a, 421b

Hessisches Staatsarchiv, Darmstadt

Landgraf Georg I., Personalia. Bestand D 4, 31/3

Hessisches Staatsarchiv, Marburg

B.1.1.4 Politische Akten nach Philipp dem Großmütigen. Bestand 4f, 17a, 17l

Hohenlohe Zentralarchiv, Neuenstein

Archiv Langeburg, La 5 Bü 400

Archiv Öhringen, Oe1 Bü 17124

Archiv Weikersheim, We Bü 2120

Houghton Library, Harvard University

Correspondence of Paolo Giovio, MS Typ. 739

Österreichische Nationalbibliothek, Vienna

Papers of Georg Handsch, MS 11183, 11200, 11204, 11205, 11206, 11207, 11240, 11247, 11251

Österreichisches Staatsarchiv, Vienna

Haus, Hof, und Staatsarchiv, RHR Grat und Feud Ärzte und Arzneiprivilegen, Boxes 1, 2, 4, 15

Sächsische Landes- und Universitätsbibliothek Dresden

Medizinischen Gutachten für dem Kurfürsten August. Mscr.Dresd. K 350

Universitätsbibliothek Heidelberg

Arzneibücher. Codices Palatini germanici, 177, 287. http://codpalgerm.uni-hd.de

Primary Sources

Abano, Pietro d'. *De venenis atque eorundem commodis remediis*. Marburg, 1537.

———. *Tractatus de venenis*. Mantua, 1473.

Acosta, Cristóbal. *Aromatum & medicamentorum in Orientali India nascentium liber*. Antwerp, 1582.

———. *Tractado de las drogas, y medicinas de las Indias Orientales*. Burgos, 1578.

Albertus Magnus. *Book of Minerals*. Trans. Dorothy Wyckoff. Oxford: Clarendon Press, 1967.

Am Wald, Georg. *Bericht und Erklerung Georgij am Wald . . . Wie vnd was gestalt das new von jm erfunden Terra Sigillata vnd vniuersal Artzeney . . . zugebrauchen sey*. St. Gallen, 1581.

———. *Gerichtsteuffel, darinn angezeigt vnd gehandlet wirt, wie vnd was massen der leidig Sathan bißweylen Unordnung vnd zerrüttung in Gerichten*. St. Gallen, 1580.

———. *Kurtzer Bericht, wie, was gestalt und warumb das Panacea am Waldina, als ein einige Medicin . . . anzuwenden seye*. Frankfurt am Main, 1591.

———. *Kurtzer Bericht, wie, was gestalt und warumb das Panacea am Waldina, als ein einige Medicin . . . anzuwenden seye*. Frankfurt am Main, 1592.

———. *Kurtzer und zum andernmal gemehrter Bericht wie, was Gestalt und warumb das Panacea am Waldina, als ein eynige Medicin . . . anzuwenden seye*. Hanau, 1594.

———. *Vortrab D. Georgen am vnd vom Wald . . . Auff Die in Truck außgefertigte Spott vnd Schmahkarten Andree Libauii*. Frankfurt am Main, 1595.

Anthony, Francis. *Panacea aurea, sive tractatus duo de insius auro potabili, nunc primum in Germania ex Londinensi exemplari excuse*. Hamburg, 1598.

———. *Panacea aurea, sive tractatus duo de ipsius auro potabili*, 1618.

Arduino, Sante. *Opus de venenis a multis hactenus desideratum*. Basel, 1562.

Avenzoar. *Liber theysir*. Venice, 1530.

Avicenna. *Liber canonis*. Hildesheim: G. Olms, 1964.

———. *Zusammengesetzte Heilmittel der Araber, nach dem fünften Buch des Canons von Ebn Sina*. Trans. Joseph von Sontheimer. Freiburg im Breisgau: Herder, 1845.

Bacci, Andrea. *L'alicorno. Discorso dell' eccellente medico, et filosofo M. Andrea Bacci*. Florence, 1573.

Bauhin, Caspar. *De lapidis bezaaris oriental. & occident. cervin. & germanici*. Basel, 1625.

Belon, Pierre. *Les observations de plusieurs singularitez et choses memorables, trouvees en Grece, Asie, Iudée, Egypte, Arabie*. Paris, 1553.

———. *Les observations de plvsievrs singvlaritez et choses memorables, trovvees en Grece, Asie, Iudée, Egypte, Arabie, & autres pays estranges*. Paris, 1588.

Benedetti, Alessandro. *De observatione in pestilentia*. Venice, 1493.

———. *Historia corporis humani, sive, Anatomice*. Ed. and trans. Giovanna Ferrari. Biblioteca della scienza italiana, 21. Florence: Giunti, 1998.

———. "History of the Human Body [1502]." In *Studies in Pre-Vesalian Anatomy: Biography, Translations, Documents*, ed. L. R. Lind. Philadelphia: American Philosophical Society, 1975.

Benivieni, Antonio. *De abditis nonnullis ac mirandis morborum e sanationum causis*. Trans. Charles Singer. Springfield, IL: Thomas, 1954.

———. *De abditis nonnullis ac mirandis morborum e sanationum causis*. Ed. Giorgio Weber. Florence: Olschki, 1994.

Beroaldo, Filippo. *Varia Philippi Beroaldi opuscula*. Basel, 1517.

Berthold, Andreas. *Außführliche Beweysung vnd Offenbarung / Wie / Wo / vnd bey weme die rechte vnd wahre Terra Sigillata in Deutschland zu finden / zu gewinnen / vnd zu bekommen sey*. Regensburg, 1594.

———. *Epistola Panegyrica / De vera et Germania Axungia Solis, sive Terra Sigillata*. N.p., 1586.

———. *Lob vnd Danckpsalmen, Auff den Frieden welcher Loeblich zu Alten Stettin*. Stettin, 1571.

———. *Newe Zeitung und Beschreibung von dem Friedeshandelstag zu Stettin*. Stettin, 1570.

———. *Terrae sigillatae nuper in Germania repertae*. Frankfurt am Main, 1583.

———. *The Vvonderfull and Strange Effect and Vertues of a New Terra Sigillata Lately Found out in Germanie*. London, 1587.

Bock, Hieronymus. *Kreüter Buch*. Strasbourg, 1546.

Boyle, Robert. *Some Considerations Touching the Usefulnesse of Experimental Naturall Philosophy*. London, 1664.

Braccesco, Giovanni. *La espositione di Geber philosopho*. Venice, 1544.

Brasavola, Antonio Musa. *De hominum aequalitate, & quare alter alterum excellat*. Venice, 1537.

———. *De medicamentis tam simplicibus, quam compositis catharticis*. Zürich, 1555.

Britannus, Bonifacius. *Antidotum contra veneficium sectarum hujus temporis*. Paris, 1565.

Brunschwig, Hieronymus. *Liber de arte distillandi de compositis*. Strasbourg, 1512.

———. *Liber de Arte Distillandi de Simplicibus*. Strasbourg: Johan Grüninger, 1500.

Calzolari, Francesco. *Francesco Calzolari da Verona e le sue lettere ad Ulisse Aldrovandi [1609]*. Rome: Tipografia Enrico Voghera, 1910.

Camerarius, Joachim. *De Theriacis et Mithridateis Commentariolus*. Nuremberg, 1533.

Cardano, Girolamo. *Hieronymi Cardani . . . De venenorum differentiis, viribus, & adversus ea remediorum praesidiis*. Basel, 1564.

Cardilucius, Johann Hiskias. *Magnalia Medico-Chymica; Oder, die Höchste Artzney- und Feuerkünstige Geheimnisse*. Nuremberg, 1676.

Castiglione, Baldassarre. *Lettere inedite e rare*. Ed. Guglielmo Gorni. Milan: Ricciardi, 1969.

———. *Lettere; ora per la prima volta date in luce e con annotazioni storiche illustrate dall'abate Pierantonio Serassi*. Ed. Pierantonio Serassi. Padua, 1769.

Catelan, Laurent. *Discours et demonstration des ingrediens de la theriaque*. Lyon, 1614.

Cellini, Benvenuto. *The Autobiography of Benvenuto Cellini*. Trans. George Bull. Harmondsworth: Penguin, 1956.

———. *My Life*. Trans. Julia Conaway Bondanella and Peter Bondanella. Oxford: Oxford University Press, 2002.

Celsus. *De Medicina*. Ed. and trans. W. G. Spencer. Cambridge: Harvard University Press, 1971. http://www.perseus.tufts.edu/hopper/text?doc=Perseus:text:1999.02.0142:book=1:chapter=pr&highlight=execution.

Ceruti, Benedetto. *Musaeum Franc. Calceolarii*. Verona, 1622.

Charas, Moyse. *New Experiments upon Vipers . . . Also a Tetter* [sic] *of Francisco Redi, Concerning Some Objections Made upon His Observations about Vipers*. London, 1677.

———. *Theriaque d'Andromacus: Avec Une Description Particuliere Des Plantes, Des Animaux & Des Mineraux Employez à Cette Grande Composition*. Nouvelle edition, Revûë & Augmentée. Paris, 1685.

Clusius, Carolus. *Exoticorvm libri decem: quibus animalium, plantarum, aromatum, aliorumq́ue peregrinorum fructuum historiae describuntur*. Antwerp, 1605.

Codronchi, Giovanni Baptista. *De morbis veneficis ac veneficiis libri quattuor*. Venice, 1595.

Collegio Medico Nuremberg. *Verneuerte Gesetz, Ordnung, und Tax . . . Deß Hey[ligen] Reichs Statt Nürnberg*. Nuremberg, 1704.

Commynes, Philippe de. *Preuves Des Mémoires de Philippe de Comines*. Paris, 1747.

Cordus, Euricus. *Von der vielfaltigen tugent vnnd waren bereitung / Deß rechten edlen Theriacs*. Marburg, 1532.

Cortese, Isabella. *I secreti de la signora Isabella Cortese*. Venice, 1565.

Crato von Krafftheim, Johannes. *Consiliorum et epistolarum medicinalium libri septem*. Vol. 2. Frankfurt am Main, 1671.

Ctesias. *Ctesias: On India, and Fragments of His Minor Works: Introduction, Translation and Commentary*. Ed. and trans. Andrew G. Nichols. London: Bristol Classical, 2011.

Dioscorides Pedanius of Anazarbos. *De medica materia libri V*. Cologne, 1529.

———. *Des Pedanios Dioskurides aus Anazarbos: Arzneimittellehre in fünf büchern*. Trans. J. Berendes. Wiesbaden: MSändig, 1970.

Dodart, Denis. *Mémoires pour servir à l'histoire des plantes*. Paris, 1676.

Drabitius, Lorenz. *Panacea, Das ist: Geistlich Recept und Heilwertige Artzney / vor allerhand Zufellen*. Erfurt, 1591.

Erasmus, Desiderius. *The Correspondence of Erasmus. Letters 1356 to 1534*. Ed. and trans. Alexander Dalzell. Toronto: University of Toronto Press, 1974.

Everaerts, Gilles. *De herba panacea quam alii tobacum*. Antwerp, 1587.

Falloppio, Gabriele. *Gabrielis Falloppii . . . Libelli duo: alter de ulceribus, alter de tumoribus praeter naturam . . .* Venice, 1563.

Ferri, Sara, ed. *Pietro Andrea Mattioli: Siena, 1501–Trento, 1578: la vita, le opere: con l'identificazione delle piante*. Ponte San Giovanni, Perugia: Quattroemme, 1997.

Fioravanti, Leonardo. *Compendium, oder, Ausszug der Secreten, Gehaymnissen und verborgenen Künsten*. Frankfurt am Main, 1604.

———. *Corona oder Kron der Artzney: . . . in vier sonderbare Bücher unterscheiden . . . Erst newlich in italiänischer Sprach . . . in Truck verfertiget. Nunmehr aber in unsere hochteutsche Sprach . . . versetzt*. Frankfurt am Main, 1618.

———. *M. Leonardi Fioravanti . . . Allgemeiner Kunst vnd Weltspiegels drey Bücher*. Trans. Lucas Jennis. Frankfurt am Main, 1619.

———. *Il tesoro della vita hvmana*. Venice, 1570.

———. *Three Exact Pieces of Leonard Phioravant Viz. His Rationall Secrets, and Chirurgery, Reviewed, and Revived*. Trans. John Hester. London, 1652.

Foligno, Gentile da. *Gentilis Fulginatis medici illustris contra pestilentiam consilium feliciter incipit*. Colle di Val d'Elsa, 1479. http://nrs.harvard.edu/urn-3:HMS.COUNT:1182903.

Fries, Lorenz. *Spiegel der Artzny*. Strasbourg, 1519.

Frummelt, Merten. *Terra Sigillata Das Ist / Kurzer Bericht von Krafft Tugent Vnd Wirckung Der Weissen Roten Vnd Grawen Erden / Welche Zur Ligenitz in Schlesien Gefunden*. Freiburg [?], 1585.

Galen. *Claudii Galeni opera omnia*. Ed. Karl Gottlob Kühn. Vol. 14. Hildesheim G. Olms, 1997.

———. *De theriaca ad Pisonem*. Ed. Enrico Coturri. Florence: L. S. Olschki, 1959.

———. *On Theriac to Piso, Attributed to Galen: A Critical Edition with Translation and Commentary*. Ed. and trans. Robert Leigh. Leiden: Brill, 2016.

Gessner, Conrad. *Epistolarium medicinalium, Conradi Gesneri, philosophi et medici Tigurini, libri 3*. Basel, 1577.

———. *Historiae animalium liber primus de quadrupedibus viuiparis*. Frankfurt am Main, 1602.

———. *Schlangenbuch: Das ist ein grundtliche und vollkom̄ne Beschreybung aller Schlangen, so im Meer, sussen Wassern und auff erden ir Wohnung haben*. Zürich, 1589.

Giovio, Paolo. *Elogia doctorum virorum ingenii monumentis illustrium*. Antwerp, 1557.

———. *Lettere inedite di Paolo Giovio tratte dall'archivio Gonzaga*. Ed. Alessandro Luzio. Mantua: Eredi Segna, 1885.

———. *Pauli Iovii Opera*. New ed. Vol. 1. Rome: Istituto Poligrafico dello Stato, Libreria dello Stato, 1956.

Glauber, Johann Rudolf. *Operis mineralis. Pars prima. . . . Item Panacea; sive Medicina universalis antimonialis, ejusque usus*. Amsterdam, 1651.

Gordon, Bernard de. "Ueber den Theriak, nach einer bisher ungedruckten Schrift des Bernhard von Gordon." Trans. J. L. Pagel. *Pharmaceutische Post* 27, Sonderabdruck (1894): 1–8.

G. R. *Panacea: Or, Select Aphorismes, Diuine and Morall*. London, 1630.

Grassi, Paride. *Il diario di Leone X di Paride de Grassi: dai volumi manoscritti degli archivi vaticani della S. Sede*. Rome: Tipografia della Pace di F. Guggiani, 1884.

Grévin, Jacques. *Devx livres des venins, ausquels il est amplement discouru des destes venimeuses, theriaques, poisons*. Antwerp, 1568.

Guaineri, Antonio. *Opera*. Venice, 1497.

Guiffrey, Jules, ed. *Inventaires de Jean Duc de Berry (1401–1416)*. Paris: Ernest Leroux, 1894.

Herring, Theodore. *Panacea Christiana. Or a Christians Soveraigne Salve for Every Soare*. London, 1624.

Hieblin, Christoph. *Tractat Von der aller furtrefflichsten vnd kräftigsten Artzney wider allerley Gifft: welches der Stein Bezaar ist*. Konstanz am Bodensee, 1598.

Hippocrates. *Opera Omnia*. Frankfurt am Main, 1595.

Hoberg, Hermann, ed. *Die Inventare des päpstlichen Schatzes in Avignon, 1314–1376*. Studi e testi (Biblioteca apostolica vaticana) 111. Vatican City: Biblioteca apostolica vaticana, 1944.

Horrox, Rosemary, ed. *The Black Death*. Manchester Medieval Sources Series. Manchester: Manchester University Press, 1994.

Isidore of Seville. *The Etymologies of Isidore of Seville*. Trans. Stephen A. Barney. Cambridge: Cambridge University Press, 2006.

Jābir ibn Ḥayyān. *Das Buch der Gifte des Ǧābir ibn Ḥayyān: arabischer Text in Faksimile*. Ed. and trans. Alfred Siggel. Wiesbaden: Franz Steiner Verlag, 1958.

Jordan, Thomas. *Pestis phaenomena*. Frankfurt, 1576.

Königshofen, Jakob Twinger von. *Die Alteste Teutsche so wol Allgemeine Als insonderheit Elsassische und Straßburgische Chronicke*. Städel, 1698.

Ein Kurzer Bericht / Wie Terra Sigillata Nützlich Kann Gebraucht Werden. 1585.

Lauterbach, Johannes. *Lauterbachius, pro Panacea Am Waldina*. Frankfurt am Main, 1592.

Lawrence O. Gostin, Cori Vanchieri, and Andrew MacPherson Pope. *Ethical Considerations for Research Involving Prisoners*. Washington, DC: National Academies Press, 2007.

Lenzi, Maria Ludovica. "Dal 'regno di Iatria' alla corte di Praga." In *Pietro Andrea Mattioli: Siena, 1501–Trento, 1578: la vita, le opere: con l'identificazione delle piante*, ed. Sara Ferri, 83–104. Ponte San Giovanni, Perugia: Quattroemme, 1997.

Libavius, Andreas. *Gegenbericht von Der Panacea Amwaldina*. Frankfurt am Main, 1595.

———. *Neoparacelsica*. Frankfurt am Main, 1594.

———. *Panacea Ambaldina victa et prostrata, das ist: Widerholter beständiger Gegenbericht von der überwundenen und todten Panaceae Ambaldina*. Frankfurt am Main, 1596.

Lindius, M. Auctorus. *Iustorum Panacea, Eine Christliche und zu Leidtmütiger Hertzen Trost gerichtete LeichPredigt*. Helmstadt, 1618.

Lonicer, Adam. *Kreuterbuch kunstliche Conterfeytunge der Bäume, Stauden, Hecken, Kreuter, Getreyde, Gewürtze*. Frankfurt am Main, 1587.

———. *Kreuterbuch, neu zugericht, künstliche Conterfeytunge der Bäume, Stauden, Hecken, Kreuter, Getreyde, Gewürtze*. Frankfurt am Main, 1557.

Ludwig, Christian Gottlieb. *Terrae Musei Regii Dresdensis*. Leipzig, 1749.

Lusitanus, Amatus. *Curationum medicinalium centuriae duae, prima & secunda*. Paris, 1554.

———. *In Dioscoridis Anazarbei De medica materia libros quinque*. Lyon, 1558. http://archive.org/details/indioscoridisan00lusigoog.

Maimonides, Moses. *On Poisons and the Protection against Lethal Drugs: A Parallel Arabic-English Edition*. Ed. Gerrit Bos. Trans. Gerrit Bos and M. R. McVaugh. 1st ed. Provo, UT: Brigham Young University Press, 2009.

Marius, Georg. *Terra sigillata: Bericht und Erforschung der kostbaren Erden*. Nuremberg, 1589.

Marra, Guglielmo de. *Opusculum quod dicitur Sertum Papale de Venenis*, Vatican Ms.Barb.Lat. 306. https://digi.vatlib.it/view/MSS_Barb.lat.306.

Mattioli, Pietro Andrea. *Les commentaires . . . sur les six livres des simples de Pedacius Dioscoride Anazarbeen: Traduit de latin en François*. Trans. Antoine Du Pinet. Lyon, 1566.

———. *Commentaires de M. Pierre André Matthiole . . . svr les six livres de Ped. Dioscoride*. Lyon, 1572.

———. *Commentarii in sex libros Pedacii Dioscoridis Anazarbei De medica materia*. Venice, 1565.

———. *De i discorsi di M. Pietro Andrea Matthioli*. Venice, 1585.

———. *Il Dioscoride: dell'eccellente dottor medico M. P. Andrea Matthioli da Siena: con i suoi discorsi*. Venice, 1548.

———. *Di Pedacio Dioscoride Anazarbeo libri cinque della historia, & materia medicinale*. Venice, 1544.

———. *I discorsi di M. Pietro Andrea Matthioli medico sanese, ne i sei libri della materia medicinale di Pedacio Dioscoride Anazarbeo*. Venice, 1555.

———. *I Discorsi . . . Ne i Sei Libri Della Materia Medicinale Di Pedacio Dioscoride Anazarbeo*. Venice, 1557.

———. *I discorsi nei sei libri di Pedacio Dioscoride*. Venice, 1568.

———. *Epistolarum medicinalium libri quinque*. Prague, 1561.

———. *Herbář: ginak Bylinář welmi vžitečný a Figůrami*. Prague, 1562.

———. *New Kreüterbuch*. Trans. Georg Handsch. Prague, 1563.

Michon, L.-A. Joseph, ed. *Documents inédits sur la grande peste de 1348: consultation de la Faculte de Paris: consultation d'un praticien de Montpellier: description de Guillaume de Machaut*. Open Collections Program at Harvard University. Paris: J.-B. Baillère et Fils; John Penington & Son, 1860. http://nrs.harvard.edu/urn-3:HMS.COUNT:3988474.

Möller, Andreas. *Theatrum Freibergense Chronicum: Beschreibung der alten löblichen BergHauptStadt Freyberg in Meissen*. Freiberg: Beuther, 1653.

Monardes, Nicolás. *De simplicibus medicamentis ex occidentali India delatis quorum in medicina usus est*. Trans. Carolus Clusius. Antwerp, 1574.

———. *Dos libros. El uno trata de todas las cosas que traen de nuestras Indias Occidentales, que sirven al uso de medicina . . . El otro libro, trata de dos medicinas maravillosas que son contra todo veneno, la piedra Bezaar, y la yerva Escuerçonera*. Seville, 1565.

———. *Joyfull Newes out of the Newfound World.* Trans. John Frampton. London, 1580.

———. *Libro qve trata dela nieve, y de suspropiedades.* Seville, 1571.

———. *Primera y segunda y tercera partes de la historia medicinal de las cosas que se traen de nuestras Indias Occidentales que siruen en medicina.* Seville, 1574.

———. *Segunda parte del libro, de las cosas que se traen de nuestras Indias Occidentales, que siruen al vso de medicina.* Seville, 1571.

———. *Simplicium medicamentorum ex novo orbe delatorum, quorum in medicina usus est.* Trans. Carolus Clusius. Antwerp, 1579.

Montanus, Johannes. *Breve, Sed Exquisitum, Vereque Philosophicum Iudicium . . . de vera, nativa, omnisq artis & fuci experte Terra Sigillata.* Breslau, 1597.

———. *Breve, sed Exquisitum, Vereque Philosophicum Iudicium . . . de vera, nativa, omnisq artis & fuci experte Terra Sigillata.* Breslau, 1610.

———. *Breve, sed exquisitum, vereque philosophicum Iudicium Doctrinis Mysterijsque varijs refertum . . . de vera, nativa, omnisque artis & fuci experte Terra Sigillata.* Nuremberg, 1585.

———. *Terrae Sigillatae Vtriusque, Sive Axungiæ Solis & Lunæ.* In Andreas Berthold, *Terrae sigillatae nuper in Germania repertae.* Frankfurt am Main, 1583.

Monte, Giovanni Baptista da. *In tertium primi epidemiorum (Hippocratis) sectionem explanationes.* Venice, 1554.

Moryson, Fynes. *Shakespeare's Europe: Unpublished Chapters of Fynes Moryson's Itinerary: Being a Survey of the Condition of Europe at the End of the 16th Century.* Ed. Charles Hughes. London: Sherratt & Hughes, 1903.

Novenianus, Philipp Michael. *Von den bösen vmbflechtenden Bauchflüssen vnd Durchlauff vrsprung.* Wittenberg, 1558.

Olivi, Gian Battista. *De reconditis et praecipuis Collectaneis ab honestissimo, et solertiss. Francisco Calceolario Veronensi in Musaeo adservatis.* Venice, 1584.

Orta, Garcia da. *Aliquot notae in Garciae Aromatum Historiam.* Trans. Carolus Clusius. Antwerp, 1582.

———. *Aromatum, et simplicium aliquot medicamentorum apud Indos nascentium.* Trans. Carolus Clusius. Antwerp, 1574.

———. *Colloquies on the Simples & Drugs of India.* Ed. and trans. Clements Markham. London: Sotheran, 1913.

———. *Coloquios dos simples, e drogas he cousas mediçinais da India e assi dalgu[m]as frutas.* Goa, 1563.

Orta, Garcia da, and Nicolás Monardes. *Dell'historia de i semplici aromati, et altre cose che vengono portate dall'Indie Orientali pertinenti all'vso della medicina.* Trans. Annibale Briganti. Venice, 1576.

Paracelsus. *Archidoxorum . . . de secretis naturae mysteriis libri decem.* Basel, 1570.

———. "The Diseases That Deprive Man of His Reason." In *Four Treatises of Theophrastus von Hohenheim,* ed. Henry E. Sigerist, trans. Gregory Zilboorg. Baltimore: Johns Hopkins Press, 1941.

———. *Das Erste Buch Der Grossen Wundarzney.* Ulm, 1536.

———. *Der Grossen Wundarzney Das Erste Buch.* Augsburg, 1537.

———. "Seven Defensiones." In *Four Treatises of Theophrastus von Hohenheim Called Paracelsus*, ed. Henry E. Sigerist, trans. C. Lilian Temkin, George Rosen, Gregory Zilboorg, and Henry E. Sigerist. Baltimore: Johns Hopkins Press, 1941.

Paré, Ambroise. *Ambroise Paré: d'après de nouveaux documents découverts aux Archives nationales et des papiers de famille*. Ed. Claude Stéphen Le Paulmier. Paris: Charavay Frères, 1884.

———. *Discours d'Ambroise Paré . . . asçavoit de la mumie, de la licorne, des venins, et de la peste*. Paris, 1582.

———. *Les Oeuvres d'Ambroise Paré, conseiller et premier chirurgien du roy, divisées en vingt-sept livres*. Paris, 1579.

———. *Les Oeuvres d'Ambroise Paré*. Lyon, 1633.

———. "Résponse de M. Ambroise Paré." In *Ambroise Paré: D'après de Nouveaux Documents Découverts Aux Archives Nationales et Des Papiers de Famille*, ed. Claude Stéphen Le Paulmier, 242–43. Paris: Charavay Frères, 1884.

Peinliche Hals-Gerichts-Ordnung des Georg Friederichen, Marggrafen zu Brandenburg: . . . Welcher Massen in J. F. G. Landen und Fürstenthumen in Peinlichen Sachen Einzuziehen. Onolzbach: Messerer, 1582.

Percival, Thomas. *Medical Ethics: Or, a Code of Institutes and Precepts, Adapted to the Professional Conduct of Physicians and Surgeons*. London, 1803.

Pliny the Elder. *Natural History*. Ed. and trans. John Bostock. London: Taylor and Francis, 1855. http://www.perseus.tufts.edu/hopper/text?doc=Plin.+Nat.+toc.

Ponzetti, Ferdinando. *Libellus de venenis*. Rome, 1521.

Poupard, Olivier. *Conseil divin touchant la malladie divine et peste en la ville de La Rochelle*, 1583. http://gallica.bnf.fr/ark:/12148/bpt6k73372t.

Raffaeli, Francesco. *Theriaca Andromachi Senioris Diuinum Inuentum: The Abouesaid Antidote Is Composed in Venice Vvith All Diligence and Careby Anthony Raffaeli Apothecary at the Signe of the Paradice at Riu'alta*. Venice, 1697.

Razi, Muhammad Ibn-Zakariya ar-. *Contenta in Hoc Volumine Liber Rasis Ad Almansorem*. Venice, 1497. http://daten.digitale-sammlungen.de/bsb00060841/image_79.

———. *Liber nonus ad Almansorem*. Venice: Ottino di Luna, 1497.

Redi, Francesco. *Francesco Redi on Vipers*. Ed. and trans. Peter K. Knoefel. Leiden: Brill, 1988.

———. *Opere di Francesco Redi*. Vol. 4. Milan, 1811.

———. *Osservazioni intorno alle vipere*. Florence, 1664.

Ristori, Renzo, and Price Zimmerman, eds. "Una lettera inedita di Paolo Giovio al cardinale Benedetto Accolti." *Archivio Storico Italiano* 122, no. 3 (1964): 505.

Rochlitz, Michael Bapst von. *Gifftjagendes Kunst vnnd Haussbuch*. Leipzig: Zacharias Berwaldt, 1591.

Royal College of Physicians of London. *Pharmacopœia Londinensis*. London, 1618.

Sala, Angelus. *Ternarius bezoarticorum, ou, Trois souverains medicaments bezoardiques, contre tous venins et empoisonnements*. Leiden, 1616.

———. *Ternarius ternariorum hemeticorum bezoardicorum laudanorum*. Erfurt, 1630.

Schmidt, Franz. *A Hangman's Diary; Being the Journal of Master Franz Schmidt, Public Executioner of Nuremberg, 1573–1617*. Ed. Albrecht Keller. New York: D. Appleton, 1928.

———. *Maister Franntzn Schmidts Nachrichters inn Nürmberg all sein Richten*. Ed. Albrecht Keller. Leipzig: W. Heim, 1913.

Schnellenberg, Tarquinius. *Experimenta von zwentzig pestilentz wurtzeln*. Frankfurt am Main, 1555.

Schönfeldt, Elias Draxdorf. *Panacea contra vim mortis, Das ist: Bewärt Kräutlein / daß den Todt kan vertreiben / Aus der grünen Awen des H. Evangelisten Johannis*. Jena, 1618.

Scot, Reginald. *The Discouerie of Witchcraft*. London: Henry Denham for William Brome, 1584.

Serres, Olivier du. *Théâtre d'agriculture et Mesnage des Champs*. Paris, 1600.

Sies, Rudolf. *Das "Pariser Pestgutachten" von 1348 in altfranzösischer Fassung*. Würzburger medizinhistorische Forschungen, vol. 7. Pattensen, Hanover: Wellm, 1977.

Silvaticus, Johannes Baptista. *De unicornu, lapide bezaar, smaragdo & margaritis: eorumq[ue] in febribus pestilen[tibus] usu tractatio*. Bergamo, 1605.

Sloane, Hans. "A Letter from Sir Hans Sloane Baronet, Late Pr. R. S. to Martin Folkes Esquire Pr. R. S. Containing Accounts of the Pretended Serpent-Stone Called Pietra de Cobra de Cabelos, and of the Pietra de Mombazza or the Rhinoceros Bezoar, Together with the Figure of a Rhinoceros with a Double Horn." *Philosophical Transactions (1683–1775)* 46, no. 491 (1749): 118–25. https://doi.org/10.1098/rstl.1749.0023.

Stein, Henri. *Charles de France, frère de Louis XI*. Mémoires et documents (Société de l'Ecole des chartes) 10. Paris: Picard, 1919.

Strotschius, Ibrahim. *De Electione et Inauguratione Maximiliani Austrii II. Rom. Regis, Francofurti Ad Moenum, Anno 1562*. Frankfurt am Main, 1563.

Testimonium de verissima ac admirabili virtute olei compositi contra pestem & omnia venena: de quo, jussu Clementis VII. Pont. Max. periculum fecere viri clariss. Rome, 1524.

Thurneisser zum Thurn, Leonhard. *Archidoxa dorin der recht war motus, lauff und gang, auch heymlikait, wirkung und krafft der planeten, gstirns und gantzen firmaments*. Münster, 1569.

———. *Historia vnnd Beschreibung influentischer, elementischer vnd natürlicher Wirckungen, aller fremden vnnd heimischen Erdgewechssen*. Berlin, 1578.

———. *Pison. Das erste Theÿl. Von kalten, warmen, minerischen und metalischen Wassern, sampt der ergleichunge der Plantarum und Erdgewechsen 10. Bücher*. Frankfurt an der Oder, 1572.

Tilburg, Cornelius. *By His Majesties License, and Special Approbation of the True Orvietan, or Antidote*. London, 1689.

Trials of War Criminals before the Nuremberg Military Tribunals under Control Council Law No. 10, October 1946–April 1949. Vol. 2. Washington, DC: US Government Printing Office, 1949.

Vigevano, Guido da. *Texarus Regis Franciae*, Bibliothèque Nationale de France, Ms. lat. 11015. https://gallica.bnf.fr/ark:/12148/btv1b520009994.image.

Villanova, Arnaldus de. *Arnaldi de Villanova Opera medica omnia*. Ed. Luis García Ballester, M. R. McVaugh, and J. A. Paniagua. 17 vols. Granada: Seminarium Historiae Medicae Granatensis, 1975.

———. "De arte cognisciendi venenis." In *Arnaldi Villanovani Opera omnia*. Basel, 1585.

———. "De venenis." In *Arnaldi Villanovani Opera omnia*. Basel, 1585.

———. "Epistola de dosi tyriacalium medicinarum." In *Arnaldi de Villanova Opera medica omnia*, ed. M. R. McVaugh, 3: 75–91. Granada: Seminarium Historiae Medicae Granatensis, 1985.

———. *Praxis medicinalis*. Leiden, 1586.

———. *Tractatus magistri Arnaldi de Villa Nova de arte cognosce[n]di venena cum quis timet sibi ea ministrari*. Rome, 1475. http://nrs.harvard.edu/urn-3:HMS.COUNT:1170267.

Wahshīya, Ibn. "The Book on Poisons of Ibn Wahshīya." Ed. Martin Levey. *Transactions of the American Philosophical Society* 56 (1966): 1–130.

Wepfer, Johann Jakob. *Cicutae aquaticae historia et noxae commentario*. Basel, 1679.

Willius, Johann Valentin. *Der nordische Bezoar, oder, Das wunderwürdige Hirsch-Horn-Saltz in leichten Reimen aus naturlichen und medicinischen Gründen*. Copenhagen, 1676.

Wittich, Johann. *Bericht von der wunderbaren bezoardischen Steinen*. Leipzig, 1592.

———. *Consilium apoplecticum*. Leipzig, 1593.

———. *Vade Mecum: Der ander Theil des kuenstlichen Newen Artzney-Buchs*. Leipzig, 1607.

———. *Von dem ligno gvayaco, Wunderbawn, res noua genandt*. Leipzig, 1592.

Wunder Artzney, vonn allerley Leibs Gebrechen, unnd zu fallenden Kranckheiten. Basel, 1573.

Zamora, Juan Gil de, and Cándida Ferrero Hernández. "Liber contra venena et animalia venenosa. Estudio preliminar, edición crítica y traducción." Universitat Autònoma de Barcelona, 2002. http://www.tdx.cat/TDX-0505103–183401.

Zimmermann, Justus. *Carmina Iusti Zimmermanni . . . pro Panacea am Waldina ad nobilissimum virum, dominum Georgium am et vom Wald*. Frankfurt am Main, 1592.

Secondary Sources

Agrimi, Jole, and Chiara Crisciani. "Per una ricerca su experimentum-experimenta: riflessione epistemologica e tradizione medica (secoli XIII–XV)." In *Presenza del lessico Greco e Latino nelle lingue contemporanee: ciclo di lezioni tenute all'Università di Macerata nell'a.a. 1987/88*, ed. Pietro Janni and Innocenzo Mazzini, 9–49. Macerata: Università degli studi, 1990.

Andretta, Elisa. "Pre- and Post-Mortem Inquiries: Assessing Poisoning in the Law Courts of Sixteenth-Century Rome." In *Pathology in Practice: Diseases and*

Dissections in Early Modern Europe, ed. Silvia De Renzi, Marco Bresadola, and Maria Conforti, 155–69. London: Routledge, 2017.

———. *Roma medica: anatomie d'un système médical au XVIe siècle*. Rome: École française de Rome, 2011.

Ardestani, Mozhgan M., Roja Rahimi, Mohammad M. Esfahani, Omar Habbal, and Mohammad Abdollahi. "The Golden Age of Medieval Islamic Toxicology." In *Toxicology in the Middle Ages and Renaissance*, ed. Philip Wexler, 11–30. London: Academic Press, 2017.

Arrizabalaga, Jon. "*Pestis Manufacta*: Plague, Poisons and Fear in Mid-Fourteenth-Century Europe." In *"It All Depends on the Dose": Poisons and Medicines in European History*, ed. Ole Peter Grell, Andrew Cunningham, and Jon Arrizabalaga, 62–80. Abingdon, UK: Routledge, 2018.

Azar, Henry. *Sage of Seville: Ibn Zuhr, His Time and His Medical Legacy*. Cairo: American University in Cairo Press, 2000.

Bakhtin, Mikhail. "Carnival and the Carnivalesque." In *Cultural Theory and Popular Culture: A Reader*, ed. John Storey, 254–59. Harlow, UK: Pearson Education, 1998.

———. *Rabelais and His World*. 1st Midland book ed. Bloomington: Indiana University Press, 1984.

Baldwin, John W. "The Intellectual Preparation for the Canon of 1215 against Ordeals." *Speculum* 36, no. 4 (1961): 613–36.

Baldwin, Martha. "The Snakestone Experiments: An Early Modern Medical Debate." *Isis* 86 (1995): 394–418.

Barkai, Ron. "Jewish Treatises on the Black Death (1350–1500): A Preliminary Study." In *Medicine from the Black Death to the French Disease*, ed. R. K. French, Jon Arrizabalaga, Andrew Cunningham, and Luis García Ballester, 6–25. Aldershot, UK: Ashgate, 1998.

Barker, Sheila. "Poisons and the Prince: Toxicology and Statecraft at the Medici Grand Ducal Court." In *Toxicology in the Middle Ages and Renaissance*, ed. Philip Wexler, 71–81. London: Academic Press, 2017.

Barrera-Osorio, Antonio. *Experiencing Nature: The Spanish American Empire and the Early Scientific Revolution*. Austin: University of Texas Press, 2006.

———. "Translating Facts: From Stories to Observations in the Work of Seventeenth-Century Dutch Translators of Spanish Books." In *Translating Knowledge in the Early Modern Low Countries*, ed. Harold John Cook and Sven Dupré, 317–32. Zürich: LIT Verlag Münster, 2012.

Bayle, Adrienne. "Thériaque et triacleurs chez Pierre-André Mathiole." In *Poison et antidote dans l'Europe des XVIe et XVIIe siècles*, ed. Sarah Voinier and Guillaume Winter. Arras: Artois presses université, 2011.

Beaty, Nancy Lee. *The Craft of Dying: A Study in the Literary Tradition of the Ars Moriendi in England*. New Haven: Yale University Press, 1970.

Bedini, Silvio A. *The Pope's Elephant*. New York: Penguin Books, 2000.

Bednarski, Steven. *A Poisoned Past: The Life and Times of Margarida de Portu, a Fourteenth-Century Accused Poisoner*. Toronto: University of Toronto Press, 2014.

Beecher, Donald. "Nicolás Monardes, John Frampton and the Medical Wonders of the New World." In *Humanismo e Ciência: Antiguidade e Renascimento*, ed. António Manuel Lopes Andrade, Carlos de Miguel Mora, and João Manuel Nunes Torrão, 141–60. Coimbra: Imprensa da Universidade de Coimbra, 2015.

Behringer, Wolfgang. *Witchcraft Persecutions in Bavaria: Popular Magic, Religious Zealotry and Reason of State in Early Modern Europe*. Cambridge: Cambridge University Press, 1997.

Bentz, Katherine M. "Ulisse Aldrovandi, Antiquities, and the Roman Inquisition." *Sixteenth Century Journal* 43 (2012): 963–88.

Bergdolt, Klaus. "Die Pest und die Juden—Mythen, Fakten, Topoi." *Aschkenas: Zeitschrift für Geschichte und Kultur der Juden* 29, no. 1 (2019): 43–62.

Bever, Edward. "Witchcraft, Female Aggression, and Power in the Early Modern Community." *Journal of Social History* 35, no. 4 (2002): 955–88.

Biagioli, Mario. *Galileo, Courtier: The Practice of Science in the Culture of Absolutism*. Chicago: University of Chicago Press, 1993.

———. "Knowledge, Freedom, and Brotherly Love: Homosociality and the Accademia dei Lincei." *Configurations* 3, no. 2 (1995): 139–66.

Bijloos, J. *Adrianus VI, de Nederlandse paus*. Haarlem: Fibula-Van Dishoeck, 1980.

Birnbaum, Marianna D. *The Long Journey of Gracia Mendes*. Budapest: Central European University Press, 2003.

Blair, Ann. "Historia in Zwinger's *Theatrum humanae vitae*." In *Historia: Empiricism and Erudition in Early Modern Europe*, ed. Gianna Pomata and Nancy G. Siraisi, 269–96. Cambridge: MIT Press, 2005.

Blécourt, Willem de, and Cornelie Usborne, eds. *Cultural Approaches to the History of Medicine: Mediating Medicine in Early Modern and Modern Europe*. New York: Palgrave Macmillan, 2004.

Bleichmar, Daniela. "Books, Bodies, and Fields: Sixteenth-Century Transatlantic Encounters with New World Materia Medica." In *Colonial Botany: Science, Commerce, and Politics in the Early Modern World*, ed. Londa L. Schiebinger and Claudia Swan, 83–99. Philadelphia: University of Pennsylvania Press, 2005.

———. "Seeing the World in a Room: Looking at Exotica in Early Modern Collections." In *Collecting across Cultures: Material Exchanges in the Early Atlantic World*, ed. Daniela Bleichmar and Peter C. Mancall. Early Modern Americas. Philadelphia: University of Pennsylvania Press, 2011.

———. *Visible Empire: Botanical Expeditions and Visual Culture in the Hispanic Enlightenment*. Chicago: University of Chicago Press, 2012.

Bleichmar, Daniela, and Peter C. Mancall, eds. *Collecting across Cultures: Material Exchanges in the Early Atlantic World*. Philadelphia: University of Pennsylvania Press, 2011.

Borschberg, Peter. "The Euro-Asian Trade in Bezoar Stones (Approx. 1500 to 1700)." In *Artistic and Cultural Exchanges between Europe and Asia, 1400–1900*, ed. Michael North, 29–43. Aldershot, UK: Ashgate, 2010.

Bouley, Bradford. *Pious Postmortems: Anatomy, Sanctity, and the Catholic Church in Early Modern Europe*. Philadelphia: University of Pennsylvania Press, 2017.

Boyden, James M. "The Worst Death Becomes a Good Death." In *The Place of the Dead: Death and Remembrance in Late Medieval and Early Modern Europe*, 240–65. Cambridge: Cambridge University Press, 2000.

Brackett, John K. *Criminal Justice and Crime in Late Renaissance Florence, 1537–1609*. Cambridge: Cambridge University Press, 1992.

Brater, D. Craig, and Walter J. Daly. "Clinical Pharmacology in the Middle Ages: Principles That Presage the 21st Century." *Clinical Pharmacology & Therapeutics* 67, no. 5 (2000): 447–50.

Breen, Benjamin. *Age of Intoxication: Origins of the Global Drug Trade*. Philadelphia: University of Pennsylvania Press, 2019.

Brettel, H. F., and D. Emrich. "Criminal Law and Forensic Medicine in the 16th Century." *Beiträge zur gerichtlichen Medizin* 49 (1991): 171–74.

Brévart, Francis B. "Between Medicine, Magic, and Religion: Wonder Drugs in German Medico-Pharmaceutical Treatises of the Thirteenth to the Sixteenth Centuries." *Speculum* 83 (2008): 1–57.

Buchrucker, Armin-Ernst. "Wer so stirbt, der stirbt wohl. Martin Luther über die Kunst zu sterben." *Luther* 64 (1993): 25.

Burgaleta, Claudio M. *José de Acosta, S.J., 1540–1600: His Life and Thought*. Chicago: Jesuit Way, 1999.

Cabanès, Augustin. *Poisons et sortilèges*. Paris: Plon-Nourrit, 1903.

Cagle, Hugh. *Assembling the Tropics: Science and Medicine in Portugal's Empire, 1450–1700*. Cambridge: Cambridge University Press, 2018.

———. "Cultures of Inquiry, Myths of Empire: Natural History in Colonial Goa." In *Medicine, Trade and Empire: Garcia de Orta's Colloquies on the Simples and Drugs of India (1563) in Context*, ed. Palmira Fontes da Costa, 107–28. Farnham, Surrey, UK: Ashgate, 2015.

Carlino, Andrea. *Books of the Body: Anatomical Ritual and Renaissance Learning*. Chicago: University of Chicago Press, 1999.

Carusi, Paola. "Il filosofo e il marinaio: alchemia islamica e medicina alle prese con la natura." In *Alchimia e medicina nel Medioevo*, ed. Chiara Crisciani and Agostino Paravicini Bagliani, 19–32. Tavarnuzze (Florence): SISMEL edizioni del Galluzzo, 2003.

Catellani, Patrizia. *L'orvietano*. Collana di studi (Accademia di scienze, lettere e arti, Modena, Italy) 25. Pisa: ETS, 2004.

Cavallo, Sandra, and Tessa Storey. *Conserving Health in Early Modern Culture: Bodies and Environments in Italy and England*. Manchester: Manchester University Press, 2017.

Černá, Jana. "A Powerful Antidote, a Strange Camel and Turkish Pepper: Iberian Science, the Discovery of the New World and the Early Modern Czech Lands." *Early Science and Medicine* 21 (2016): 214–31.

Chiaramonte, Enrica, and Silvia Tozzi. "Un medico umanista fra dottrina e pratica." In *Pietro Andrea Mattioli: Siena, 1501–Trento, 1578: la vita, le opere: con l'identificazione delle piante*, ed. Sara Ferri, 61–81. Ponte San Giovanni, Perugia: Quattroemme, 1997.

Christen, Arden G., and Joan A. Christen. "The Unicorn and the Narwhal: A Tale of the Tooth." *Journal of the History of Dentistry* 59, no. 3 (2011): 135–42.

Cicogna, Emmanuele Antonio. *Delle inscrizioni Veneziane*. Venice: Orlandelli, 1830.

Cilliers, Louise, and Francois Retief. "Poisons, Poisoners, and Poisoning in Ancient Rome." In *History of Toxicology and Environmental Health*, ed. Philip Wexler, 1: 127–37. Boston: Academic Press, 2014.

Clayton, Martin. "Leonardo's Anatomy Years." *Nature* 484, no. 7394 (2012): 314–16.

Clayton, Martin, and Ron Philo, eds. *Leonardo da Vinci, Anatomist*. London: Royal Collection Publications, 2012.

Clouse, Michele L. *Medicine, Government and Public Health in Philip II's Spain: Shared Interests, Competing Authorities*. London: Taylor and Francis, 2011.

Cobb, Cathy. "The Case against the Borgias: Motive, Opportunity, and Means." In *Toxicology in the Middle Ages and Renaissance*, ed. Philip Wexler, 53–62. London: Academic Press, 2017.

Cohen, Thomas V. *Love and Death in Renaissance Italy*. Chicago: University of Chicago Press, 2004.

Cohn, Samuel K. "The Black Death and the Burning of Jews." *Past & Present*, no. 196 (2007): 3–36.

Collard, Franck. *The Crime of Poison in the Middle Ages*. Trans. Deborah Nelson-Campbell. Westport, CT: Praeger, 2008.

———. "Ouvrir pour découvrir. Réflexions sur les expertises de cadavres empoisonnés à l'époque médiévale." In *Le corps à l'épreuve: poisons, remèdes et chirurgie: aspects des pratiques médicales dans l'antiquité et au Moyen Age*, ed. Franck Collard and Evelyne Samama, 177–90. Langres: D. Guéniot, 2002.

———. "Pharmaca aut venena, pharmaca sive venena. Réflexions sur les perceptions d'une contiguïté durant le Moyen Âge latin." In *Le corps à l'épreuve: poisons, remèdes et chirurgie: aspects des pratiques médicales dans l'antiquité et au Moyen Age*, ed. Franck Collard and Evelyne Samama, 105–22. Langres: D. Guéniot, 2002.

———. "Le poison et le sang dans la culture médiévale." *Médiévales*, no. 60 (2011): 129–55.

———. *Pouvoir et poison: histoire d'un crime politique de l'antiquité à nos jours*. Paris: Seuil, 2007.

Collard, Franck, and Evelyne Samama, eds. *Le corps à l'épreuve: poisons, remèdes et chirurgie: aspects des pratiques médicales dans l'antiquité et au Moyen Age*. Langres: D. Guéniot, 2002.

Collins, Harry M. *The Golem: What You Should Know about Science*. Cambridge: Cambridge University Press, 1998.

———. "Studies of Expertise and Experience." *Topoi* 37 (2018): 67–77.

Collins, Harry M., and T. J. Pinch. *Dr. Golem: How to Think about Medicine*. Chicago: University of Chicago Press, 2005.

Conforti, Maria. "Visible Signs, Invisible Processes: Explaining Poison in the Late Seventeenth Century." In *Pathology in Practice: Diseases and Dissections in Early*

Modern Europe, ed. Silvia De Renzi, Marco Bresadola, and Maria Conforti, 116–34. London: Routledge, 2017.

Cook, Harold J. *The Decline of the Old Medical Regime in Stuart London*. Ithaca: Cornell University Press, 1986.

———. "Global Economies and Local Knowledge in the East Indies: Jacobus Bontius Learns the Facts of Nature." In *Colonial Botany: Science, Commerce, and Politics in the Early Modern World*, ed. Londa L. Schiebinger and Claudia Swan, 100–118. Philadelphia: University of Pennsylvania Press, 2005.

———. *Matters of Exchange: Commerce, Medicine, and Science in the Dutch Golden Age*. New Haven: Yale University Press, 2007.

———. "The New Philosophy and Medicine in Seventeenth-Century England." In *Reappraisals of the Scientific Revolution*, ed. David C. Lindberg and Robert S. Westman, 397–436. Cambridge: Cambridge University Press, 1990.

———. "Sir John Colbatch and Augustan Medicine: Experimentalism, Character and Entrepreneurialism." *Annals of Science* 47 (1990): 475–505.

———. "Victories for Empiricism, Failures for Theory: Medicine and Science in the Seventeenth Century." In *The Body as Object and Instrument of Knowledge*, ed. Charles T. Wolfe and Ofer Gal, 9–32. Dordrecht: Springer, 2010.

Cook, Harold J., and Sven Dupré. *Translating Knowledge in the Early Modern Low Countries*. Zürich: LIT Verlag Münster, 2012.

Cook, Harold J., and Timothy D. Walker. "Circulation of Medicine in the Early Modern Atlantic World." *Social History of Medicine* 26, no. 3 (2013): 337–51.

Cooper, Alix. *Inventing the Indigenous: Local Knowledge and Natural History in Early Modern Europe*. Cambridge: Cambridge University Press, 2007.

Corradi, Alfonso. "Degli esperimenti tossocologici in anima nobili nel cinquecento." *Annali universali di medicina e chirurgica* 277 (1886): 73–100.

———. "Degli esperimenti tossocologici in anima nobili nel cinquecento." In *Memorie del R. Istituto Lombardo di scienze e lettere. Classe di Scienze matematiche e naturali*, 31–53. Milan: Tip. Bernadoni di C. Rebeschini, 1891.

Costa, Palmira Fontes da. "Geographical Expansion and the Reconfiguration of Medical Authority: Garcia de Orta's Colloquies on the Simples and Drugs of India (1563)." *Studies in History and Philosophy of Science Part A*, 43, no. 1 (March 2012): 74–81.

———, ed. *Medicine, Trade and Empire: Garcia de Orta's Colloquies on the Simples and Drugs of India (1563) in Context*. Farnham, Surrey, UK: Ashgate, 2015.

Crawford, Matthew James. *The Andean Wonder Drug: Cinchona Bark and Imperial Science in the Spanish Atlantic, 1630–1800*. Pittsburgh, PA: University of Pittsburgh Press, 2016.

Crawford, Matthew James, and Joseph M. Gabriel, eds. *Drugs on the Page: Pharmacopoeias and Healing Knowledge in the Early Modern Atlantic World*. Pittsburgh: University of Pittsburgh Press, 2019.

Creager, Angela N. *Life Atomic: A History of Radioisotopes in Science and Medicine*. Chicago: University of Chicago Press, 2013.

Crombie, A. C. "Avicenna's Influence on the Mediaeval Scientific Tradition." In *Avicenna: Scientist & Philosopher: A Millenary Symposium*, 84–107. London: Luzac, 1952.

———. *Science, Optics, and Music in Medieval and Early Modern Thought.* London: Hambledon Press, 1990.

Crystal, Malcolm Lee. "Medicine in Vienna in the Sixteenth and Seventeenth Centuries." PhD diss., University of Virginia, 1994.

Cunningham, Andrew. *The Anatomical Renaissance.* Brookfield, VT: Scholar Press, 1997.

———. *The Anatomist Anatomis'd: An Experimental Discipline in Enlightenment Europe.* Farnham, Surrey, UK: Ashgate, 2010.

———. Introduction. In *"It All Depends on the Dose": Poisons and Medicines in European History*, ed. Ole Peter Grell, Andrew Cunningham, and Jon Arrizabalaga, 1–8. Abingdon, UK: Routledge, 2018.

Curry, H. A., Nicholas Jardine, James A. Secord, and E. C. Spary, eds. *Worlds of Natural History.* Cambridge: Cambridge University Press, 2018.

Dackerman, Susan. "Dürer's Indexical Fantasy: The Rhinoceros and Printmaking." In *Prints and the Pursuit of Knowledge in Early Modern Europe*, ed. Susan Dackerman, 164–83. Cambridge, MA: Harvard Art Museums/Yale University Press, 2011.

Daly, Walter J., and D. Craig Brater. "Medieval Contributions to the Search for Truth in Clinical Medicine." *Perspectives in Biology and Medicine* 43, no. 4 (2000): 530–40.

Dann, Georg Edmund. *Einführung in die Pharmaziegeschichte.* Stuttgart: Wissenschaftliche Verlagsgesellschaft, 1975.

Dannenfeldt, Karl H. "The Introduction of a New Sixteenth-Century Drug, Terra Silesiaca." *Medical History* 28 (1984): 174–88.

Daston, Lorraine. "Observation." In *Prints and the Pursuit of Knowledge in Early Modern Europe*, ed. Susan Dackerman, 126–33. Cambridge, MA: Harvard Art Museums/Yale University Press, 2011.

Daston, Lorraine, and Elizabeth Lunbeck, eds. *Histories of Scientific Observation.* Chicago: University of Chicago Press, 2011.

Daston, Lorraine, and Katharine Park. *Wonders and the Order of Nature, 1150–1750.* New York: Zone Books, 1998.

Davies, Surekha. *Renaissance Ethnography and the Invention of the Human: New Worlds, Maps and Monsters.* Cambridge: Cambridge University Press, 2016.

Dean, Trevor. "Getting out of Jail: Suicide, Escape and Release in Late Medieval and Renaissance Bologna." *Historical Research* 90, no. 249 (2017): 449–64.

Dear, Peter. *Discipline and Experience: The Mathematical Way in the Scientific Revolution.* Chicago: University of Chicago Press, 1995.

———. "The Meanings of Experience." In *The Cambridge History of Science*, vol. 3: *Early Modern Science*, ed. Katharine Park and Lorraine Daston, 106–31. Cambridge: Cambridge University Press, 2006.

———. "Narratives, Anecdotes, and Experiments: Turning Experience into Science in the Seventeenth Century." In *The Literary Structure of Scientific Argument*, 135–63. Philadelphia: University of Pennsylvania Press, 1991.

Debru, Armelle, ed. *Galen on Pharmacology: Philosophy, History, and Medicine*. Vol. 16. Studies in Ancient Medicine, 0925–1421. Leiden: Brill, 1997.

Delbourgo, James. *Collecting the World: Hans Sloane and the Origins of the British Museum*. Cambridge: Belknap Press of Harvard University Press, 2017.

Delisle, Candice. "'The Spices of Our Art': Medical Observation in Conrad Gesner's Letters." In *Communicating Observations in Early Modern Letters (1500–1675): Epistolography and Epistemology in the Age of the Scientific Revolution*, ed. Dirk van Miert, 27–42. London: Warburg Institute, 2013.

Demaitre, Luke E. *Doctor Bernard de Gordon: Professor and Practitioner*. Toronto: Pontifical Institute of Mediaeval Studies, 1980.

De Renzi, Silvia. "Seats and Series: Dissecting Disease in the Seventeenth Century." In *Pathology in Practice: Diseases and Dissections in Early Modern Europe*, ed. Silvia De Renzi, Marco Bresadola, and Maria Conforti, 96–115. London: Routledge, 2017.

De Renzi, Silvia, Marco Bresadola, and Maria Conforti, eds. *Pathology in Practice: Diseases and Dissections in Early Modern Europe*. London: Routledge, 2017.

DeVun, Leah. *Prophecy, Alchemy, and the End of Time: John of Rupescissa in the Late Middle Ages*. New York: Columbia University Press, 2009.

Dickson, Sarah. *Panacea or Precious Bane: Tobacco in Sixteenth-Century Literature*. New York: Public Library, 1954.

DiMeo, Michelle, and Sara Pennell, eds. *Reading and Writing Recipe Books, 1550–1800*. Manchester: Manchester University Press, 2013.

Distelberger, Rudolf. "The Hapsburg Collection in Vienna during the Seventeenth Century." In *The Origins of Museums: The Cabinet of Curiosities in Sixteenth- and Seventeenth-Century Europe*, ed. O. R. Impey and Arthur MacGregor, 39–46. Oxford: Ashmolean, 2017.

Dols, Michael W. *The Black Death in the Middle East*. Princeton: Princeton University Press, 1977.

Döring, Eduard. *Handbuch der Münz-, Wechsel-, Mass- und Gewichtskunde*. 1854.

Dousset, Jean-Claude. *Histoire des médicaments: des origines à nos jours*. Nice: Ovadia, 2009.

Duffin, Christopher J. "Bezoar Stones and Their Mounts." *Jewellery History Today*, no. 16 (2013): 3–4.

———. "Porcupine Stones." *Pharmaceutical Historian* 43, no. 1 (2013): 13–22.

Dülmen, Richard van. *Theatre of Horror: Crime and Punishment in Early Modern Germany*. Cambridge, UK: Polity Press, 1990.

Dyck, Erika, and Larry Stewart, eds. *The Uses of Humans in Experiment: Perspectives from the 17th to the 20th Century*. Clio Medica, vol. 95. Leiden: Brill, 2016.

Eamon, William. "How to Read a Book of Secrets." In *Secrets and Knowledge in Medicine and Science, 1500–1800*, ed. Elaine Leong and Alisha Rankin, 23–46. Farnham, Surrey, UK: Ashgate, 2011.

———. "Masters of Fire: Italian Alchemists at the Court of King Philip II." In *Chymia: Science and Nature in Medieval and Early Modern Europe*, ed. Miguel López Pérez, Didier Kahn, and Mar Rey Bueno, 138–56. Newcastle upon Tyne: Cambridge Scholars, 2010.

———. *The Professor of Secrets: Mystery, Medicine, and Alchemy in Renaissance Italy*. Washington, DC: National Geographic, 2010.

———. *Science and the Secrets of Nature: Books of Secrets in Medieval and Early Modern Culture*. Princeton: Princeton University Press, 1994.

Edelstein, Bruce. "Leone Leoni, Benvenuto Cellini and Francesco Vinta, a Medici Agent in Milan." *Sculpture Journal* 4 (2000): 35–45.

Edgerton, Samuel Y. *Pictures and Punishment: Art and Criminal Prosecution during the Florentine Renaissance*. Ithaca: Cornell University Press, 1985.

Egmond, Florike. "Figuring Exotic Nature in Sixteenth-Century Europe: Garcia Da Orta and Carolus Clusius." In *Medicine, Trade and Empire: Garcia de Orta's Colloquies on the Simples and Drugs of India (1563) in Context*, ed. Palmira Fontes da Costa, 167–94. Farnham, Surrey, UK: Ashgate, 2015.

———. *The World of Carolus Clusius: Natural History in the Making, 1550–1610*. London: Pickering & Chatto, 2010.

Egmond, Florike, and Sachiko Kusukawa. "Circulation of Images and Graphic Practices in Renaissance Natural History: The Example of Conrad Gessner." *Gesnerus* 73 (2016): 29–72.

Eichhorn, Gertraud. "Der Paracelsist Georg am Wald (1554–1616), Ein Gebürtiger Passauer: Ein Bericht." *Ostbairische Grenzmarken* 38 (1996): 35–46.

Fabbri, Christiane Nockels. "Treating Medieval Plague: The Wonderful Virtues of Theriac." *Early Science and Medicine* 12, no. 3 (2007): 247–83.

Fabiani, Giuseppe. *La vita di Pietro Andrea Mattioli: Raccolta dalle sue Opere*. Siena: Bargellini, 1872. http://hdl.handle.net/2027/hvd.hn4uer.

Feinberg, Larry J. "Imagination All Compact: Tavolette and Confraternity Rituals for the Condemned in Renaissance Italy." *Apollo* 161, no. 519 (May 1, 2005): 48–57.

Ferrari, A. de. "Matteo Corti." In *Dizionario biografico degli Italiani*, ed. Alberto Maria Ghisalberti, 29: 795–97. Rome: Istituto della Enciclopedia italiana, 1960.

Ferrari, Giovanna. *L'esperienza del passato: Alessandro Benedetti filologo e medico umanista*. Florence: L. S. Olschki, 1996.

———. "Public Anatomy Lessons and the Carnival: The Anatomy Theatre of Bologna." *Past & Present*, no. 117 (1987): 50–106.

Ferri, Sara. "Il 'Dioscoride,' i 'Discorsi,' i 'Commentarii': gli amici e i nemici." In *Pietro Andrea Mattioli: Siena, 1501–Trento, 1578: la vita, le opere: con l'identificazione delle piante*, ed. Ferri, 15–48. Ponte San Giovanni, Perugia: Quattroemme, 1997.

———, ed. *Pietro Andrea Mattioli: Siena, 1501–Trento, 1578: la vita, le opere: con l'identificazione delle piante*. Ponte San Giovanni, Perugia: Quattroemme, 1997.

Figueroa, Luis Millones. "The Bezoar Stone: A Natural Wonder in the New World." *Hispanófila* 171 (2014): 139–56.

Findlen, Paula. "Controlling the Experiment: Rhetoric, Court Patronage, and the Experimental Method of Francesco Redi." *History of Science* 31 (1993): 35–64.

———. "The Formation of a Scientific Community in Sixteenth-Century Italy." In *Natural Particulars: Nature and the Disciplines in Renaissance Europe*, ed. Anthony Grafton and Nancy G. Siraisi, 369–400. Cambridge: MIT Press, 1999.

———. *Possessing Nature: Museums, Collecting, and Scientific Culture in Early Modern Italy*. Studies on the History of Society and Culture 20. Berkeley: University of California Press, 1994.

———. "Possessing the Past: The Material World of the Italian Renaissance." *American Historical Review* 103, no. 1 (1998): 83–114.

Fischer, Adolf. "Die älteste evangelische Kirchenordnung und die frühesten Kirchenvisitationen in Hohenlohe." *Zeitschrift fur Kirchenrecht* 6 (1880).

———. *Geschichte des Hauses Hohenlohe*. Vol. 2. Stuttgart: Alfred Müller, 1868.

Fischer, Louis-Paul, and Véronique Cossu Ferra Fischer. "La licorne et la corne de licorne chez les apothicaires et les médecins." *Histoire des sciences medicales* 45 (2011): 265–74.

Fischer-Homberger, Esther. *Medizin vor Gericht: Gerichtsmedizin von der Renaissance bis zur Aufklärung*. Bern: Huber, 1983.

Fortuna, Stefania. "The Latin Editions of Galen's *Opera Omnia* (1490–1625) and Their Prefaces." *Early Science and Medicine* 17, no. 4 (September 1, 2012): 391–412.

Fosi, Irene. *Papal Justice: Subjects and Courts in the Papal State, 1500–1750*. Washington, DC: Catholic University of America Press, 2011.

———. "Signori e tribunali. Criminalità nobiliare e giustizia pontificia nella Roma del Cinquecento." In *Signori, patrizi, cavalieri in Italia centro-meridionale nell'età moderna*, ed. Maria Antonietta Visceglia, 214–30. Rome: Laterza, 1992.

Foucault, Michel. *The Birth of the Clinic: An Archaeology of Medical Perception*. New York: Pantheon Books, 1973.

———. *Discipline and Punish: The Birth of the Prison*. 1st American ed. New York: Pantheon Books, 1977.

———. *The Order of Things: An Archaeology of the Human Sciences*. New York: Pantheon Books, 1971.

———. "La vérité et les formes juridiques." In *Dits et écrits: 1954–1988*, vol. 2. Paris: Gallimard, 1994.

French, Roger K. "The Medical Ethics of Gabriele de Zerbi." In *Doctors and Ethics: The Earlier Historical Setting of Professional Ethics*, ed. Johanna Geyer-Kordesch, Andrew Wear, and Roger K. French, 72–97. Amsterdam: Rodopi, 1993.

———. *William Harvey's Natural Philosophy*. Cambridge: Cambridge University Press, 1994.

French, Roger K., Jon Arrizabalaga, Andrew Cunningham, and Luis García Ballester, eds. *Medicine from the Black Death to the French Disease*. Aldershot, UK: Ashgate, 1998.

Frisch, Andrea. *The Invention of the Eyewitness: Witnessing and Testimony in Early Modern France*. Chapel Hill: University of North Carolina Press, 2017.

Fuiková, Eliska. "The Collection of Rudolf II at Prague: Cabinet of Curiosities or Scientific Museum?" In *The Origins of Museums: The Cabinet of Curiosities in Sixteenth- and Seventeenth-Century Europe*, ed. O. R. Impey and Arthur MacGregor, 47–53. Oxford: Ashmolean, 2017.

Gabriel, Joseph M. *Medical Monopoly: Intellectual Property Rights and the Origins of the Modern Pharmaceutical Industry*. Chicago: University of Chicago Press, 2014.

Gaisser, Julia Haig. "Seeking Patronage under the Medici Popes: A Tale of Two Humanists." In *The Pontificate of Clement VII: History, Politics, Culture*, ed. Kenneth Gouwens and Sheryl E. Reiss, 293–312. Aldershot, UK: Ashgate, 2005.

Gambaccini, Piero. *Mountebanks and Medicasters: A History of Italian Charlatans from the Middle Ages to the Present*. Jefferson, NC: McFarland, 2004.

Gantenbein, Urs Leo. "Poison and Its Dose: Paracelsus on Toxicology." In *Toxicology in the Middle Ages and Renaissance*, ed. Philip Wexler, 1–10. London: Academic Press, 2017.

Gantenbein, Urs Leo, and Pia Holenstein, eds. *Nova ACTA Paracelsica: Beitraege zur Paracelsus-Forschung*. Bern: Peter Lang, 2005.

Geary, Patrick. *Living with the Dead in the Middle Ages*. Ithaca, NY: Cornell University Press, 1994.

Gentilcore, David. *Food and Health in Early Modern Europe: Diet, Medicine and Society, 1450–1800*. London: Bloomsbury Academic, 2016.

———. *Healers and Healing in Early Modern Italy*. New York: St. Martin's Press, 1998.

———. *Medical Charlatanism in Early Modern Italy*. Oxford: Oxford University Press, 2006.

Gere, Cathy. *Pain, Pleasure, and the Greater Good: From the Panopticon to the Skinner Box and Beyond*. Chicago: University of Chicago Press, 2017.

Geyer-Kordesch, Johanna, Andrew Wear, and Roger K. French, eds. *Doctors and Ethics: The Earlier Historical Setting of Professional Ethics*. Amsterdam: Rodopi, 1993.

Gibbs, Frederick W. *Poison, Medicine, and Disease in Late Medieval and Early Modern Europe*. Milton Park, Abingdon, UK: Routledge, 2019.

———. "Poisonous Properties, Bodies, and Forms in the Fifteenth Century." *Preternature* 2 (2013): 19–46.

Gómez, Pablo F. *The Experiential Caribbean: Creating Knowledge and Healing in the Early Modern Atlantic*. Chapel Hill: University of North Carolina Press, 2017.

Gooding, David, Trevor Pinch, and Simon Schaffer, eds. *The Uses of Experiment: Studies in the Natural Sciences*. Cambridge: Cambridge University Press, 1989.

Gordon, Bruce, and Peter Marshall. *The Place of the Dead: Death and Remembrance in Late Medieval and Early Modern Europe*. Cambridge: Cambridge University Press, 2000.

Grafton, Anthony, and Nancy G. Siraisi, eds. *Natural Particulars: Nature and the Disciplines in Renaissance Europe*. Cambridge: MIT Press, 1999.

Granados, Dionisio Ollero. "New Light on Celsus' 'De Medicina.'" *Sudhoffs Archiv* 62, no. 4 (1978): 359–77.

Green, Monica H., ed. *Pandemic Disease in the Medieval World: Rethinking the Black Death*. Kalamazoo: Arc Medieval Press, 2015.

Greene, Jeremy A. "Therapeutic Proofs and Medical Truths: The Enduring Legacy of Early Modern Drug Trials." *Bulletin of the History of Medicine* 91 (2017): 420–29.

Grell, Ole Peter, Andrew Cunningham, and Jon Arrizabalaga, eds. *"It All Depends on the Dose": Poisons and Medicines in European History*. Abingdon, UK: Routledge, 2018.

Guerrini, Anita. *The Courtiers' Anatomists: Animals and Humans in Louis XIV's Paris*. Chicago: University of Chicago Press, 2015.

———. "The Ethics of Animal Experimentation in Seventeenth-Century England." *Journal of the History of Ideas* 50 (1989): 391–407.

———. *Experimenting with Humans and Animals: From Galen to Animal Rights*. Johns Hopkins Introductory Studies in the History of Science. Baltimore: Johns Hopkins University Press, 2003.

Haage, Bernhard D. *Das gereimte Pestregimen des Cod. Sang. 1164 und seine Sippe: Metamorphosen eines Pestgedichtes*. Würzburger medizinhistorische Forschungen, vol. 8. Pattensen, Hanover: HWellm, 1977.

Hafemann, Klaus. *Magister Johann Wittich (1537–1596)*. Würzburg: Universität Würzburg Georg-Sticker-Institut für Geschichte der Medizin, 1956.

Hall, Bert. "Guido Da Vigevano's Texaurus Regis Franciae." In *Studies on Medieval Fachliteratur: Proceedings of the Special Session on Medieval Fachliteratur of the Sixteenth International Congress on Medieval Studies, Kalamazoo, Michigan (U.S.A.), May 10, 1981*, ed. William Eamon, 33–44. Brussels: Omirel, UFSAL, 1982.

Hamby, Wallace B. *Ambroise Paré, Surgeon of the Renaissance*. St. Louis: W. H. Green, 1967.

Harkness, Deborah. "'Strange' Ideas and 'English' Knowledge: Natural Science Exchange in Elizabethan London." In *Merchants and Marvels: Commerce, Science, and Art in Early Modern Europe*, ed. Pamela H. Smith and Paula Findlen, 137–60. New York: Routledge, 2002.

Harrington, Joel F. *The Faithful Executioner: Life and Death, Honor and Shame in the Turbulent Sixteenth Century*. New York: Farrar, Straus and Giroux, 2013.

Hasluck, F. W. *Christianity and Islam under the Sultans*. Ed. Margaret Masson Hardie Hasluck. Oxford: Clarendon Press, 1929.

Hedesan, Georgiana D. *An Alchemical Quest for Universal Knowledge: The "Christian Philosophy" of Jan Baptist Van Helmont (1579–1644)*. Universal Reform, Studies in Intellectual History, 1550–1770. London: Routledge, 2016.

———. "Alchemy, Potency, Imagination: Paracelsus's Theories of Poison." In *"It All Depends on the Dose": Poisons and Medicines in European History*, ed. Ole Peter Grell, Andrew Cunningham, and Jon Arrizabalaga, 81–102. Abingdon, UK: Routledge, 2018.

———. "Jan Baptist Van Helmont and the Medical-Alchemical Perspectives of Poison." In *Toxicology in the Middle Ages and Renaissance*, ed. Philip Wexler, 91–100. London: Academic Press, 2017.

Heinrichs, Erik A. "The Live Chicken Treatment for Buboes: Trying a Plague Cure in Medieval and Early Modern Europe." *Bulletin of the History of Medicine* 91, no. 2 (July 25, 2017): 210–32.

———. "The Plague Cures of Caspar Kegler: Print, Alchemy, and Medical Marketing in Sixteenth-Century Germany." *Sixteenth-Century Journal* 43 (2012): 417–40.

———. *Plague, Print, and the Reformation: The German Reform of Healing, 1473–1573.* London: Routledge, 2018.

Herlihy, David. *The Black Death and the Transformation of the West.* Cambridge: Harvard University Press, 1997.

Hildebrandt, Mireille. "The Trial of the Expert: *Épreuve* and *Preuve*." *New Criminal Law Review* 10 (2007): 78–101.

Hirsch, August. "Georg Amwald auf Durnhoff." In *Allgemeine Deutsche Biographie*, 1: 418–19. Leipzig, 1875.

———. "Thomas Jordan." In *Allgemeine Deutsche Biographie*, 14: 520. Leipzig, 1881.

Höfele, Andreas, and Werner von Koppenfels. *Renaissance Go-Betweens: Cultural Exchange in Early Modern Europe*. Spectrum Literaturwissenschaft 2. Berlin: Walter de Gruyter, 2005.

Holman, Beth L. "For 'Honor and Profit': Benvenuto Cellini's Medal of Clement VII and His Competition with Giovanni Bernardi." *Renaissance Quarterly* 58, no. 2 (2005): 512–75.

Holste, Thomas. *Der Theriakkrämer: Ein Beitrag zur Frühgeschichte der Arzneimittelwerbung*. Vol. 5. Würzburger medizinhistorische Forschungen. Pattensen, Hanover: Wellm, 1976.

Impey, O. R., and Arthur MacGregor, eds. *The Origins of Museums: The Cabinet of Curiosities in Sixteenth- and Seventeenth-Century Europe*. Oxford: Ashmolean, 2017.

Jankrift, Kay Peter. "Judenpogrome in rheinischen und westfälischen Städten im Umfeld des Schwarzen Todes." *Aschkenas* 16, no. 2 (2008): 545–60.

Jenner, Mark S. R., and Patrick Wallis. *Medicine and the Market in England and Its Colonies, c. 1450–c. 1850*. Basingstoke, UK: Palgrave Macmillan, 2007.

Johnson, Christine R. *The German Discovery of the World: Renaissance Encounters with the Strange and Marvelous*. Charlottesville: University of Virginia Press, 2008.

Jones, Claire. "Formula and Formulation: 'Efficacy Phrases' in Medieval English Medical Manuscripts." *Neuphilologische Mitteilungen* 99 (1998): 199–209.

Kappel, Jutta. "Turned Serpentine Works." In *Princely Splendor: The Dresden Court, 1580–1620*, ed. Dirk Syndram and Antje Scherner, 198–205. Milan: Electa, 2004.

Karamanou, Marianna, and George Androutsos. "Theriaca Magna: The Glorious Cure-All Remedy." In *History of Toxicology and Environmental Health*, ed. Philip Wexler, 1: 35–43. Boston: Academic Press, 2014.

Klairmont Lingo, Alison. "Empirics and Charlatans in Early Modern France: The Genesis of the Classification of the 'Other' in Medical Practice." *Journal of Social History* 19 (1986): 583–603.

Klein, Joel A. "Chymical Medicine, Corpuscularism, and Controversy: A Study of Daniel Sennert's Works and Letters." PhD diss., Indiana University, 2014.

———. "Daniel Sennert, the Philosophical Hen, and the Epistolary Quest for a (Nearly) Universal Medicine." *Ambix* 62 (2015): 29–49.

Klestinec, Cynthia. *Theaters of Anatomy: Students, Teachers, and Traditions of Dissection in Renaissance Venice*. Baltimore: Johns Hopkins University Press, 2011.

———. "Translating Learned Surgery." *Journal of the History of Medicine and Allied Sciences* 72 (May 2, 2017): 34–50.

Koeppe, Wolfram. "Exotica and the Kunstkammer: 'Snake Stones, Iridescent Sea Snails, and Eggs of the Giant Iron-Devouring Bird.'" In *Princely Splendor: The Dresden Court, 1580–1620*, ed. Dirk Syndram and Antje Scherner, 80–89. Milan: Electa, 2004.

Kopp, Hermann. *Geschichte der chemie*. Vol. 1. Braunschweig: F. Vieweg, 1843.

Kunz, George Frederick. *The Magic of Jewels and Charms*. London: Lippincott, 1915.

Kusukawa, Sachiko. *Picturing the Book of Nature: Image, Text, and Argument in Sixteenth-Century Human Anatomy and Medical Botany*. Chicago: University of Chicago Press, 2012.

Langbein, John H. *Prosecuting Crime in the Renaissance: England, Germany, France*. Cambridge: Harvard University Press, 1974.

Lavers, Chris. *The Natural History of Unicorns*. New York: Morrow, 2009.

Lawrence O. Gostin, Cori Vanchieri, and Andrew MacPherson Pope. *Ethical Considerations for Research Involving Prisoners*. Washington, DC: National Academies Press, 2007.

Lederer, Susan E. *Subjected to Science: Human Experimentation in America before the Second World War*. Baltimore: Johns Hopkins University Press, 1995.

Lehfeldt, Elizabeth A. *The Black Death*. Boston: Houghton Mifflin, 2005.

Leigh, Robert. Introduction. *On Theriac to Piso, Attributed to Galen: A Critical Edition with Translation and Commentary*. Trans. Robert Leigh. Leiden: Brill, 2016.

Lentsch, Roberte. "La proba. L'épreuve des poisons à la cour des papes d'Avignon." In *Les Prélats, l'Eglise et la société: XIe–XVe siècles: hommage à Bernard Guillemain*, ed. Anne-Marie Dom and Françoise Bériac. Talence: Université Michel de Montaigne, Bordeaux III, 1994.

Lenzi, Maria Ludovica. "Dal 'regno di Iatria' alla corte di Praga." In *Pietro Andrea Mattioli: Siena, 1501–Trento, 1578: la vita, le opere: con l'identificazione delle piante*, ed. Sara Ferri, 83–104. Ponte San Giovanni, Perugia: Quattroemme, 1997.

Leong, Elaine. *Recipes and Everyday Knowledge: Medicine, Science, and the Household in Early Modern England*. Chicago: University of Chicago Press, 2018.

Leong, Elaine, and Sara Pennell. "Recipe Collections and the Currency of Medical Knowledge in the Early Modern 'Medical Marketplace." In *Medicine and the Market in England and Its Colonies, c.1450–c.1850*, ed. Mark S. R. Jenner and Patrick Wallis, 133–52. Basingstoke, UK: Palgrave Macmillan, 2007.

Leong, Elaine, and Alisha Rankin, eds. *Secrets and Knowledge in Medicine and Science, 1500–1800*. Farnham, Surrey, UK: Ashgate, 2011.

———. "Testing Drugs and Trying Cures: Experiment and Medicine in Medieval and Early Modern Europe." *Bulletin of the History of Medicine* 91 (2017): 157–82.

Le Paulmier, Claude Stéphen. *L'orviétan: histoire d'une famille de charlatans du Pont-Neuf aux XVIIe et XVIIIe siècles*. Paris: Librillustrée, 1893.

Lesage, Claire. "La littérature des 'secrets' e 'I secreti d'Isabella Cortese." *Chroniques italiennes* 36 (1993): 145–78.

Levey, Martin. *Early Arabic Pharmacology: An Introduction Based on Ancient and Medieval Sources*. Leiden: Brill, 1973.

———. "Medieval Arabic Toxicology: The Book on Poisons of Ibn Wahshīya and Its Relation to Early Indian and Greek Texts." *Transactions of the American Philosophical Society* 56 (1966): 1–130.

Liboni, Gionata. "Humanist Post-Mortems: Philology and Therapy." In *Pathology in Practice: Diseases and Dissections in Early Modern Europe*, ed. Silvia De Renzi, Marco Bresadola, and Maria Conforti. London: Routledge, 2017.

Lindemann, Mary. *Health and Healing in Eighteenth-Century Germany*. Baltimore: Johns Hopkins University Press, 1996.

———. *Medicine and Society in Early Modern Europe*. 2nd ed. Cambridge: Cambridge University Press, 2010.

Long, Pamela O. *Artisan/Practitioners and the Rise of the New Sciences, 1400–1600*. Corvallis: Oregon State University Press, 2011.

Lopes Andrade, António Manuel. "Garcia de Orta and Amato Lusitano's Views on Materia Medica: A Comparative Perspective." In *Medicine, Trade and Empire: Garcia de Orta's Colloquies on the Simples and Drugs of India (1563) in Context*, ed. Palmira Fontes da Costa. Farnham, Surrey, UK: Ashgate, 2015.

López Pérez, Miguel, Didier Kahn, and Mar Rey Bueno, eds. *Chymia: Science and Nature in Medieval and Early Modern Europe*. Newcastle upon Tyne: Cambridge Scholars, 2010.

Lo Presti, Roberto. "Anatomy as Epistemology: The Body of Man and the Body of Medicine in Vesalius and His Ancient Sources (Celsus, Galen)." *Renaissance and Reformation / Renaissance et Réforme* 33 (2010): 27–60.

Lovejoy, Arthur O. *The Great Chain of Being: A Study of the History of an Idea*. Cambridge: Harvard University Press, 1936.

Macgregor, Arthur. "Medicinal Terra Sigillata: A Historical, Geographical and Typological Review." *Geological Society, London, Special Publications* 375, no. 1 (January 1, 2013): 113–36.

Maclean, Ian. *Learning and the Market Place: Essays in the History of the Early Modern Book*. Leiden: Brill, 2009.

———. *Logic, Signs, and Nature in the Renaissance: The Case of Learned Medicine*. Cambridge: Cambridge University Press, 2002.

Maehle, Andreas-Holger. *Drugs on Trial: Experimental Pharmacology and Therapeutic Innovation in the Eighteenth Century*. Clio Medica. Vol. 53. Amsterdam: Rodopi, 1999.

———. "The Ethical Discourse on Animal Experimentation, 1650–1900." In *Doctors and Ethics: The Earlier Historical Setting of Professional Ethics*, ed. Johanna

Geyer-Kordesch, Andrew Wear, and Roger K. French, 203–51. Amsterdam: Rodopi, 1993.

———. *Johann Jakob Wepfer (1620–1695) als Toxikologe: die Fallstudien und Tierexperimente aus seiner Abhandlung über den Wasserschierling (1679)*. Aarau: Sauerländer, 1987.

———. "Zur wissenschaftlichen und moralischen Rechtfertigung toxikologischer Tierversuche im 17. Jahrhundert: Johann Jakob Wepfer und Johann Jakob Härder." *Gesnerus* 43 (1986): 213–21.

Marcus, Jacob Rader, and Marc Saperstein, eds. "The Black Death and the Jews: 1348–1349." In *The Jews in Christian Europe*, 153–59. Pittsburgh: University of Pittsburgh Press, 2015.

Marks, Harry M. *The Progress of Experiment: Science and Therapeutic Reform in the United States, 1900–1990*. Cambridge: Cambridge University Press, 1987.

Marland, Hilary, and Margaret Pelling, eds. *Task of Healing: Medicine, Religion, and Gender in England and the Netherlands, 1450–1800*. Rotterdam: Erasmus, 1996.

Martin, Craig. *Subverting Aristotle: Religion, History, and Philosophy in Early Modern Science*. Baltimore: Johns Hopkins University Press, 2014.

Martin, Randall. *Women, Murder, and Equity in Early Modern England*. New York: Routledge, 2008.

Mattern, Susan P. *The Prince of Medicine: Galen in the Roman Empire*. New York: Oxford University Press, 2013.

Matthee, Rudi. "Exotic Substances: The Introduction and Global Spread of Tobacco, Coffee, Cocoa, Tea, and Distilled Liquor, Sixteenth to Eighteenth Centuries." In *Drugs and Narcotics in History*, ed. Roy Porter and Mikulás Teich. Cambridge: Cambridge University Press, 1995.

Matus, Zachary A. *Franciscans and the Elixir of Life: Religion and Science in the Later Middle Ages*. Middle Ages Series. Philadelphia: University of Pennsylvania Press, 2017.

Mayor, Adrienne. "Alexander the Great: A Questionable Death." In *History of Toxicology and Environmental Health*, ed. Philip Wexler, 1: 52–59. Boston: Academic Press, 2014.

———. "Mithradates of Pontus and His Universal Antidote." In *History of Toxicology and Environmental Health*, ed. Philip Wexler, 1: 21–34. Boston: Academic Press, 2014.

———. *The Poison King: The Life and Legend of Mithradates, Rome's Deadliest Enemy*. Princeton: Princeton University Press, 2010.

McVaugh, Michael. "Determining a Drug's Properties: Medieval Experimental Protocols." *Bulletin of the History of Medicine* 91, no. 2 (July 25, 2017): 183–209.

———. "The Experience-Based Medicine of the Thirteenth Century." *Early Science and Medicine* 14 (2009): 105–30.

———. "The Experimenta of Arnald of Villanova." *Journal of Medieval and Renaissance Studies* 1 (1971): 107–18.

———. "Introduction to the 'Epistola de dosi tyriacalium medicinarum.'" In *Arnaldi de Villanova Opera Medica Omnia*, 3: 57–73. Granada: Seminarium Historiae Medicae Granatensis, 1985.

———. "Quantified Medical Theory and Practice at Fourteenth-Century Montpellier." *Bulletin of the History of Medicine* 43 (1969): 397–413.

———. "Theriac at Montpellier 1285–1325." *Sudhoffs Archiv* 56 (1972): 113–44.

Meadow, Mark A. "Merchants and Marvels: Hans Jacob Fugger and the Origins of the Wunderkammer." In *Merchants and Marvels: Commerce, Science, and Art in Early Modern Europe*, ed. Pamela H. Smith and Paula Findlen, 182–200. New York: Routledge, 2002.

Menzhausen, Joachim. "Elector August's Kunstkammer: An Analysis of the Inventory of 1587." In *The Origins of Museums: The Cabinet of Curiosities in Sixteenth- and Seventeenth-Century Europe*, ed. O. R. Impey and Arthur MacGregor, 69–75. Oxford: Ashmolean, 2017.

Merback, Mitchell B. *The Thief, the Cross, and the Wheel: Pain and the Spectacle of Punishment in Medieval and Renaissance Europe*. Chicago: University of Chicago Press, 1999.

Meyer, G. J. *The Borgias: The Hidden History*. 1st ed. New York: Bantam Books, 2013.

Mez-Mangold, Lydia. *A History of Drugs*. Carnforth, Lancs., UK: Parthenon, 1986.

Miert, Dirk van, ed. *Communicating Observations in Early Modern Letters (1500–1675): Epistolography and Epistemology in the Age of the Scientific Revolution*. London: Warburg Institute, 2013.

Möhsen, Johann Karl Wilhelm. *Beiträge zur Geschichte der Wissenschaften in der Mark Brandenburg*. Berlin, 1783.

Mollenauer, Lynn Wood. *Strange Revelations: Magic, Poison, and Sacrilege in Louis XIV's France*. Magic in History. University Park: Pennsylvania State University Press, 2007.

Moran, Bruce T. *The Alchemical World of the German Court: Occult Philosophy and Chemical Medicine in the Circle of Moritz of Hessen, 1572–1632*. Sudhoffs Archiv. Beihefte, Heft 29. Stuttgart: Franz Steiner Verlag, 1991.

———. *Andreas Libavius and the Transformation of Alchemy: Separating Chemical Cultures with Polemical Fire*. Sagamore Beach, MA: Science History Publications, 2007.

———. "Art and Artisanship in Early Modern Alchemy." *Getty Research Journal* (2013): 1–14.

———. "German Prince-Practitioners: Aspects in the Development of Courtly Science, Technology, and Procedures in the Renaissance." *Technology and Culture* 22 (1981): 253–74.

———. "The 'Herbarius' of Paracelsus." *Pharmacy in History* 35, no. 3 (1993): 99–127.

———. *Paracelsus: An Alchemical Life*. London: Reaktion Books, 2019.

———. "Preserving the Cutting Edge: Traveling Woodblocks, Material Networks, and Visualizing Plants in Early Modern Europe." In *The Structures of Practical*

Knowledge, ed. Matteo Valleriani, 393–420. Cham: Springer International, 2017.

———. "Wilhelm IV of Hesse-Kassel: Informal Communication and the Aristocratic Context of Discovery." In *Scientific Discovery: Case Studies*, ed. Thomas Nickles, 67–96. Dordrecht: D. Riedel, 1983.

Müller-Jahncke, Wolf-Dieter. "Georg am Wald (1554–1616): Arzt und Unternehmer." In *Analecta paracelsica: Studien zum Nachleben Theophrast von Hohenheims im deutschen Kulturgebiet der frühen Neuzeit*, ed. Joachim Telle, 213–304. Stuttgart: Franz Steiner Verlag, 1994.

Mullini, Roberta. "'I Cornelius à Tilbourn': Hotchpotches, Poisons, Antidotes and Royal Gifts. The Career of a London Irregular Physician at the End of the Seventeenth Century." In *Chlorophyll Killers: Pozioni, Veleni, Narcotici tra Letteratura Noir e Scienza*, ed. Jan Martin Ivo Klaver and Giuseppe Puntarello, 119–48. Urbino: Aras, 2016.

Murphy, Hannah. *A New Order of Medicine: The Rise of Physicians in Reformation Nuremberg*. Pittsburgh: University of Pittsburgh Press, 2019.

Naphy, William G. *Plagues, Poisons, and Potions: Plague-Spreading Conspiracies in the Western Alps, c. 1530–1640*. Manchester: Manchester University Press, 2002.

Nappi, Carla. "Bolatu's Pharmacy: Theriac in Early Modern China." *Early Science and Medicine* 14, no. 6 (2009): 737–64.

Nasser, Mona, Aida Tibi, and Emilie Savage-Smith. "Ibn Sina's Canon of Medicine: Eleventh Century Rules for Assessing the Effects of Drugs." *Journal of the Royal Society of Medicine* 102 (2009): 78–80.

Newman, William R. *Newton the Alchemist: Science, Enigma, and the Quest for Nature's "Secret Fire."* Princeton: Princeton University Press, 2019.

———. *Promethean Ambitions: Alchemy and the Quest to Perfect Nature*. Chicago: University of Chicago Press, 2004.

Newman, William R., and Lawrence Principe. *Alchemy Tried in the Fire: Starkey, Boyle, and the Fate of Helmontian Chymistry*. Chicago: University of Chicago Press, 2002.

Noack, Winfried. *Landgraf Georg I. von Hessen und die Obergrafschaft Katzenelnbogen (1567–1596)*. Darmstadt: Historischer Verein für Hessen, 1966.

Noel, Gerard. *The Renaissance Popes: Statesmen, Warriors, and the Great Borgia Myth*. New York: Carrol & Graf, 2006.

Norton, Marcy. *Sacred Gifts, Profane Pleasures: A History of Tobacco and Chocolate in the Atlantic World*. Ithaca: Cornell University Press, 2008.

———. "Tasting Empire: Chocolate and the European Internalization of Mesoamerican Aesthetics." *American Historical Review* 111, no. 3 (2006): 660–91.

Novoa, James Nelson. "Unicorns and Bezoars in a Portuguese House in Rome: António Da Fonseca's Portuguese Inventories." *Ágora: Estudos Clássicos em Debate*, no. 14 (2012): 91–112.

Nummedal, Tara E. "Alchemical Reproduction and the Career of Anna Maria Zieglerin." *Ambix* 48 (2001): 56–68.

———. *Alchemy and Authority in the Holy Roman Empire*. Chicago: University of Chicago Press, 2007.

———. *Anna Zieglerin and the Lion's Blood: Alchemy and End Times in Reformation Germany*. Philadelphia: University of Pennsylvania Press, 2019.

———. "Anna Zieglerin's Alchemical Revelations." In *Secrets and Knowledge in Medicine and Science, 1500–1800*, ed. Elaine Leong and Alisha Rankin, 125–42. Farnham, Surrey, UK: Ashgate, 2011.

Nutton, Vivian. "Galen on Theriac: Problems of Authenticity." In *Galen on Pharmacology: Philosophy, History, and Medicine*, ed. Armelle Debru, 133–51. Leiden: Brill, 1997.

———. "Humanist Surgery." In *The Medical Renaissance of the Sixteenth Century*, ed. Andrew Wear, Roger K. French, and I. M. Lonie, 75–99. Cambridge: Cambridge University Press, 1985.

———. "'Qui magni Galeni doctrinam in re medica primus revocavit': Matteo Corti und der Galenismus im medizinischen Unterricht der Renaissance." In *Der Humanismus und die oberen Fakultäten*, ed. Gundolf Keil, Bernd Moeller, and Winfried Trusen. Weinheim: Acta humaniora, VCH, 1987.

———. "Wittenberg Anatomy." In *Medicine and the Reformation*, ed. Ole Peter Grell and Andrew Cunningham, 11–32. London: Routledge, 1993.

O'Connor, Mary Catharine. *The Art of Dying Well: The Development of the Ars Moriendi*. New York: Columbia University Press, 1942.

Ogilvie, Brian W. "The Many Books of Nature: Renaissance Naturalists and Information Overload." *Journal of the History of Ideas* 64 (2003): 29–40.

———. *The Science of Describing: Natural History in Renaissance Europe*. Chicago: University of Chicago Press, 2006.

Olivieri, Luigi Pompili. *Il senato romano nella sette epoche di svariato governo da Romolo fine a noi*. Rome: Contedini, 1840.

Olson, Kelly. "Cosmetics in Roman Antiquity: Substance, Remedy, Poison." *Classical World* 102, no. 3 (2009): 291–310.

Packard, Francis R. *The Life and Times of Ambroise Paré (1510–1590): With a New Translation of His Apology and an Account of His Journeys in Divers Places*. New York: P. B. Hoeber, 1921.

Pagel, Walter. *Joan Baptista van Helmont: Reformer of Science and Medicine*. Cambridge Monographs on the History of Medicine. Cambridge: Cambridge University Press, 1982.

Palmer, Richard. "Pharmacy in the Republic of Venice." In *The Medical Renaissance of the Sixteenth Century*, ed. Andrew Wear, Roger K. French, and I. M. Lonie, 100–117. Cambridge: Cambridge University Press, 1985.

Paravicini Bagliani, Agostino. *The Pope's Body*. Trans. David S. Peterson. Chicago: University of Chicago Press, 2000.

———. "Ruggero Bacone e l'alchimia di lunga vita: Riflessioni sui testi." In *Alchimia e medicina nel Medioevo*, ed. Chiara Crisciani and Agostino Paravicini Bagliani, 33–54. Tavarnuzze (Florence): SISMEL edizioni del Galluzzo, 2003.

Pardo-Tomás, José. "East Indies, West Indies: Garcia da Orta and the Spanish Treatises on Exotic Materia Medica." In *Medicine, Trade and Empire: Garcia de Orta's Colloquies on the Simples and Drugs of India (1563) in Context*, ed. Palmira Fontes da Costa, 195–212. Farnham, Surrey, UK: Ashgate, 2015.

———. "Obras españolas sobre historia natural y materia médica americanas en la Italia del siglo XVI." *Asclepio, archivo iberoamericano de historia de la medicina y antropología médica* 43 (1991): 51–94.

———. "Two Glimpses of America from a Distance: Carolus Clusius and Nicolás Monardes." In *Carolus Clusius. Towards a Cultural History of a Renaissance Naturalist*, ed. Florike Egmond, Paul Hoftijzer, and Robert Visser, 173–93. Amsterdam: Koninklijke Nederlandse Akademie van Wetenschappen, 2007.

Park, Katharine. "Country Medicine in the City Marketplace: Snakehandlers as Itinerant Healers." *Renaissance Studies* 15 (2001): 104–20.

———. "The Criminal and the Saintly Body: Autopsy and Dissection in Renaissance Italy." *Renaissance Quarterly* 47, no. 1 (1994): 1–33.

———. *Doctors and Medicine in Early Renaissance Florence*. Princeton: Princeton University Press, 1985.

———. "The Life of the Corpse: Division and Dissection in the Late Middle Ages." *Bulletin of the History of Medicine* 50 (1985): 111–32.

———. "Natural Particulars: Medical Epistemology, Practice, and the Literature of Healing Springs." In *Natural Particulars: Nature and the Disciplines in Renaissance Europe*, ed. Anthony Grafton and Nancy G. Siraisi, 347–68. Cambridge: MIT Press, 1999.

———. "Observation in the Margins, 500–1500." In *Histories of Scientific Observation*, ed. Lorraine Daston and Elizabeth Lunbeck, 15–44. Chicago: University of Chicago Press, 2011.

———. *Secrets of Women: Gender, Generation, and the Origins of Human Dissection*. New York: Zone Books, 2006.

Parrish, Sean David. "Marketing Nature: Apothecaries, Medicinal Retailing, and Scientific Culture in Early Modern Venice, 1565–1730." PhD diss., Duke University, 2015.

Partner, Peter. *Renaissance Rome, 1500–1559: A Portrait of a Society*. Berkeley: University of California Press, 1976.

Partsch, Josef. *Schlesien: eine Landeskunde für das deutsche Volk auf wissenschaftlicher Grundlage*. Breslau: F. Hirt, 1896.

Pasolini, Pier Desiderio. *Caterina Sforza*. Vol. 3. Rome: E. Loescher, 1893.

Pastor, Ludwig. *History of the Popes, from the Close of the Middle Ages, Drawn from the Secret Archives of the Vatican and Other Original Sources*. Vol. 9. St. Louis: Herder, 1902.

Pastore, Alessandro. "Médecine légale et investigation judiciaire: expérimenter le poison sur les animaux en Italie à l'époque moderne." *Revue d'Histoire des Sciences Humaines* 22 (2010): 17–35.

———. *Il medico in tribunale: la perizia medica nella procedura penale d'antico regime: secoli XVI–XVIII*. Vol. 2. Bellinzona: Casagrande, 1998.

———. *Veleno: credenze, crimini, saperi nell'Italia moderna*. Bologna: Il mulino, 2010.

Paul, Serge. *Les Grandes Empoisonneuses*. Paris: l'Édition, 1909.

Payne, Lynda. "'A Speedie Reformation': Barber-Surgeons, Anatomization, and the Reformation of Medicine in Tudor London." In *Paracelsian Moments: Science, Medicine, and Astrology in Early Modern Europe*, ed. Gerhild Scholz Williams and Charles D. Gunnoe, 71–92. Kirksville, MO: Truman State University Press, 2002.

Pelling, Margaret. *The Common Lot Sickness, Medical Occupations, and the Urban Poor in Early Modern England*. London: Longman, 1998.

———. *Medical Conflicts in Early Modern London: Patronage, Physicians, and Irregular Practitioners, 1550–1640*. Oxford: Clarendon Press, 2003.

Pender, Stephen. "Examples and Experience: On the Uncertainty of Medicine." *British Journal for the History of Science* 39 (March 2006): 1–28.

Pereira, Michela. *The Alchemical Corpus Attributed to Raymond Lull*. London: Warburg Institute, University of London, 1989.

———. "L'alchimista come medico perfetto nel *Testamentum* pseudolulliano." In *Alchimia e medicina nel Medioevo*, ed. Chiara Crisciani and Agostino Paravicini Bagliani. Tavarnuzze (Florence): SISMEL edizioni del Galluzzo, 2003.

———. *Arcana sapienza: l'alchimia dalle origini a Jung*. Rome: Carocci, 2001.

———. "Heavens on Earth. From the Tabula Smaragdina to the Alchemical Fifth Essence." *Early Science and Medicine* 5 (2000): 131–44.

———. "*Mater Medicinarum*: English Physicians and the Alchemical Elixir in the Fifteenth Century." In *Medicine from the Black Death to the French Disease*, ed. R. K. French, Jon Arrizabalaga, Andrew Cunningham, and Luis García Ballester, 26–52. Aldershot, UK: Ashgate, 1998.

———. "Medicina in the Alchemical Writings Attributed to Raimond Lull (14th–17th Centuries)." In *Alchemy and Chemistry in the 16th and 17th Centuries*, ed. Piyo Rattansi and Antonio Clericuzio. Dordrecht: Kluwer Academic Publishers, 1994.

———. *L'oro dei filosofi: saggio sulle idee di un alchimista del Trecento*. Spoleto: Centro italiano di studi sull'alto Medioevo, 1992.

———. "Utili segreti. Strumenti per lambiccare e acque distillate nei 'Discorsi.'" In *Pietro Andrea Mattioli: Siena, 1501–Trento, 1578: la vita, le opere: con l'identificazione delle piante*, ed. Sara Ferri, 281–305. Ponte San Giovanni, Perugia: Quattroemme, 1997.

Peters, Julia Stone. *Theatre of the Book, 1480–1880: Print, Text, and Performance in Europe*. Oxford: Oxford University Press, 2000.

Photos-Jones, E., C. Edwards, F. Häner, L. Lawton, C. Keane, A. Leanord, and V. Perdikatsis. "Archaeological Medicinal Earths as Antibacterial Agents: The Case of the Basel Lemnian Sphragides." *Geological Society, London, Special Publications* 452, no. 1 (January 1, 2017): 141–53.

Pihlajamaki, H. "'Swimming the Witch, Pricking for the Devil's Mark': Ordeals in the Early Modern Witchcraft Trials." *Journal of Legal History* 21 (2000): 35–58.

Pimentel, Juan, and Isabel Soler. "Painting Naked Truth: The Colóquios of Garcia da Orta (1563)." *Journal of Early Modern History* 18, no. 1–2 (2014): 101–20.

Pinon, Laurent. "Gessner and the Historical Depth of Renaissance Natural History." In *Historia: Empiricism and Erudition in Early Modern Europe*, ed. Gianna Pomata and Nancy G. Siraisi, 241–67. Cambridge: MIT Press, 2005.

Pinto, Antônio Guimarães. "Preconceito e ciência: o ataque de Pietro Andrea Mattioli a Amato Lusitano." *Humanitas* 65 (August 30, 2016): 161–86.

Pomata, Gianna. *Contracting a Cure: Patients, Healers, and the Law in Early Modern Bologna*. Baltimore: Johns Hopkins University Press, 1998.

———. "Observation Rising: Birth of an Epistemic Genre." In *Histories of Scientific Observation*, ed. Lorraine Daston and Elizabeth Lunbeck, 45–80. Chicago: University of Chicago Press, 2011.

———. "Praxis Historialis: The Uses of Historia in Early Modern Medicine." In *Historia: Empiricism and Erudition in Early Modern Europe*, 125–37. Cambridge: MIT Press, 2005.

———. *La promessa di guarigione: malati e curatori in antico regime: Bologna XVI–XVIII secolo*. Rome: Laterza, 1994.

———. "Sharing Cases: The Observationes in Early Modern Medicine." *Early Science and Medicine* 15 (2010): 193–236.

———. "A Word of the Empirics: The Ancient Concept of Observation and Its Recovery in Early Modern Medicine." *Annals of Science* 68 (January 2011): 1–25.

Pomata, Gianna, and Nancy G. Siraisi, eds. *Historia: Empiricism and Erudition in Early Modern Europe*. Cambridge: MIT Press, 2005.

Pop, Ioan-Aurel. "Religiones and Nationes in Transylvania during the 16th Century: Between Acceptance and Exclusion." *Journal for the Study of Religions and Ideologies* 12 (2013): 209–36.

Principe, Lawrence. *The Aspiring Adept: Robert Boyle and His Alchemical Quest: Including Boyle's "Lost" Dialogue on the Transmutation of Metals*. Princeton: Princeton University Press, 1998.

———. *The Secrets of Alchemy*. Chicago: University of Chicago Press, 2013.

Prosperi, Adriano. "Consolation or Condemnation: The Debates on Withholding Sacraments from Prisoners." In *The Art of Executing Well: Rituals of Execution in Renaissance Italy*, ed. Nicholas Terpstra. Kirksville, MO: Truman State University Press, 2008.

Pugliano, Valentina. "Natural History in the Apothecary's Shop." In *Worlds of Natural History*, ed. H. A. Curry et al., 44–60. Cambridge: Cambridge University Press, 2018.

———. "Pharmacy, Testing, and the Language of Truth in Renaissance Italy." *Bulletin of the History of Medicine* 91, no. 2 (July 25, 2017): 233–73.

Ragland, Evan R. "Experimental Clinical Medicine and Drug Action in Mid-Seventeenth-Century Leiden." *Bulletin of the History of Medicine* 91, no. 2 (July 25, 2017): 331–61.

———. "Making Trials in Sixteenth-Century European Academic Medicine." *Isis* 108 (2017): 503–28.

Ramisch, Hans Karlmann. *Landkreis Feuchtwangen*. Munich: Deutscher Kunstverlag, 1964.

Rankin, Alisha. "Becoming an Expert Practitioner: Court Experimentalism and the Medical Skills of Anna of Saxony (1532–1585)." *Isis* 98 (2007): 23–53.

———. "Empirics, Physicians, and Wonder Drugs in Early Modern Germany: The Case of the 'Panacea Amwaldina.'" *Early Science and Medicine* 14 (2009): 680–710.

———. "Exotic Materials and Treasured Knowledge: The Valuable Legacy of Noblewomen's Pharmacy in Early Modern Germany." *Renaissance Studies* 28 (2014): 533–55.

———. "Gender, Poison, and Antidotes in Early Modern Europe." In *"It All Depends on the Dose": Poisons and Medicines in European History*, ed. Ole Peter Grell, Andrew Cunningham, and Jon Arrizabalaga, 132–49. History of Medicine in Context. Abingdon, UK: Routledge, 2018.

———. "How to Cure the Golden Vein: Medical Remedies as Wissenschaft in Early Modern Germany." In *Ways of Making and Knowing: The Material Culture of Empirical Knowledge*, ed. Pamela H. Smith, Amy R. W. Meyers, and Harold J. Cook, 109–33. Ann Arbor: University of Michigan Press, 2014.

———. "On Anecdote and Antidotes: Poison Trials in Sixteenth-Century Europe." *Bulletin of the History of Medicine* 91 (2017): 274–302.

———. *Panaceia's Daughters: Noblewomen as Healers in Early Modern Germany*. Chicago: University of Chicago Press, 2013.

Rankin, Alisha, and Justin Rivest. "Medicine, Monopoly, and the Premodern State—Early Clinical Trials." *New England Journal of Medicine* 375, no. 2 (July 14, 2016): 106–9.

Ray, Meredith K. *Daughters of Alchemy: Women and Scientific Culture in Early Modern Italy*. Cambridge: Harvard University Press, 2015.

Rebecchini, Guido. "Rituals of Justice and the Construction of Space in Sixteenth-Century Rome." *I Tatti Studies in the Italian Renaissance* 16, no. 1/2 (2013): 153–79.

Reinis, Austra. *Reforming the Art of Dying: The Ars Moriendi in the German Reformation*. Aldershot, UK: Ashgate, 2007.

Ricordel, Joëlle. "Le traité sur la thériaque d'Ibn Rushd (Averroes)." *Revue d'Histoire de la Pharmacie* 88 (2000): 81–90.

Rivest, Justin. "Beyond the Pharmacopoeia? Secret Remedies, Exclusive Privileges, and Trademarks in Early Modern France." In *Drugs on the Page: Pharmacopoeias and Healing Knowledge in the Early Modern Atlantic World*, ed. Matthew James Crawford and Joseph M. Gabriel, 81–100. Pittsburgh: University of Pittsburgh Press, 2019.

———. "Testing Drugs and Attesting Cures: Pharmaceutical Monopolies and Military Contracts in Eighteenth-Century France." *Bulletin of the History of Medicine* 91, no. 2 (July 25, 2017): 362–90.

Robisheaux, Thomas Willard. *The Last Witch of Langenburg: Murder in a German Village*. New York: W. W. Norton, 2009.

———. *Rural Society and the Search for Order in Early Modern Germany*. Cambridge: Cambridge University Press, 2002.

Rosner, Fred. *The Medical Legacy of Moses Maimonides*. Hoboken, NJ: KTAV, 1998.

Rowling, J. K. *Harry Potter and the Sorcerer's Stone*. New York: Scholastic, 1997.

Rublack, Ulinka. *The Astronomer and the Witch: Johannes Kepler's Fight for His Mother*. Oxford: Oxford University Press, 2015.

———. *The Crimes of Women in Early Modern Germany*. Oxford: Clarendon Press, 1999.

Sameiro Barroso, Maria do. "The Bezoar Stone: A Princely Antidote, the Távora Sequeira Pinto Collection." *Acta Medico-Historica Adriatica* 12 (2014): 77–98.

Sawday, Jonathan. *The Body Emblazoned: Dissection and the Human Body in Renaissance Culture*. London: Routledge, 1995.

Schabel, C., and F. Pedersen. "Miraculous, Natural, or Jewish Conspiracy? Pierre Ceffons' Question on the Black Death, with Astrological Predictions by Gersonides and Jean de Murs / Firmin de Beuval." *Recherches de théologie et philosophie médiévales* 81 (2014): 137–79.

Scheicher, Elizabeth. "The Collection of Archduke Ferdinand II at Schloss Ambras: Its Purpose, Composition, and Evolution." In *The Origins of Museums: The Cabinet of Curiosities in Sixteenth- and Seventeenth-Century Europe*, ed. O. R. Impey and Arthur MacGregor, 29–38. Oxford: Ashmolean, 2017.

Schelenz, Hermann. *Geschichte der Pharmazie*. Berlin: J. Springer, 1904.

Schickore, Jutta. *About Method: Experimenters, Snake Venom, and the History of Writing Scientifically*. Chicago: University of Chicago Press, 2017.

———. "The Significance of Re-doing Experiments: A Contribution to Historically Informed Methodology." *Erkenn* 75 (2011): 325–47.

———. "Trying Again and Again: Multiple Repetitions in Early Modern Reports of Experiments on Snake Bites." *Early Science and Medicine* 15 (2010): 567–617.

Schiebinger, Londa L. "Human Experimentation in the Eighteenth Century: Natural Boundaries and Valid Testing." In *The Moral Authority of Nature*, ed. Lorraine Daston and Fernando Vidal, 384–408. Chicago: University of Chicago Press, 2004.

———. "Medical Experimentation and Race in the Eighteenth-Century Atlantic World." *Social History of Medicine* 26 (2013): 364–82.

———. *The Mind Has No Sex?: Women in the Origins of Modern Science*. Cambridge: Harvard University Press, 1989.

———. "Prospecting for Drugs: European Naturalists in the West Indies." In *Colonial Botany: Science, Commerce, and Politics in the Early Modern World*, ed. Londa L. Schiebinger and Claudia Swan, 119–33. Philadelphia: University of Pennsylvania Press, 2005.

———. *Secret Cures of Slaves: People, Plants, and Medicine in the Eighteenth-Century Atlantic World*. Stanford: Stanford University Press, 2017.

Schiebinger, Londa L., and Claudia Swan, eds. *Colonial Botany: Science, Commerce, and Politics in the Early Modern World*. Philadelphia: University of Pennsylvania Press, 2005.

Schmitt, Charles B. "Experience and Experiment: A Comparison of Zabarella's View with Galileo's in *De Motu*." *Studies in the Renaissance* 16 (1969): 80–138.

Scholz Williams, Gerhild, and Charles D. Gunnoe, eds. *Paracelsian Moments: Science, Medicine, and Astrology in Early Modern Europe*. Kirksville, MO: Truman State University Press, 2002.

Schröder, Ingrid. "Wunderdrogentraktate—Texte zwischen Humoralpathologie, Magie und Religion." *Jahrbuch für germanistische Sprachgeschichte* 3, no. 1 (2012): 345–60.

Schultze, Jürgen, and Hermann Fillitz, eds. *Prag um 1600: Kunst und Kultur am Hofe Rudolfs II*. Vols. 1–2. Freren: Kunsthistorisches Museum Wien Luca Verlag, 1988.

Schutz, Bernard. *Art and Anatomy in Renaissance Italy*. Ann Arbor: UMI Research Press, 1982.

Scully, Terence. *The Art of Cookery in the Middle Ages*. Woodbridge, Suffolk, UK: Boydell Press, 1995.

Seelig, Lorenz. "The Munich Kunstkammer, 1565–1807." In *The Origins of Museums: The Cabinet of Curiosities in Sixteenth- and Seventeenth-Century Europe*, ed. O. R. Impey and Arthur MacGregor, 76–89. Oxford: Ashmolean, 2017.

Shapin, Steven. *A Social History of Truth: Civility and Science in Seventeenth-Century England*. Chicago: University of Chicago Press, 1994.

Shapin, Steven, and Simon Schaffer. *Leviathan and the Air-Pump: Hobbes, Boyle, and the Experimental Life*. Princeton: Princeton University Press, 1985.

Sharpe, J. A. *Crime in Early Modern England, 1550–1750*. 2nd ed. New York: Longman, 1999.

Shaw, James E., and Evelyn S. Welch. *Making and Marketing Medicine in Renaissance Florence*. Clio Medica, 89. Amsterdam: Rodopi, 2011.

Siraisi, Nancy G. "Anatomizing the Past: Physicians and History in Renaissance Culture." *Renaissance Quarterly* 53, no. 1 (2000): 1–30.

———. *Communities of Learned Experience: Epistolary Medicine in the Renaissance*. Baltimore: Johns Hopkins University Press, 2013.

———. *History, Medicine, and the Traditions of Renaissance Learning*. Ann Arbor: University of Michigan Press, 2007.

———. *Medieval and Early Renaissance Medicine: An Introduction to Knowledge and Practice*. Chicago: University of Chicago Press, 1990.

———. "Pietro d'Abano and Taddeo Alderotti: Two Models of Medical Culture." *Medioevo* 11 (1985): 139.

———. "Reflections on Italian Medical Writings of the Fourteenth and Fifteenth Centuries." *Annals of the New York Academy of Sciences* 412, no. 1 (October 1, 1983): 155–68.

Smith, Pamela H. *The Body of the Artisan: Art and Experience in the Scientific Revolution*. Chicago: University of Chicago Press, 2004.

Smith, Pamela H., and Paula Findlen, eds. *Merchants and Marvels: Commerce, Science, and Art in Early Modern Europe*. New York: Routledge, 2002.

Smith, Pamela H., Amy R. W. Meyers, and Harold J. Cook, eds. *Ways of Making and Knowing: The Material Culture of Empirical Knowledge*. Ann Arbor: University of Michigan Press, 2014.

Smith, Pamela H., and Benjamin Schmidt, eds. *Making Knowledge in Early Modern Europe: Practices, Objects, and Texts, 1400–1800*. Chicago: University of Chicago Press, 2007.

Spitzer, Gabriele. *—und die Spree führt Gold: Leonhard Thurneysser zum Thurn: Astrologe, Alchimist, Arzt und Drucker im Berlin des 16. Jahrhunderts*. Berlin: Staatsbibliothek zu Berlin, Preussischer Kulturbesitz, 1996.

Staden, Heinrich von. *Herophilus: The Art of Medicine in Early Alexandria*. Cambridge: Cambridge University Press, 1989.

Stannard, Jerry. "Dioscorides and Renaissance Materia Medica." In *Materia Medica in the XVIth Century: Proceedings of a Symposium of the International Academy of the History of Medicine, Held at the University of Basel, 7th September 1964*, ed. Marcel Florkin, 1–21. Oxford: Symposium Publications Division, Pergamon Press, 1966.

Stark, Laura. *Behind Closed Doors: IRBs and the Making of Ethical Research*. Chicago: University of Chicago Press, 2012.

———. "Work, Welfare, and the Values of Voluntarism: Rethinking Anscombe's 'Action under a Description' in Postwar Markets for Human Subjects." *American Journal of Cultural Sociology* 5 (2017): 181–224.

Stark, Marnie P. "Mounted Bezoar Stones, Seychelles Nuts, and Rhinoceros Horns: Decorative Objects as Antidotes in Early Modern Europe." *Studies in the Decorative Arts* 11 (2003): 69–94.

Stephenson, Marcia. "From Marvelous Antidote to the Poison of Idolatry: The Transatlantic Role of Andean Bezoar Stones during the Late Sixteenth and Early Seventeenth Centuries." *Hispanic American Historical Review* 90 (2009): 3–39.

Stille, Günther. *Krankheit und Arznei: Die Geschichte der Medikamente*. Berlin: Springer-Verlag, 1994.

Stinger, Charles L. *The Renaissance in Rome*. Bloomington: Indiana University Press, 1985.

Stolberg, Michael. "Empiricism in Sixteenth-Century Medical Practice: The Notebooks of Georg Handsch." *Early Science and Medicine* 18 (2013): 487–516.

———. *Experiencing Illness and the Sick Body in Early Modern Europe*. Basingstoke, UK: Palgrave Macmillan, 2011.

———. *Homo patiens: Krankheits- und Körpererfahrung in der Frühen Neuzeit*. Cologne: Böhlau Verlag, 2003.

———. "Learning from the Common Folks: Academic Physicians and Medical Lay Culture in the Sixteenth Century." *Social History of Medicine* 27 (2014): 649–67.

———. "The Many Uses of Writing: A Humanist Physician in Sixteenth-Century Prague." In *Civic Medicine: Physician, Polity, and Pen in Early Modern Europe*, ed. J. Andrew Mendelsohn, Annemarie Kinzelbach, and Ruth Schilling, 66–87. London: Routledge, 2020.

———. "'Mein Äsculapisches Orakel!' Patientenbriefe als Quelle einer Kulturgeschichte der Krankheitserfahrung im 18. Jahrhundert." *Österreichischen Zeitschrift für Geschichtswissenschaften* 7 (1996): 385–404.

———. "Post-Mortems, Anatomical Dissections, and Humoral Pathology in the Sixteenth and Early Seventeenth Centuries." In *Pathology in Practice: Diseases and Dissections in Early Modern Europe*, ed. Silvia De Renzi, Marco Bresadola, and Maria Conforti, 79–95. London: Routledge, 2017.

———. "A Sixteenth-Century Physician and His Patients: The Practice Journal of Hiob Finzel, 1565–1589." *Social History of Medicine*, 2018.

Stuart, Kathy. *Defiled Trades and Social Outcasts: Honor and Ritual Pollution in Early Modern Germany*. Cambridge: Cambridge University Press, 1999.

Taape, Tillman. "Hieronymus Brunschwig and the Making of Vernacular Medical Knowledge in Early German Print." PhD diss., Pembroke College, University of Cambridge, 2017.

Teigen, Philip M. "Taste and Quality in 15th- and 16th-Century Galenic Pharmacology." *Pharmacy in History* 29 (1987): 60–68.

Telle, Joachim, ed. *Analecta paracelsica: Studien zum Nachleben Theophrast von Hohenheims im deutschen Kulturgebiet der frühen Neuzeit*. Stuttgart: Franz Steiner Verlag, 1994.

Terpstra, Nicholas, ed. *The Art of Executing Well: Rituals of Execution in Renaissance Italy*. Kirksville, MO: Truman State University Press, 2008.

———. "Confraternal Prison Charity and Political Consolidation in Sixteenth-Century Bologna." *Journal of Modern History* 66 (1994): 217–48.

———. "Theory into Practice: Executions, Comforting, and Comforters in Renaissance Italy." In *The Art of Executing Well: Rituals of Execution in Renaissance Italy*, ed. Nicholas Terpstra, 118–58. Kirksville, MO: Truman State University Press, 2008.

Thompson, C. J. S. *Poison Mysteries in History, Romance and Crime*. London: Scientific Press, 1923.

Thorndike, Lynn. *A History of Magic and Experimental Science*. 8 vols. Vol. 2. New York: Macmillan, 1923.

———. *A History of Magic and Experimental Science*. 8 vols. Vol. 3. New York: Columbia University Press, 1934.

———. *A History of Magic and Experimental Science*. 8 vols. Vol. 5. New York: Columbia University Press, 1941.

Timmermann, Achim. "'Locus Calvariae': Walking and Hanging with Christ and the Good Thief, c. 1350–c. 1700." *Artibus et Historiae* 35 (2014): 137–61.

Tiraboschi, Girolamo. *Storia della letteratura Italiana*. 2nd ed. Vol. 7, part 3. Modena, 1791. http://hdl.handle.net/2027/ucm.5324325886.

Totelin, Laurence M. V. "Mithradates' Antidote: A Pharmacological Ghost." *Early Science and Medicine* 9 (2004): 1–19.

Touwaide, Alain. "Harmful Botanicals." In *History of Toxicology and Environmental Health*, ed. Philip Wexler, 1: 60–68. Boston: Academic Press, 2014.

———. "Nicander, Thêriaka, and Alexipharmaka: Venoms, Poisons, and Literature." In *History of Toxicology and Environmental Health*, ed. Philip Wexler, 44–51. Boston: Academic Press, 2014.

Tribby, Jay. "Cooking (with) Clio and Cleo: Eloquence and Experiment in Seventeenth-Century Florence." *Journal of the History of Ideas* 52 (1991): 417–39.

Truitt, E. R. *Medieval Robots: Mechanism, Magic, Nature, and Art*. Philadelphia: University of Pennsylvania Press, 2015.

———. "The Virtues of Balm in Late Medieval Literature." *Early Science and Medicine* 14 (2009): 711–36.

Vicchi, Leone. *Villa Borghese nella storia e nella tradizione del popolo romano: ricerche monografiche*. Rome: Forzani, 1886.

Visceglia, Maria Antonietta. *Morte e elezione del papa: norme, riti e conflitti*, vol. 2: *L'Eta moderna*. Rome: Viella, 2013.

Voinier, Sarah, and Guillaume Winter, eds. *Poison et antidote dans l'Europe des XVIe et XVIIe siècles*. Arras: Artois presses université, 2011.

Wagner, Harold. *Ars Moriendi: Erwägungen zur Kunst des Sterbens*. Ed. Torsten Kruse. Freiburg in Breisgau: Herder, 1989.

Walker, Garthine. *Crime, Gender and Social Order in Early Modern England*. Cambridge: Cambridge University Press, 2003.

Wallis, Faith, ed. *Medieval Medicine: A Reader*. Toronto: University of Toronto Press, 2010.

Wallis, Patrick. "Exotic Drugs and English Medicine: England's Drug Trade, 1550–1800." *Social History of Medicine* 25 (2011): 20–46.

Wallis, Patrick, and Catherine Wright. "Evidence, Artisan Experience, and Authority in Early Modern England." In *Ways of Making and Knowing: The Material Culture of Empirical Knowledge*, ed. Pamela H. Smith, Amy R. W. Meyers, and Harold J. Cook, 138–63. Ann Arbor: University of Michigan Press, 2014.

Watson, Gilbert. *Theriac and Mithridatium: A Study in Therapeutics*. London: Wellcome Historical Medical Library, 1966.

Wear, Andrew, Roger K. French, and I. M. Lonie, eds. *The Medical Renaissance of the Sixteenth Century*. Cambridge: Cambridge University Press, 1985.

Webster, Charles. "Paracelsus: Medicine as Popular Protest." In *Medicine and the Reformation*, ed. Ole Peter Grell and Andrew Cunningham, 57–78. London: Routledge, 1993.

———. *Paracelsus: Medicine, Magic and Mission at the End of Time*. New Haven: Yale University Press, 2008.

Weiler, Inge. *Giftmordwissen und Giftmörderinnen: Eine diskursgeschichtliche Studie*. Tübingen: M. Niemeyer, 1998.

Werrett, Simon. *Thrifty Science: Making the Most of Materials in the History of Experiment*. Chicago: University of Chicago Press, 2019.

Weyer, Jost. *Graf Wolfgang II. von Hohenlohe und die Alchemie: Alchemistische Studien in Schloß Weikersheim, 1587–1610*. Sigmaringen: Jan Thorbecke, 1992.

Wheeler, Jo. *Renaissance Secrets, Recipes and Formulas*. London: Victoria and Albert Museum, 2009.

Wibel, Johann Christian. *Hohenlohische Kyrchen- und Reformationshistorie: aus bewährten Urkunden und Schriften verfasset*. Ansbach, 1755.

Wickens, G. M. *Avicenna: Scientist and Philosopher: A Millenary Symposium*. Vol. OP 29,079. [University Microfilms. Publications.] London: Luzac, 1952.

Wickersheimer, Ernest. *Les accusations d'empoisonnement portées pendant la première moitié du XIVe siècle contre les lépreux et les Juifs*. Anvers, 1927.

Wiese, Erich. *Biedermeierreise durch Schlesien*. Darmstadt: Bläschke, 1966.

Wilson, Miranda. *Poison's Dark Works in Renaissance England*. Lanham, MD: Bucknell University Press, 2013.

Wilson, Roslynne Valerie. "Collaborations in Art and Medicine, 1491–1543: The Development of Anatomical Studies in Italian Medical Treatises." PhD diss., Case Western Reserve University, 1988.

Wiltenburg, Joy. *Crime and Culture in Early Modern Germany*. Charlottesville: University of Virginia Press, 2012.

Winkler, Lutz. *Galens Schrift "De antidotis": ein Beitrag zur Geschichte von Antidot und Theriak*. Marburg, 1980.

Winnerling, Tobias. "Der Fürsten-Theriak: Die wundersamen Knoblauchkräfte der Frühen Neuzeit." *Würzburger medizinhistorische Mitteilungen* 29 (2010): 235–362.

Winterbottom, Anna E. *Hybrid Knowledge in the Early East India Company World*. London: Palgrave Macmillan, 2016.

Yale, Elizabeth. *Sociable Knowledge: Natural History and the Nation in Early Modern Britain*. Philadelphia: University of Pennsylvania Press, 2016.

Zamora, Juan Gil de, and Cándida Ferrero Hernández. "Liber contra venena et animalia venenosa. Estudio preliminar, edición crítica y traducción." Universitat Autònoma de Barcelona, 2002. http://www.tdx.cat/TDX-0505103-183401.

Zedler, Johann Heinrich. "Wald, Georg an und vom, order Anwaldus." In *Grosses Vollständiges Universal-Lexicon aller Wissenschafften und Künste*, 52: 1197–98. Leipzig and Halle, 1747.

———. "Johann Montanus." In *Grosses Vollständiges Universal-Lexicon aller Wissenschafften und Künste*, 21: 1256–57. Leipzig and Halle, 1739.

Ziegler, Philip. *The Black Death*. New York: Harper & Row, 1971.

Zimmermann, T. C. Price. *Paolo Giovio: The Historian and the Crisis of Sixteenth-Century Italy*. Princeton: Princeton University Press, 1995.

Zimmermann, Volker. "Der Rosmarin als Heilpflanze und Wunderdroge. Ein Beitrag zu den mittelalterlichen Drogenmonographien." *Sudhoffs Archiv* 64 (1980): 351.

Ziser, Michael. "Sovereign Remedies: Natural Authority and the 'Counterblaste to Tobacco.'" *William and Mary Quarterly* 62, no. 4 (2005): 719–44.

Index

Page numbers in italics refer to figures.